MIXING
in Inland
and Coastal Waters

MIXING
in Inland
and Coastal Waters

HUGO B. FISCHER

Department of Civil Engineering
University of California
Berkeley, California

E. JOHN LIST
ROBERT C. Y. KOH

W. M. Keck Laboratory of Hydraulics
and Water Resources
Division of Engineering and Applied Science
California Institute of Technology
Pasadena, California

JÖRG IMBERGER

Department of Civil Engineering
University of California
Berkeley, California

NORMAN H. BROOKS

W. M. Keck Laboratory of Hydraulics
and Water Resources
Division of Engineering and Applied Science
California Institute of Technology
Pasadena, California

ACADEMIC PRESS, INC.
A DIVISION OF
HARCOURT BRACE & COMPANY
San Diego   New York   Boston
London   Sydney   Tokyo   Toronto

COPYRIGHT © 1979, BY ACADEMIC PRESS, INC.
ALL RIGHTS RESERVED.
NO PART OF THIS PUBLICATION MAY BE REPRODUCED OR
TRANSMITTED IN ANY FORM OR BY ANY MEANS, ELECTRONIC
OR MECHANICAL, INCLUDING PHOTOCOPY, RECORDING, OR ANY
INFORMATION STORAGE AND RETRIEVAL SYSTEM, WITHOUT
PERMISSION IN WRITING FROM THE PUBLISHER.

ACADEMIC PRESS, INC.
1250 Sixth Avenue
San Diego, California 92101

United Kingdom Edition published by
ACADEMIC PRESS, INC. (LONDON) LTD.
24/28 Oval Road, London NW1 7DX

Library of Congress Cataloging in Publication Data
Main entry under title:

Mixing in inland and coastal waters.

    Bibliography: p.
    1. Mixing. 2. Hydrodynamics. I. Fischer,
Hugo B.
TC171.M57   628'.39   78–22524
ISBN 0–12–258150–4

PRINTED IN THE UNITED STATES OF AMERICA
93 94 95 96 97 QW 15 14 13 12 11 10

Contents

Chapter 2   Fickian Diffusion

Chapter 3   Turbulent Diffusion

Chapter 4   Shear Flow Dispersion

Chapter 5   Mixing in Rivers

Chapter 6   Mixing in Reservoirs

Chapter 7   Mixing in Estuaries

Chapter 8   River and Estuary Models

Chapter 9   **Turbulent Jets and Plumes**

Chapter 10   **Design of Ocean Wastewater Discharge Systems**

Preface

This book is an outgrowth of research contributions and teaching experiences by all of the authors in applying modern fluid mechanics to problems of pollutant transport and mixing in the water environment. It should be suitable for use in first year graduate level courses for engineering and science students, although more material is contained than can reasonably be taught in a one-year course, and most instructors will probably wish to cover only selected portions. The book should also be useful as a reference for practicing hydraulic and environmental engineers, as well as anyone involved in engineering studies for disposal of wastes into the environment. The practicing consulting or design engineer will find a thorough explanation of the fundamental processes, as well as many references to the current technical literature which may be followed for greater detail. Besides being led to the current literature, the student should gain a deep enough understanding of basics to be able to read with understanding the future technical literature in this evolving field.

Chapter 1 discusses the relevance of this book to overall environmental management and explains certain basic concepts which apply throughout, such as dimensional reasoning. Chapter 2 presents the classical theory of diffusion in the context of molecular diffusion, primarily as an introduction to the equations and concepts used in later chapters (e.g., Chapter 3 on turbulent mixing). Chapter 4 describes shear flow dispersion—that phenomenon that describes the stretching and mixing of pollutant clouds caused by the combined action of shear and lateral mixing—and completes the presentation of background material necessary to the study of mixing in the environment. Chapters 5–7 apply the material in the earlier chapters to the cases of mixing in rivers, in reservoirs, and in estuaries, and Chapter 8 completes the discussion of rivers and estuaries by describing the use of physical and numerical models.

Whereas the previous chapters deal with natural flow conditions, Chapter 9 treats buoyant jets and plumes, strong man-induced flow patterns used to achieve rapid initial dilutions for water quality control. Chapter 10 gives a design-oriented discussion of outfall diffusers and includes sections on the internal hydraulics of outfalls and the techniques of hydraulic modeling of outfall flows with density differences.

The level of treatment presumes that the reader has an understanding of calculus up to some introductory work on partial differential equations, and an undergraduate level background in hydraulics or fluid mechanics; some knowledge of the science of hydrology and the basic characteristics of rivers, lakes, estuaries, and the coastal ocean, and a general orientation in environmental engineering, including water supply and waste treatment will also be helpful. The senior author has taught most of the material (Chapters 6 and 9 excepted) to a class of beginning and advanced graduate students in hydraulic and sanitary engineering at Berkeley for the past 10 years. Those students who brought only the usual undergraduate background in Civil Engineering have been able to understand the course, but it is certainly the case that the practitioner of mixing studies in the environment will benefit from advanced courses in fluid mechanics and mathematics, and the better the background the student brings to this book the more easily he or she will understand the material.

A note is in order on the writing of the various chapters. Although this is a jointly authored book, each chapter had a primary author or authors. Chapter 1 is by NHB; Chapters 2 and 3 are jointly by EJL, JI, and HBF. Chapters 4, 5, 7, and 8 are primarily by HBF, Chapter 6 by JI, Chapter 9 by EJL, and Chapter 10 by RCYK.

Acknowledgments

This book was begun in 1970 during the first author's sabbatical leave at the University of Cambridge, and much of it was completed during the same author's second sabbatical at the Woods Hole Oceanographic Institution in 1977. Most sincere thanks are due for the gracious hospitality and assistance at both institutions and particularly to hosts Professor George Batchelor and Dr. Gabriel Csanady. The authors would particularly like to express their appreciation for the stimulus and valuable research contributions made by their students and colleagues, and for comments received on portions of the manuscript from Jack Kennedy, William Sayre, Ian Wood, Frederick Sherman, and Lloyd Townley. J. C. Patterson, Ian Loh, and R. H. B. Hebbert assisted in the writing and computations of Section 6.8.2 and collaborated in the collection of much of the data from the Wellington Reservoir. Karen Ray at Berkeley and Theresa Fall and David Byrum at Caltech assisted in typing and drafting. Finally, we are particularly appreciative of the efforts of Mabel Iwamoto, secretary of the Hydraulics Group at Berkeley, and Joan Mathews, secretary of the Keck Hydraulics Laboratory.

We also gratefully acknowledge the sponsorship over many years of the research that led to this book by the National Science Foundation, the U.S. Public Health Service, the Environmental Protection Agency, the Ford Energy Program at Caltech, the Southern California Edison Company, the Pacific Gas and Electric Company, the U.S. Geological Survey, the Australian Water Resources Council, the Australian Research Grants Committee, the Public Works Department and the Metropolitan Water Board of Western Australia, and last but most certainly not least, the Water Resources Center of the University of California. We particularly note the support of the California Institute of Technology and the University of California, Berkeley, in the final preparation of the manuscript; without the support of these institutions and ultimately of the people of the state of California there would be no book.

Finally, for the support and patience of our five families we are truly thankful.

MIXING
in Inland
and Coastal Waters

Chapter 1

Concepts and Definitions

In recent years hydraulic engineers have frequently been asked to analyze and predict mixing in natural bodies of water. It is no longer sufficient to deal only with water *quantities* because of the growing concern over water *quality*. Many pollutants enter the hydrologic cycle both intentionally and unintentionally; downstream water quality depends on both the hydrodynamics of transport and mixing, and the chemistry and biology of natural water systems. The purpose of this book is to deal with the hydrodynamic aspects of water quality management in natural bodies of water.

1.1 THE ROLE OF HYDROLOGY AND HYDRAULIC ENGINEERING IN ENVIRONMENTAL MANAGEMENT

The hydrosphere, like the atmosphere, is of extraordinary importance to mankind because water and air are fluids. By their motions they transport and disperse many essential elements for life and productivity. They are also absolutely essential for management of the quality of the natural environment, including the disposal of residuals. Examples range all the way from ordinary breathing (you exhale extra carbon dioxide to be dispersed) to massive discharges of wastewater into rivers and oceans. Regardless of the size of the problem, if these vital fluids were utterly motionless we would suffocate or be inundated in our liquid wastes!

1

1.1.1 Overall Framework for Environmental Management

Hydrologic mixing processes are but a small part of the overall process of environmental management. How the overall environmental system works is shown by the diagram in Fig. 1.1, applying to all environmental media—air, water, land. It applies to the management of hundreds of different pollutant substances, which do not act independently because there are many synergisms. For example, an increase in the temperature of a river, due to heated water discharge, may decrease its assimilative capacities for oxygen-demanding wastes such as sewage effluents. Note that the diagram is not limited to emission-response relations for pollutants, but also implicitly covers the environmental effects of all kinds of activities—e.g., building dams or dredging harbors, two typical hydraulic activities.

For problems dealing with the water environment, the hydraulic engineer works mainly on the box called "Transport, Transformation, Accumulation." This box is shared with chemists and biologists, who are also concerned with the processes that take place between the point where a pollutant is discharged into the water environment and some other sites where the ambient water quality is observed. The hydraulic engineer has a special challenge, however, because the design of hydraulic structures, such as outfalls, may have a profound effect on the water quality observed. Therefore, he must at times be the designer who takes the ambient water quality requirements and works backwards to develop the concepts and dimensions for an engineering system to achieve those results.

A good environmental control system is one which skillfully optimizes, depending upon the pollutants being handled, the combination of: (a) control of pollutants at the source (often called "pretreatment" for sewer systems), (b) wastewater treatment, and (c) dispersal in the environment. In summary, hydraulic engineers are often called upon to make the interfaces between man's activities, involving water and wastewater, and the natural environment. For a long time (all the way back to the Roman Empire!), they have had the responsibility for drawing our water supply from natural water bodies; now we must pay equal attention to how it is returned—in diminished amounts and of poorer quality.

1.1.2 Using the Water Environment for Waste Assimilation

From the geologic point of view, the fresh waters of the earth have always carried away the residuals from the land, usually delivering them to the ocean. This includes weathered mineral matter either in solution or as sedimentary particles. Much organic debris is also swept along by the rivers. The ocean has always been the sink for all the material washed off the land. Therefore, it is a

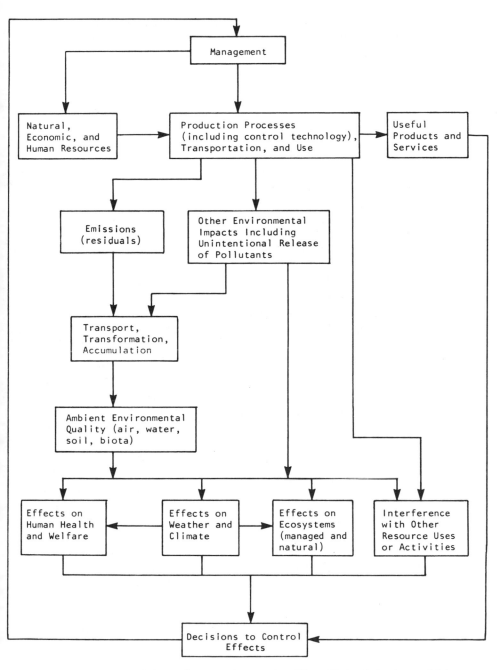

Figure 1.1   Framework for environmental protection. [From National Academy of Sciences (1977).]

perfectly natural thing for man also to use, in a limited way, the hydrologic cycle to transport his residuals. In fact, some of the dissolved salts, which must be flushed out to maintain mass balance of salts, initially enter the water supply systems by diversions from natural streams or groundwater.

If we think of using the water environment for receiving wastewater or residuals, it is important to consider the wide range of typical waste materials. We can arrange a list starting with the least dangerous types of pollutant and proceeding to the most hazardous.

(1)   *Natural Inorganic Salts and Sediments.*   These materials are not toxic and only become possible pollutants in excessive doses, such as increases in turbidity due to dredge-spoil dumping.

(2)   *Waste Heat.*   Once-through cooling systems for electric generating plants use water for carrying away large quantities of low-grade waste heat. When the body of water is large enough, it serves as the cold reservoir of the plant's thermodynamic cycle.

(3)   *Organic Wastes.*   Domestic sewage containing ecosystem materials (like carbon, nitrogen, and phosphorous) can cause bad stenches and nuisances, but if adequately treated and dispersed, these ecosystem materials can be safely assimilated into large water bodies. The biochemical oxygen demand may be sufficiently reduced by dilution so that it can be satisfied by the natural dissolved oxygen in the water body.

(4)   *Trace Metals.*   Examples are lead, mercury, and cadmium which are naturally present in the environment in very small amounts, but man's waste-waters often have very much higher concentrations which can be toxic.

(5)   *Synthetic Organic Chemicals.*   These substances are slow to degrade in the environment and are often bioaccumulated in the food chain. Even though wastewaters may be subjected to high initial dilution, the food chain is capable in some instances of multiplying the concentrations by a factor of 10^5 in successive food chain steps. It is remarkable how biological processes can do just the opposite of the physical process of turbulent mixing which reduces concentration.

(6)   *Radioactive Materials.*   The necessity of long-term storage of radioactive wastes without leakage or contamination of natural waters is causing grave concern because of the high toxicity of these materials (e.g., plutonium).

(7)   *Chemical and Biological Warfare Agents.*   Clearly these can not be dispersed in the environment without great danger to human beings as they are designed to be exceedingly toxic in very small doses.

With an extreme range of types of water pollutants from something as ubiquitous as waste heat to deathly chemicals, environmental strategies must obviously relate to the substance to be disposed of! For example, the strategy of *wide dispersal* ("dilution is the solution to pollution") is suitable only for heat and natural organic materials which are to be reassimilated in the global ecosystem. Trace metals in small amounts and nontoxic compounds can be dispersed in large bodies of water if the resulting increases in the background concentrations are minimal. But the strategy of *containment* or prevention of the release of the pollutant is by far preferable for persistent organic chemicals as well as for trace metals. Nonproduction and nonuse is certainly the safest strategy for really lethal chemicals; as for radioactive wastes, many argue that they are so dangerous as to warrant banning of nuclear power plants.

If all kinds of waste materials are allowed to be mixed together, then we have a very difficult problem of devising a suitable strategy. Treatment systems are not efficient in separating toxic materials from nontoxics; for example, trace metals tend to be preferentially attached to sewage particles. The metals are then preferentially removed by treatment along with particles that become part of the sludges. We then have the problem of what to do with sewage sludge which is "enriched" with trace contaminants, even though the liquid effluent is improved.

If we try to contain *all* of our wastes, including organic carbon, we simply have an impossible storage problem. Therefore, effective management necessitates keeping materials which can be reasonably dispersed in the environment separate from those which, because of their characteristics, should not be released.

The purpose of this book is to present the fundamental hydrodynamics for predictions of the dispersal of wastewater, both for intentional releases and for accidental spills.

1.1.3   Mass Balance Concepts in Residuals Management, Involving the Hydrologic Cycle

A fundamental and inescapable law of environmental management is conservation of mass. Usually we deal with the flux of a substance starting from its source of release into the environment; the fluxes for subsequent transport and diffusion must balance the source flux with adjustments for chemical and biological conversions and sinks, such as deposition on the seafloor. The fluxes of some substances may follow multiple pathways in air, land, or water. An example is lead in gasoline which is mostly emitted as aerosols from the exhaust pipes of automobiles. Subsequently, these particles may fall into the street and be washed by the storm runoff to the nearest natural body of water; or they may be carried by the wind and deposited in nearby or faraway land masses or oceans. For such substances, clearly there is no point in trying to devise a management strategy without considering the complete mass balance.

Sometimes it is useful to take an even larger view. By drawing a circle around a city, for example, we can say that the total influx of substance X must equal total efflux if we expect to have a steady state. Similarly, an irrigation tract cannot have a long-term salt balance unless the total dissolved solids removed by drainage balance the incoming salts in the water supply.

The regulatory system generally distinguishes between point and nonpoint sources of pollutants. A point source is the discharge from a structure which is specifically designed for the outflow of wastewater from some industrial process or municipal sewerage system. Point sources have been the target of most of the laws and regulations for water pollution control. The accidental spill of oil from a ship and the release of radioactive wastes from a power plant can also be considered as point sources.

Nonpoint sources, on the other hand, are defined as widely distributed points where pollutants are introduced into the hydrologic cycle. In such cases, water treatment is usually not feasible. Examples are the runoff of salts used for deicing highways in winter, soil erosion, acid rainfall, and street drainage. These examples illustrate that there is a wide variety of man's activities impacting the water environment for which careful analysis is needed in order to devise effective control strategies.

1.1.4   Impacts of Some Traditional Activities of Hydraulic Engineers

The analyses given in this book are intended not only to help the hydraulic engineering profession to design works specifically required for water quality control, such as outfalls, but also to be useful in the study and control of the adverse effects of traditional hydraulic works. Some examples are enumerated below.

Man-made reservoirs may cause deterioration in water quality because of summertime thermal stratification associated with oxygen depletion in the lower layers. Diversion of water for various consumptive uses or to other watersheds reduces river flow and its ability to provide natural flushing and repel salinity intrusion in estuaries. Various conveyances, like canals, can transport huge amounts of dissolved salts, sediment, nutrients and parasites to places that otherwise would not receive such doses of these materials. Agricultural drainage systems may greatly accelerate the leaching of nutrients and salts from the land into natural hydrologic systems. Breakwaters for harbors interfere with natural nearshore circulation which could otherwise carry away pollutants. Estuarine modifications or barriers can radically change the circulation patterns with dire consequences for flushing of pollutants.

The message is clear. Water quality has become a key issue of practically every hydraulic engineering endeavor, whether constructed specifically for water quality control or for other purposes.

1.2   ENVIRONMENTAL HYDRAULICS

Doing environmental analyses of various hydraulic works requires more than the traditional hydraulic subjects, as suggested in the following paragraphs.

1.2.1 Hydrologic Transport Processes

This subject refers to the physical processes of flow of natural water bodies which cause pollutants or natural substances to be transported and mixed, or exchanged, with other media. It is similar to what the chemical engineers call "transport processes," but here we apply it to a natural water environment rather than to man-made unit processes. Included among hydrologic transport processes are the following:

Advection.   Transport by an imposed current system, as in a river or coastal waters.

Convection.   Vertical transport induced by hydrostatic instability, such as the flow over a heated plate, or below a chilled water surface in a lake.

Diffusion (Molecular).   The scattering of particles by random molecular motions, which may be described by Fick's law and the classical diffusion equation.

Diffusion (Turbulent).   The random scattering of particles by turbulent motion, considered roughly analogous to molecular diffusion, but with "eddy" diffusion coefficients (which are much larger than molecular diffusion coefficients).

Shear.   The advection of fluid at different velocities at different positions; this may be simply the normal velocity profile for a turbulent flow where the water flows faster with increasing elevation above the bed of the stream; or shear may be the changes in both magnitude and direction of the velocity vector with depth in complex flows such as in estuaries or coastal waters.

Dispersion.   The scattering of particles or a cloud of contaminants by the combined effects of shear and transverse diffusion.

Mixing.   Diffusion or dispersion as described above; turbulent diffusion in buoyant jets and plumes (see following section); any process which causes one parcel of water to be mingled with or diluted by another.

Evaporation.   The transport of water vapor from a water or soil surface to the atmosphere.

Radiation.   The flux of radiant energy, such as at a water surface.

Particle Settling.   The sinking (or rising) of particles having densities different from the ambient fluid, such as sand grains or dead plankton. (In lakes and oceans the latter may be the dominant mechanism for downward transport of nutrients, often all the way to the bottom.)

Particle Entrainment.   The picking up of particles, such as sand or organic detritus, from the bed of a water body by turbulent flow past the bed.

In varying ways these processes apply to various types of water bodies—lakes, rivers, estuaries, coastal oceans, and groundwater. This book concentrates on mixing in open bodies of water (i.e., excluding ground water), and does not treat the last two items on the above list.

When we calculate the flow of water in a channel, we need only know the *mean* velocity. However, for the analysis of pollutant transport and dispersal, the velocity variations in both time and space, and the irregularities of channel geometry, are all very important in determining what happens. It is important to remember the general principle that the *fluctuations and irregularities in hydrologic systems are just as important as the mean flows for pollutant analysis.*

1.2.2 Buoyant Jets and Plumes

To increase the dilution of an effluent discharge with the surrounding waters, engineers use structures which produce submerged jets, or if the discharge fluid is lighter or heavier the flows are called submerged buoyant jets. If the initial momentum is of no consequence then the flow pattern is called a plume. The analysis of buoyant jets or plumes depends not only on the jet parameters but also on the ambient conditions represented by the ambient density stratification and the current profile. Since the jet flows of hydrologic interest are practically all turbulent, their behavior has to be described in a suitable statistical sense.

For large discharges a multiple jet arrangement is used, which can often produce immediate initial dilutions of several hundred to one (volume of mixture divided by initial effluent volume). This subject is treated in detail in Chapter 9, and applications are described in Chapter 10.

1.2.3 Density-Stratified Flows in a Natural Environment and Geophysical Fluid Mechanics

Many problems in mixing in the natural environment are often complicated by density stratifications due to temperature variation in lakes and reservoirs, or salinity profiles in estuaries. This internal structure has a very great effect on both mean flow fields and the turbulent mixing and dispersion. But since often stratification results from transport processes, we have a strong feedback system, i.e., mixing depends on the flow field and the density stratification, while on the other hand, the flow and stratification depend on the mixing.

In a broader sense, the entire flow patterns in the ocean are related to density stratification and rotation of the earth, as well as surface exchanges with the atmosphere, tidal and wave effects, and fresh water inflows at the continental margins. However, these subjects are beyond the scope of the current book.

Analysis and effects of density-stratified flows in lakes and estuaries are presented in Chapters 6 and 7, respectively.

1.2.4 Sedimentation and Erosion

As previously mentioned, this book will not treat particle settling and entrainment, even though they are sometimes important hydrologic transport processes. Also not covered is the broader subject of stream and shoreline morphology, concerned with the erosion, transport, and deposition of sediments by rivers and by littoral (longshore) currents. The reader is urged to consult several recent books on this subject [e.g., Vanoni (1975), Raudkivi (1976), and Graf (1971)].

1.2.5 Interdisciplinary Modeling

Since the fate of pollutants is governed by a combination of physical, chemical, and biological processes, there is now a lot of interest and research in developing improved predictive models for combining all these effects. Fortunately, for some problems there is a separation of time scales between the dominance of the various processes; for example, the strong hydrodynamic mixing that occurs in the jets from an outfall diffuser lasts only for a very few minutes, whereas many biochemical reactions have characteristic times of many minutes, hours, or even days; finally, biological or ecological effects (barring episodes of acute toxicity) usually are the integrated effects of many weeks and months. Sometimes, the relative time scales of the dominant processes may be reversed from the case of outfall jet mixing; for example, in slow coastal upwelling of nutrient-rich water, the rate of biological activity is limited by how fast the new upwelled water mixes with the nutrient-poor surface water.

To some extent then, as a first approximation, we may piece together submodels which focus on different disciplines. However, the reader should be cautious because some problems, like the settling of organic particles in the ocean, cannot be reliably handled by separate submodels for different effects. Much progress is still needed in interdisciplinary models; they could well be the subject of a whole separate book.

1.3 STRATEGIES AND APPROACHES FOR PROBLEM SOLVING

When solving a problem of mixing in the water environment, the investigator must first devise an overall strategy for breaking it into logical component parts (or submodels), using reasonable approximations, and then decide what combination of approaches to use to develop the best prediction. Typical approaches would include computer modeling, hydraulic modeling, and field experimentation.

There is no "best" way for any problem—a lot of judgement is essential. For example, which is more reliable for predicting temperature patterns around a heated water discharge: a computer model or a physical hydraulic model? Such a choice is sometimes a multimillion dollar question. Suppose, for instance, that a regulatory agency uses computer modeling (which they believe to be best) and determines that a proposed thermal discharge from a power plant will violate ambient thermal requirements; the agency would then refuse to grant an operating permit. On the other hand, the utility company's engineers may insist that their hydraulic model study of the discharges *does* show compliance. The latter might believe the laboratory model is more reliable because it is able to reproduce complex three-dimensional stratified flows. But the math modelers might point to their advantages of being able to include meteorological variables (wind, surface cooling), and to avoid scaling errors inevitably made in the laboratory model where viscous effects are relatively too strong.

As the reader might perceive, the best approach is often a mixed one, whereby the engineers will work different parts of a problem by whatever procedure is best, or in case of doubt, use two or more approaches to help bracket the uncertainties. Consequently it is most unwise to generalize on which approach is best.

1.3.1 Strategies

The first necessity in problem solving is to be sure you know exactly what the question is! For instance, do you seek to predict maximum instantaneous point concentrations of a pollutant, or changes in monthly averages over a broad area? The first question might be important for acute toxic effects, whereas the latter might pertain to subtle long-term ecological or climatological effects. Even if you want to know the answers to both short-term and long-term mixing, different modeling approaches should be used.

At different scales of length and time, different processes will be important. Table 1.1 gives an example of the progression through length and time scales for an ocean discharge of treated sewage effluent through a submarine outfall. Various pollutants have different relative importance at different time scales; for example, toxicity of ammonia might occur only at times $< 10^4$ sec, and biochemical oxygen demand (BOD) is important only at time scales $< 10^6$ sec, whereas chronic toxicity from persistent trace contaminants may occur at times $> 10^8$ sec (~ 3 years).

1.3.1.1 Definition of Submodels

It is inconvenient to develop an omnibus model (either physical or computer) to cover all steps. Rather the problem is broken into pieces for different length and time scales (e.g., Table 1.1), such that some simplifying assumptions can readily be made. For example, for a thermal discharge we may neglect heat loss from the water surface in the "near field," where the initial jet and plume mixing

TABLE 1.1

The Effluent Flow from a Sewer Outfall Passes through a Succession of Physical Processes at Scales from Small-to-Large

Phase	Phenomenon	Length scale[a] (m)	Time scale[a] (sec)
(1)	*Initial jet mixing* (rise of buoyant jets over an outfall diffuser in a stratified fluid).	$< 10^2$	$< 10^3$
(2)	*Establishment of sewage field or cloud*, travelling with the mean current; lateral gravitational spreading.	$10^1 - 10^3$	$10^2 - 10^3$
(3)	*Natural lateral diffusion* and/or dispersion.	$10^2 - 10^4$	$10^3 - 10^5$
(4)	*Advection* by currents (including scales of water motion too large compared to sewage plume to be called turbulence).	$10^3 - 10^5$	$10^3 - 10^6$
(5)	*Large scale flushing* (advection integrated over many tidal cycles); upwelling or down-welling; sedimentation.	$10^4 - 10^6$	$10^6 - 10^8$

[a] Approximate orders of magnitude.

occurs; hydraulic modeling might be used here. But in the "far field", where heat loss is a dominant factor, we would use a computer model, with the results from the near-field hydraulic model used as input.

Models of any kind are of necessity idealized representations. They tend to concentrate on the dominant processes and the important features of the environment (e.g., currents, stratification, topography). The process still omits many secondary details or interactions for lack of knowledge, or lack of ability to handle them (e.g., vertical mixing caused by internal gravity waves in a water reservoir). Current research is actively striving to improve our models. However, progress in recent years has been great and it is now possible to get reasonably good predictions of mixing. The various sections of this book will indicate how well we can do this and what the uncertainties are. In general, probable errors in predictions of $\pm 25\%$, or even $\pm 50\%$, should not be considered unusual in this business.

1.3.1.2 Variability of Discharge Rate and Environmental Parameters

Modeling of pollutant mixing and dispersal in water cannot be based on fixed parameters, but must provide for a simulation over representative periods of time including time-varying parameters. Such parameters include quantity and quality of discharge (often showing a diurnal cycle), density stratification (with strong seasonal variation), and currents (diurnal and seasonal changes). These varying parameters are not just put in as uncorrelated random variables following certain probability distributions, but must reflect actual time sequences which are only partly random. For example, the temperature increment of cooling water from a power plant will be highest when the power plant load is highest, usually in the late afternoon; or in some places (like near San Francisco's Golden Gate) the current is heavily dominated by the back-and-forth tidal motion. Often simulations use historic data series (like hourly current measurements at an outfall site for a particular month) or else stochastically generated synthetic sequences with a proper deterministic part.

The predictions of simulations with variable inputs come out as time series which can be plotted as time graphs or reduced to probability distributions. Most regulatory agencies now recognize the variability of outcomes and use probabilities. For instance, the California bathing water standard for total coliform bacteria is 10/mliter which may be exceeded in up to 20 % of any month's samples. Conversely, because of this statistical variability of the environment, it is troublesome if the environmental regulations are written as absolute upper limits (even 99-percentile values are better!).

1.3.2   Approaches

By "approaches", we mean what kinds of problem-solving tools to use, either on the various parts or on the whole problem. To be considered below are order-of-magnitude analysis, computer techniques, hydraulic models, and field studies.

1.3.2.1   Order-of-Magnitude Analysis

For any mixing problem a skillful analyst should be able to work out a rough approximation for the solution within a fraction of an hour! The process of developing quick approximate solutions is called "order-of-magnitude analysis" or sometimes "scaling." (The term "order of magnitude" sometimes refers to the exponent of a variable, but we use it in the sense of powers of ten.) Results of order-of-magnitude analysis can be expected to show the correct dependence on the most important parameters, and yield numerical answers within factor of three to five (often very much better). This margin of error may seem huge, but as a first step approximate answers are worth much more than no answers at all!

In spite of elaborate differential equations and modeling procedures, it is essential for the student to develop skill in order-of-magnitude analysis. Using this skill, one can develop good judgment for formulating reasonable approximations, and breaking problems up into good submodels. The engineer can then plan the detailed tasks he and his staff must tackle in order to solve a comprehensive problem. Without some preview by order-of-magnitude analysis, much time and effort can be wasted. For example, if a field program is being planned to measure density stratification and currents near a proposed ocean outfall discharge site, there is a dilemma. One cannot do a careful analysis to determine a proper location without the data, but one cannot get ocean data in approximately the right place without some preliminary estimate of the location of the discharge point! An order-of-magnitude analysis, coupled with experience, can greatly reduce the range of locations immediately, thereby saving much time, effort, and expense in a field program.

Such analyses are based on dimensional analysis, with some coefficients learned from experiments or from previous solutions of differential equations. The following example, although drawing on material appearing later in the text, will help to illustrate the procedure. Approximately how long does it take a pollutant suddenly introduced at the surface of a river to become fully mixed over the depth? This problem is analogous to heat conduction in a finite rod with a sudden addition of heat at one end; otherwise the rod is fully insulated from its surroundings. The eddy diffusivity in a river is the analog of the thermal diffusivity in a solid, both having units of (length)^2/time. By dimensional analysis (Section 1.5) the mixing time (similar to the time for the heat to be conducted throughout the rod) is

$$T \propto d^2/\varepsilon_v \quad \text{or} \quad T = \alpha d^2/\varepsilon_v \tag{1.1}$$

where d is the depth and ε_v the vertical eddy diffusivity.† By solving the unsteady diffusion equation [see the solutions given as Eqs. (2.47) and (5.9)] we can find the coefficient for whatever definition we select for "fully mixed." For less than 10% overall variation in the mean concentration profile over depth, $\alpha > 0.35$ as shown in Section 5.1.3. Next we use $\bar{\varepsilon} = 0.07u^*d$ [from Chapter 5, Eq. (5.3)] and $u^* = \bar{u}\sqrt{f/8}$, where u^* is the shear velocity, $\bar{u}$ the cross-sectional mean velocity, and f the Darcy–Weisbach friction factor [see also Eqs. (1.22)–(1.26)]; the result is

$$T = \alpha \frac{d^2}{0.07\bar{u}d} \sqrt{\frac{8}{f}}. \tag{1.2}$$

Since a typical value of $\sqrt{8/f}$ might be 15, and using $\alpha = 0.35$

$$T \approx 75d/\bar{u}. \tag{1.3}$$

In terms of distance $x = \bar{u}T$, we find

$$x/d \approx 75. \tag{1.4}$$

The point of this example was not to get a precise answer but to determine that

---

† The quantity d^2/ε is called the "time scale" of a mixing problem, i.e., it is a characteristic quantity with units of time, which is divided into real time to get a dimensionless time variable to use in analysis.

the order of magnitude is likely to be about one hundred depths downstream and not one or ten times the depth nor thousands times the depth.

Dimensional reasoning is the key to doing order-of-magnitude estimates and it will be emphasized extensively throughout the book. A review of dimensional analysis is given in Section 1.5.

1.3.2.2 Computer Techniques

A computer calculation can be no better than the validity of the underlying approximations made when representing a complex process by mathematical equations. Even though computers now have a huge capacity, it is still necessary to make skillful assumptions, and leave out altogether those processes which have little effect on the results. Order-of-magnitude analyses and dimensional reasoning again play a very important part (see Chapter 8).

A wide variety of computer techniques may be used on mixing problems. There are computer models which essentially solve differential equations by finite difference or finite element techniques, or simulate diffusion as a stochastic (or "Monte Carlo") process with superimposed transport by currents. Some computer programs analyze observed environmental data (such as ocean currents) to extract key parameters for use as inputs to other submodels. A computer can be used to calculate some result (like initial plume dilution) at fixed time intervals (say every hour), assuming a quasi-steady process (no coupling forward or backward in time). Complex simulation models integrate various physical, chemical, and/or biological effects by doing sequential step calculations for each of the different processes one at a time in a prescribed order. This is equivalent to uncoupling the processes over short time periods, when it is *not* valid to do so for long periods. This brief enumeration of computer techniques, although not intended as a complete listing or classification, does indicate the wide scope and usefulness of such techniques.

1.3.2.3 Hydraulic Models

Although hydraulic models may be more expensive than computer models, they have a very great advantage for some situations, especially for three-dimensional density-stratified flows. In a sense a hydraulic model can educate the investigator about various phenomena (such as large scale vortices, internal waves and hydraulic jumps, multilayer shear flows, blocking, induced upwelling, and gravitational spreading), whereas these have to be anticipated in developing an adequate computer model. Besides, we are not yet able to build a computer model to handle all these large scale phenomena in three dimensions, often with transients.

But hydraulic models have their limitations too. The scaling is done by Froude law because gravity effects including buoyancy are usually important (see Chapters 6, 8, and 10). Consequently, the Reynolds numbers are much reduced from the prototype, altering turbulence, and resistance characteristics. Big models, such as those for estuaries, have to be distorted to obtain adequate Reynolds numbers and to avoid excessive model resistance; a careful calibration against measured field data is necessary.

At the boundaries of a model it is often difficult to provide the proper boundary conditions, especially for stratified flows. Meteorological factors (the surface boundary conditions) are also usually impossible to scale properly.

1.3.2.4   Field Studies

In addition to Eulerian-type measurements made at a *fixed* location in a body of water, it is often useful to do *Lagrangian* experiments. These are characterized by following drogues or drifters to track flow trajectories and dispersion. In Chapter 3, Taylor's classical theory is developed from the Lagrangian view. No theoretical basis has been found for converting Eulerian current data (time series at a fixed point) to Lagrangian (time series for a given parcel of water), which is necessary for a proper analysis of diffusion.

However, a fixed-location recording current meter gives vast quantities of data at reasonable cost. Tracking drogues is very tedious, and gives distressingly little data in relation to costs. A mixed approach is often chosen—use of fixed recording current meters supplemented by a few drogue or dye studies. Furthermore, floating drift cards are helpful in determining wind effects, i.e., what fraction of them dropped near a proposed discharge site get washed onto the beach.

Another field situation is often useful for study. A similar existing discharge may be studied at full scale in order to verify or adjust numerical models. Then, since the new project is likely to be somewhat different from the old, the adjusted numerical model is run for the new design parameters.

1.3.2.5   Mixed Approaches

For large complicated situations, a careful interweaving of all of the above approaches is the most practical. Each piece of a problem should be done in the most practical way. The final synthesis of this varied approach will undoubtedly be better than depending only on computer models, or on hydraulic models, or on field experiments. Each problem becomes a special case study, and even though we provide many tools in this book, it is still up to the judgment of the user to develop the best strategies and approaches for his particular analysis.

Finally, the reader is cautioned not to expect high precision from the analyses in this book even though they are carefully and rigorously developed. They all describe results in a statistical sense, and our observations of water bodies often do not average over large enough time and space to conform to the statistical assumptions.

There is a further complication—the elusive baseline condition. Regulations for ambient water quality are often expressed as changes from a natural or undisturbed condition. But in some cases, such as for heated water discharges, the allowable increment may be comparable to the natural variability of the same water quality parameter (measured as a standard deviation σ). In other words, the noisy background obscures the signal! A good example is the site of the San Onofre Nuclear Generating Station on the Southern California coast where the allowed ambient temperature increment, 4°F (2.2°C), is comparable to the range of natural short term temperature fluctuations [approximately $4\sigma \approx 5.5$°F (3°C) in summer; see List and Koh (1976)].

1.4 BASIC DEFINITIONS AND CONCEPTS

Certain definitions and concepts which arise frequently in the book are presented in this section.

1.4.1 Concentration

Let C be the concentration in units of mass of tracer or contaminant per unit volume. It may be defined at a point at a given time by the instantaneous limit:

$$C = \lim_{\Delta V \to 0} \Delta M / \Delta V, \tag{1.5}$$

where ΔM is the tracer mass in elemental volume ΔV. In taking the limit we cannot actually let $\Delta V \to 0$, but must stop at a size which is still *large* compared to molecules or the particles comprising ΔM. But this size, $(\Delta V)^{1/3}$, must also be *small* enough so that continuous concentration gradients (e.g., dc/dx) can be defined and used in differential equations.

The temperature of water may be considered a measure of the concentration of heat (C_h) (i.e., heat content per unit volume) according to the relation

$$\rho c_p T = C_h$$

where ρ is the mass density, c_p the specific heat at constant pressure, and T the temperature (absolute or relative to any reference value). Similarly ρu (u is the velocity) may be thought of as the concentration of momentum.

As defined (1.5), $C = C(x, y, z, t)$. But $C(x, y, z, t)$ will still include turbulent fluctuations and may not be convenient for our purposes. We can speak of several kinds of averages as follows:

Time average of C

$$\bar{C}_t(x, y, z, t_0) = \frac{1}{T} \int_{t_0}^{t_0 + T} C(x, y, z, t) \, dt. \tag{1.6}$$

This average is still a function of position, the averaging time T, and the initial time t_0. However, if the turbulence is of time scales much less than T, then the resulting $\bar{C}_t$ is a slowly varying function, reflecting only the change of flow rate and ambient water conditions (over hours), rather than short-duration turbulence-induced fluctuations (of the order of minutes or seconds).

Spatial average of C

$$\bar{C}_V(x_0, y_0, z_0, t) = \frac{1}{V} \iiint_{\Delta V} C(x, y, z, t)\, dV. \tag{1.7}$$

We generally consider the averaging volume to be relatively small, such as the size of a grab sample, with coordinates (x_0, y_0, z_0) indicating the location of the center of the volume V. This kind of average wipes out turbulent fluctuations occurring on scales smaller than $V^{1/3}$.

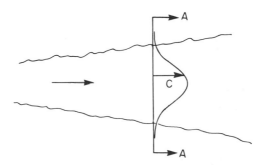

Figure 1.2   Definition sketch of cross-section of plume or wastewater field.

Flux average of C.   For a plume or wastewater field passing a given cross section AA (Fig. 1.2), a flux average $(\bar{C}_f)$ may be defined as

Flux of contaminant mass through AA $= \bar{C}_f \cdot$ (flux of water through AA):

$$\int_A Cu\, dA = \bar{C}_f \int_A u\, dA = \bar{C}_f Q \quad \text{or} \quad \bar{C}_f(t) = \int_A Cu\, dA / Q. \tag{1.8}$$

The area of integration A is the area occupied by the plume in the cross section, with the edges defined to occur at some small threshold level of C. Basically, the flux average of C is the value to be multiplied by the total flow rate to get total mass flux. The flux average may also be thought of as a spatial average where V is the volume of plume water that passes a given cross section in a unit of time.

Average for a group of samples.   If N bottle samples are taken, each with volume $V_i, i = 1, \ldots, N$, and measured C_i, then

$$\bar{C} = \sum_{1}^{N} V_i C_i \bigg/ \sum_{1}^{N} V_i = \text{total mass/total volume}. \qquad (1.9)$$

This is the same value that would be obtained if all samples were mixed together in a large container.

1.4.2   Dilution

Dilution usually is defined as

$$S = \frac{\text{total volume of a sample}}{\text{volume of effluent contained in the sample}}. \qquad (1.10)$$

The reciprocal of S is thus equal to the volume fraction of effluent in a sample, designated p,

$$p = 1/S = \text{volume fraction of effluent}$$
$$= \text{relative concentration}. \qquad (1.11)$$

The term *relative* concentration is used to indicate that $p = 1$ for undiluted effluent ($S = 1$) and $p = 0$ for pure ambient water ($S = \infty$). In between these limits, the mixture contains p parts of effluent and $(1 - p)$ parts of ambient water.

Alternatively, some writers and agencies use a dilution D defined as

$$D = \frac{1 - p}{p} = \frac{\text{volume of ambient water in the sample}}{\text{volume of effluent in the sample}}. \qquad (1.12)$$

By definition, also

$$D = S - 1. \qquad (1.13)$$

By rearranging (1.13) we find

$$p = 1/(D + 1). \qquad (1.14)$$

Because hydrodynamic models yield relative concentrations (p), it is more convenient to use S rather than D.

Furthermore, ecological effects are all related to concentration C of a particular contaminant X. Defining

$$C_s = \text{background concentration of substance } X \text{ in ambient water,}$$

and

$$C_d = \text{concentration of } X \text{ in the effluent discharge,}$$

it follows that

$$S = \frac{C_d - C_s}{C - C_s} \quad \text{or} \quad p = \frac{C - C_s}{C_d - C_s}. \tag{1.15}$$

$$C = C_s + \frac{1}{S}(C_d - C_s) = C_s + p(C_d - C_s). \tag{1.16}$$

In simple terms, the increment of concentration above background is reduced by the dilution factor S (or multiplied by p) from the point of discharge to the point of measurement of C. The ambient concentration is noted to be a simple linear function of p for any given discharge.

1.4.3 Average Dilution

Because the dilution is defined basically as the reciprocal of the relative concentration p, then the dilution of a composite sample is, according to Eq. (1.10),

$$\bar{S} = \frac{\text{total volume}}{\text{total effluent volume}} = \frac{1}{p}. \tag{1.17}$$

Since the total effluent volume in a composite of N samples is $\sum V_i p_i = \sum V_i(1/S_i)$, we find

$$\bar{S} = \sum_{1}^{N} V_i \bigg/ \sum_{1}^{N} V_i \frac{1}{S_i}. \tag{1.18}$$

Equation (1.18) shows that dilutions should be averaged *harmonically* in order to get the proper value to describe a composite. Simply put, first we average the relative concentrations, $p_i = 1/S_i$ to get $\bar{p}$, then find $\bar{S} = 1/\bar{p}$. To average p, use the same definitions and formulas as for C [Eqs. (1.6)–(1.9)]. It is not appropriate to average dilutions directly because $S \to \infty$ at the edge of a plume, and all arithmetic averaging procedures will give unreasonable or infinite results.

1.4.4 Density

The *weight density* (weight per unit volume) of water plays an important role in mixing in a water body. If ρ is the *mass* density, g is the gravitational acceleration, then ρg is the *weight* density (also called unit weight). The variations of practical interest are usually less than 3%, which is unimportant for *fluid acceleration* ($\rho + \Delta\rho \approx \rho$); but for buoyancy of discharges or stability of density-stratified lakes, estuaries, and oceans it is the weight or buoyancy *differences*

$(g \, \Delta\rho)$ which are controlling. The buoyancy per unit mass $(g \, \Delta\rho/\rho)$ may be considered as a modified gravitational acceleration

$$g' = g \, \Delta\rho/\rho. \tag{1.19}$$

A convenient way to represent water densities close to 1 g/cm³ is to use σ_t units defined by oceanographers as:

$$\rho = 1 + (\sigma_t/1000) \quad (g/cm^3) \qquad or \qquad \rho = 1000 + \sigma_t \quad (kg/m^3), \tag{1.20}$$

when the pressure is 1 atm. For example, if seawater has $\sigma_t = 26.35$ then $\rho = 1.02635$ g/cm³. For oceans and estuaries, σ_t is a function of salinity and temperature (for tables see Appendix A).

The density in a deep lake or ocean also increases noticeably with depth due to the increased hydrostatic pressure. This compression does not affect buoyancy or stability because all water masses moved up and down are similarly compressed (to first-order accuracy). Therefore, the convention has been adopted to reduce all densities to σ_t (at 1 atm pressure), and to neglect compressibility in the equations of motion. (This procedure is like using potential temperatures in the atmosphere.)

For analyses of estuaries or nearshore ocean waters, it is usually necessary to work to accuracies of at least $\pm 0.1 \sigma_t$ units; for mixing analyses in deep lakes and oceans, much higher accuracy (± 0.01 to $\pm 0.001 \sigma_t$ units) is often needed.

1.4.5 Density Stratification

Density stratification in the ambient fluid is described by the density profile $\rho_a(z)$, where z is the vertical coordinate. It is almost universally stable (in sharp contrast with the atmosphere), or $d\rho_a/dz < 0$, when z is positive in the upward sense. For expressing the unit weight gradient, which is the relevant quantity, we need to use $-g \, d\rho_a/dz$, where the minus sign is used for convenience to make the term positive.

Linear density stratification, which is an approximation which is useful for analysis, is defined by

$$-g \, d\rho_a/dz = constant \text{ (i.e., a straight line profile).} \tag{1.21}$$

1.4.6 Dynamically Active versus Passive Substances

Some substances being mixed in the water environment may significantly affect the water density and hence the flow patterns, while some do not. For example, the mixing of a heated water discharge may modify the density distribution so much that the ambient velocities are affected through the equations of motion. Another example is the distribution of salinity in an estuary.

These substances are called *dynamically active* because the strong coupling between the equations for mixing and for the flow dynamics necessitates solving the problem as a whole.

A *dynamically passive* substance is one which does not cause density changes sufficient to affect the flow dynamics. These are easier to analyze because the flow pattern may be separately determined and used as given input to the mixing equation.

1.4.7 Velocity Distribution in Turbulent Shear Flow

For turbulent shear flow in a long pipe or channel with a constant flow cross section ("uniform flow"), the boundary layer will be fully developed, and a flow pattern will be established which is independent of the distance along the pipe or channel. The slowest velocities are near the walls, and the fastest are the furthest from the walls. The shear stresses at the walls are counterbalanced by the driving forces—pressure gradients and gravity. If the time-average velocity is measured at various points in the cross section, a velocity contour map can be drawn for the cross section (see for example, Fig. 5.10). This is often called the velocity distribution, $u(y, z)$, where u is in the x direction and (y, z) are coordinates in the cross section. If the flow is nonuniform in the x direction, then $u = u(x, y, z)$.

The mean velocity in the cross section $\bar{u}$ is related to the *mean* wall shear stress τ_0 by the relation

$$\tau_0 = \tfrac{1}{8} f \rho \bar{u}^2, \tag{1.22}$$

where f is the familiar Darcy–Weisbach friction factor. For convenience, we define the *shear velocity* u^* to be

$$u^* = \sqrt{\tau_0/\rho} \tag{1.23}$$

which has units of velocity. Then Eq. (1.22) becomes

$$\bar{u}/u^* = \sqrt{8/f}. \tag{1.24}$$

For circular pipes, f may be estimated from the familiar pipe friction (or Moody) diagram,† giving f as a function of Reynolds number and relative roughness. For wide open channels and other nearly symmetric shapes, the Moody diagram may also be used when the pipe diameter D is replaced by $4r_h$ (r_h is the hydraulic radius = cross-sectional area/wetted perimeter).

The mean shear stress τ_0 may be found directly from the balance of forces as

$$\tau_0 = \rho g r_h S, \tag{1.25}$$

or

$$u^* = \sqrt{g r_h S}, \tag{1.26}$$

† See any standard fluid mechanics or hydraulics text.

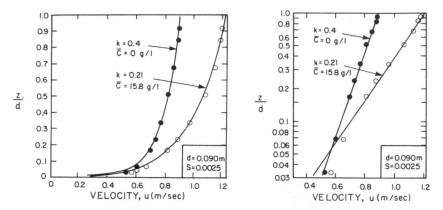

Figure 1.3 Linear and semilogarithmic graphs of velocity profiles in a flow 0.090 m deep and 0.85 m wide with clear water and with a heavy load of suspended 0.1 mm sand. The sand reduces the von Karmen constant k from 0.40 to 0.21. [From Vanoni and Brooks (1957).]

wherein S is the hydraulic gradient (slope of the energy grade line, or open channel slope for uniform flow).

In the special cases of circular pipes or wide two-dimensional channels, the velocity distribution is observed to be excellently approximated by a logarithmic function as follows†:

Pipe

$$u = \bar{u} + \frac{3}{2}\frac{u^*}{k} + \frac{2.30}{k} u^* \log_{10} \frac{z}{R} \qquad \left(u = \bar{u} \quad \text{at} \quad \frac{z}{R} = 0.223\right), \quad (1.27)$$

Wide Channel

$$u = \bar{u} + \frac{u^*}{k} + \frac{2.30}{k} u^* \log_{10} \frac{z}{d} \qquad \left(u = \bar{u} \quad \text{at} \quad \frac{z}{d} = 0.368\right), \quad (1.28)$$

where z is the distance from the wall, R the pipe radius, d the channel depth, and k the von Karman constant ≈ 0.4 for nonstratified flow and no suspended sediment. Figure 1.3 illustrates the logarithmic profile for a wide channel [eq. (1.28)], compared to a measured velocity profile in a flume experiment. From Eqs. (1.27) and (1.28) it may be noted that the variability in the velocity profile (i.e., slope on semilog graph) is proportional to u^* but is independent of $\bar{u}$. We can immediately predict then that longitudinal dispersion in a shear flow (Chapter 4) will be found to depend on the shear velocity, but not on the mean velocity. This is just one example of why the shear velocity is a very useful flow characteristic.

† For further discussion of logarithmic profiles, the law of the wall, and the relationships to the Moody diagram and Nikuradse's experiments, see Rouse, *Fluid Mechanics for Hydraulic Engineers,* or other basic textbooks.

1.5 DIMENSIONAL ANALYSIS

The theory of dimensional analysis is covered in many texts and courses, but little attention is given to the art of using it to maximum advantage. The use of dimensional analysis does not require that one forget everything in pursuit of blind formalisms. Rather when one skillfully blends his knowledge and experience in fluid mechanics with dimensional analysis, it becomes a very powerful tool for mixing problems.

1.5.1 Buckingham π-Theorem

The basic purpose of dimensional analysis is to find useful dimensionless groups of variables to describe a process and to provide the basis for similarity between physical models and prototypes. The Buckingham π-theorem tells us how many dimensionless groups are necessary for a given list of variables.

The theorem may be stated as follows:

Given:
   (1) Variables $q_1, q_2, \ldots, q_n$, each representing the magnitude of a physical quantity;
   (2) q_1 is the dependent variable and $q_2, q_3, \ldots, q_n$ are independent, i.e.,

$$q_1 = f(q_2, q_3, \ldots, q_n); \qquad (1.29)$$

   (3) They involve k different physical dimensions (length, time, mass, temperature, heat, ...).
Then:
   (1) There are $(n - k)$ dimensionless groups, $\pi_1, \pi_2, \ldots, \pi_{n-k}$, which can be independently formed from $q_1, q_2, \ldots, q_n$.
   (2) The functional relationship of the qs [Eq. (1.29)] reduces to $\pi_1 = F(\pi_2, \pi_3, \ldots, \pi_{n-k})$.
   (3) No single group need depend on more than $k + 1$ of the qs.

Most textbooks show the formal way in which to find the exponents (α_{ij}) to formulate the πs, i.e.,

$$\pi_i = \prod_{j=1}^{n} q_j^{\alpha_{ij}}, \qquad 1 \le i \le n - k$$

where α_{ij} is a matrix of exponents to be chosen to make all the units cancel out, so each π_i is dimensionless. For each i, $(n - k - 1)$ of the α_{ij}'s can be set equal to zero according to the π-theorem. With rare exception this type of formal approach is a waste of time. Instead some judgement and a little trial and error works nicely, as will be demonstrated here.

1.5.2 Suggestions for Using Dimensional Analysis

Some suggestions for this art are given in the following paragraphs. The experienced person soon comes to use dimensional arguments almost unconsciously.

(1)   It is all important to select a proper list of variables. Obviously a wrong list will give the wrong answer. Less obvious is the need to make the list as *short* as possible—rather than making the longest most complete list possible. Too long a list of qs means that the list of πs is also too long, thus losing the advantage of dimensional analysis.

(2)   Be sure to put only one dependent variable on the list, because only *one* functional relationship is being described at a time. If there is a problem with several dependent variables of interest (as in the example below), then apply the π-theorem in turn for each one.

(3)   Combine multiple variables into functionally important single variables. For example, in the case of a buoyant discharge (for which $\Delta\rho \ll \rho$), the list could include ρ, $\Delta\rho$, and g. But since $\Delta\rho$ and g are only relevant as they occur in the product $g\Delta\rho$, the buoyant force per unit volume, only the single variable $g\Delta\rho$ is needed in place of two, g and $\Delta\rho$.

(4)   Make tentative assumptions based on your knowledge of fluid mechanics, such as omitting viscosity for high Reynolds number flows. (This sometimes works, not always!)

(5)   In considering the number of physical dimensions (k), you may count mass of a particular trace contaminant carried in solution in addition to mass of the fluid itself. This is a useful trick to find dimensionless groups for dilution in jets and plumes (see example below).

(6)   In forming the dimensionless groups, look for familiar dimensionless numbers in fluid mechanics (Reynolds numbers, Froude numbers, Richardson numbers, friction factors, drag coefficients, discharge coefficients, length ratios, etc.). Remember there is no *unique* way to form the groups, so one might as well stick to the familiar dimensionless numbers for which we already have some experience and intuition. Be sure your groups are independent of each other (i.e., one should *not* be just a combination of two others). Use all the variables, unless one kind of unit appears in only one physical quantity. In that case you may strike that variable from the list and start again.

(7)   For convenience keep the dependent variable in only one group (q_1 only in π_1) so that the functional relationship in dimensionless form can be explicit rather than implicit if possible.

(8)   If you suspect that a variable might drop out when you examine experimental results, then be sure it appears only in one group. In that case, dropping one variable (like viscosity) clearly drops one group, and what is left is already the solution to the problem with one less variable. If you let an

unimportant variable show up in two groups you may not even notice that it could be neglected.

1.5.3 Example of the Application of Dimensional Analysis

As an example of the application of dimensional analysis, consider the process of sludge dumping in a linearly stratified ocean (gradient $= -d\rho_a/dz$), as shown in Fig. 1.4. A volume V_s at density ρ_s is released suddenly from the bottom of a barge into surface water of density ρ_0. Find the maximum depth of penetration (d_{max}), the minimum dilution at that depth, and the time of descent. Because the heavy fluid mixes initially with some of the lighter surface water as it sinks, its density approaches the ambient value, until a position of neutral buoyancy is reached. After a short "overshoot," the cloud will subsequently spread laterally without sinking further.

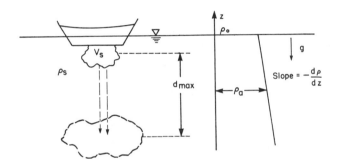

Figure 1.4 Schematic diagram of instantaneous sludge dump in a linearly stratified ocean.

Without much thought one would write the following list of variables:

		Units
(1)	d_{max}	L
(2)	V_s	L^3
(3)	g	LT^{-2}
(4)	ρ_0	ML^{-3}
(5)	ρ_s	ML^{-3}
(6)	$-d\rho_a/dz$	ML^{-4}

By the π-theorem, $n - k = 6 - 3 = 3$ dimensionless groups. But this result is not useful because the list is too long. Rather we should define $\Delta\rho = \rho_s - \rho_0$, and note that the submerged weight per unit volume is $g\Delta\rho$. The whole flow field is driven by the total submerged weight which is $V_s g\Delta\rho$. Similarly the gradient is only interesting as it affects submerged weight, so $-d\rho_a/dz$ is replaced by $-g\, d\rho_a/dz$. The new list is as follows:

		Units
(1)	d_{max}	L
(2)	$V_s g\Delta\rho$	$(L^4 T^{-2})(ML^{-3})$
(3)	$-g\, d\rho_a/dz$	$(T^{-2})(ML^{-3})$
(4)	ρ_0	ML^{-3}

Now we find only $n - k = 4 - 3 = 1$ dimensionless group, which is

$$\pi_1 = d_{max}^4 \left(-\frac{g}{\rho_0}\frac{d\rho_a}{dz} \right) \Big/ V_s g \frac{\Delta\rho}{\rho_0}.$$

This was easily obtained by inspection, by first dividing (2) and (3) by ρ_0 to cancel all the mass terms; then the quotient of (2) divided by (3) has units of L^4. Note that g and ρ_0 are left in, rather than canceled, to remind us of the dynamic nature of the problem; in some other problems g and ρ_0 would not happen to cancel out.

Another way of looking at this problem is to use the list obtained after mass terms are canceled:

			Units
(1)	d_{max}		L
(2)	$V_s g(\Delta\rho/\rho_0) = V_s g'$		$L^4 T^{-2}$
(3)	$-(g/\rho_0)(d\rho_a/dz) = g\varepsilon$	(defines ε)	T^{-2}

Since there is only one group, it must be a constant, so we can write

$$d_{max} = A(V_s g')^{1/4}/(g\varepsilon)^{1/4}. \qquad (1.30)$$

We have thus derived the basic dependence of d_{max} on the source and stratification parameters, with only the value of the constant A to be found by experiment. By experiments, the value of A has been found to be $A \approx 2.66$ (Morton et al., 1956). This sludge dumping problem is analogous to a buoyant "thermal"— the flow pattern due to an instantaneous release of heat in a stratified atmosphere.

The above list is certainly as short as possible, but we may have omitted some important effects, such as viscosity or the effect of the initial volume per se.

Again we revise the list, using kinematic viscosity v because we have already divided the other terms by ρ_0:

		Units
(1)	d_{max}	L
(2)	$V_s g'$	$L^4 T^{-2}$
(3)	$g\varepsilon$	T^{-2}
(4)	v	$L^2 T^{-1}$
(5)	V_s	L^3

Now $n - k = 5 - 2 = 3$ groups. The strategy is to make one of the two additional groups be a Reynolds number, and the other one include V_s, as follows:

$$\pi_1 = d_{max}^4 (g\varepsilon)/(V_s g') \qquad \text{(same as before)}$$

$$\pi_2 = \sqrt{V_s g'}/v = \text{Re} \qquad \text{(Reynolds number)}$$

$$\pi_3 = V_s/d_{max}^3$$

Now we would expect Re to drop out when it becomes large (fully developed turbulence in the sinking cloud), and π_3 to disappear when $\pi_3 = V_s/d_{max}^3 \ll 1$. If π_3 is of the order of one, then the whole problem is different and we must retain π_3, as a variable parameter because the dimension of the cloud $V_s^{1/3}$ is comparable to the sinking depth d_{max}, which is not the way it is shown in Fig. 1.4.

The use of d_{max} in π_3 may be awkward because d_{max} is the dependent variable. We can modify π_3 by multiplying it by $\pi_1^{3/4}$, to get a revised group π_3':

$$\pi_3' = \pi_3 \cdot \pi_1^{3/4} = V_s(g\varepsilon)^{3/4}/(V_s g')^{3/4} \sim V_s^{1/4}.$$

Finally we write

$$d_{max} = A[(V_s g')^{1/4}/(g\varepsilon)^{1/4}]f(\text{Re}, \pi_3'). \qquad (1.31)$$

This expression is arranged so that asymptotically the function f approaches unity for $\text{Re} \to \infty$ and $\pi_3' \to 0$. By experiments we could determine how large Re and how small π_3' need to be for $f \approx 1$.

As an illustration, consider the following numerical values:

$$V_s = 10^3 \quad m^3,$$

$$\rho_s = 1.045 \text{ g/mliter},$$

$$\rho_0 = 1.027 \text{ g/mliter}$$

$$g = 9.8 \text{ m/sec},$$

$$\varepsilon = -\frac{1}{\rho_0}\frac{d\rho_a}{dz} = 2 \times 10^{-5} = 2\sigma_t \quad \text{units per 100 m}.$$

The basic quantities are

$$g' = g\frac{\Delta \rho}{\rho_0} = 9.8\frac{0.018}{1.027} = 0.172 \text{ m/sec}^2,$$

$$V_s g' = 172 \text{ m}^4/\text{sec}^2,$$

$$g\varepsilon = 9.8 (2 \times 10^{-5}) = 1.96 \times 10^{-4} \text{ sec}^{-2}.$$

By Eq. (1.30)

$$d_{\max} = 2.66\left(\frac{172}{1.96 \times 10^{-4}}\right)^{1/4} = 81 \text{ m}.$$

The minimum dilution in the mixed cloud after descent to its equilibrium levels can be derived from the maximum concentration of sludge particles (dry weight/volume), assuming that the particle settling in the suspension is negligible during the time of fall. With the initial sludge mass being $C_s V_s$, the list of variables, by reasoning similar to the above, is as follows:

		Units
(1)	$C_{\max}$	M/L^3
(2)	$C_s V_s$	M
(3)	$V_s g'$	L^4T^{-2}
(4)	$g\varepsilon$	T^{-2}

We keep the same independent variables (3) and (4) as before. Note that the use of M for sludge mass is different from the water mass used previously [now suppressed by using the kinematic form of variables (3) and (4)]. Also the product $V_s C_s$ is used as *one* quantity rather than two, on the presumption that it is the total mass only that counts. (Also the reader should be cautioned that $C_s \neq \Delta\rho$, because the mass of sludge particles is not simply additive to the water mass inasmuch as the particles displace water.)

With this list, $n - k = 4 - 3 = 1$, and the single group is easily determined by inspection to be

$$\pi_1 = \frac{C_{\max}(V_s g')^{3/4}}{V_s C_s (g\varepsilon)^{3/4}}.$$

Rearranging, setting $\pi_1 = \text{constant} = B$, and using the definition of dilution

$$\frac{C_s}{C_{\max}} = S_{\min} = \frac{B}{V_s}\left(\frac{V_s g'}{g\varepsilon}\right)^{3/4}, \qquad S_{\min} = \frac{Bg'^{3/4}}{V_s^{1/4}(g\varepsilon)^{3/4}}.$$

Or combining with Eq. (1.30)

$$S_{\min} = \frac{B}{A^3}\frac{d_{\max}^3}{V_s}.$$

In other words the dilution increases as d_{max}^3. The reader may prove that the volume of the cloud after it stops sinking is also proportional to d_{max}^3. The value of B must be determined by experiment, but once found, it should apply to all cases of this flow situation within the limits of the Reynolds number and the size of the cloud described above.

Finally, the time t_0 it takes to sink must be proportional to $(g\varepsilon)^{-1/2}$ and independent of the size of the initial dump $(V_s g')$. This is deduced from the list of variables:

		Units
(1)	t_0	T
(2)	$V_s g'$	$L^4 T^{-2}$
(3)	$g\varepsilon$	T^{-2}

Here $n - k = 3 - 2 = 1$ group. But since L occurs in only one term $(V_s g')$, that term must be dropped, and $\pi_1 = t_0(g\varepsilon)^{1/2}$. The result is intuitively reasonable because a heavier cloud not only sinks faster [velocity $\propto (V_s g')^{1/4}$], but also further [d_{max} also $\propto (V_s g')^{1/4}$], so the time it takes is independent of $V_s g'$. This example has shown the power and usefulness of dimensional analysis for mixing problems. The reader is urged to keep thinking in terms of dimensionless groups throughout the book and to note that many results, except for the dimensionless coefficients or functions, can be quickly deduced by dimensional analysis.

Chapter 2

Fickian Diffusion

In this chapter we present some of the equations and concepts underlying molecular diffusion processes, first in a fluid at rest and then in a moving fluid. Molecular diffusion by itself is not of great direct consequence in environmental problems, except on the microscopic scale of chemical and biological reactions, but in many cases environmental dispersion problems can be described by processes that are strongly analogous to molecular diffusion but on a grander scale. We will therefore present two different rationalizations for the molecular diffusion equation and discuss its more important solutions.

2.1 FICK'S LAW OF DIFFUSION

As in many physical processes, observation leads to an empirical description followed by a physical argument for its validity. Fourier's law of heat flow (1822) is a classic example.† It was used as the basis for the development of a large body of theory governing the flow of heat long before the actual physics of the heat flow process was understood. Similarly for diffusion, Adolph Fick, a German physiologist, published a paper in 1855 entitled "Über Diffusion" in

† The time rate of flow of heat per unit area in a given direction is proportional to the temperature gradient in that direction.

which he described how Fourier's heat flow led to a hypothesis to describe the molecular diffusion process. He restated the idea in English (1855):

> It is quite natural to suppose that this law for the diffusion of salt in its solvent must be identical with that according to which the diffusion of heat in a conducting body takes place; upon this law Fourier founded his celebrated theory of heat, and it is the same which Ohm applied with much extraordinary success to the diffusion of electricity in a conductor.

Restated, Fick's law says that the flux of solute mass, that is, the mass of a solute crossing a unit area per unit time in a given direction, is proportional to the gradient of solute concentration in that direction.

For a one-dimensional diffusion process, Fick's law can be stated mathematically as

$$q = -D\, \partial C/\partial x, \tag{2.1}$$

where q is the solute mass flux, C the mass concentration of diffusing solute, D the coefficient of proportionality, and the minus sign indicates transport is from high to low concentrations. D has dimensions of $(\text{length})^2/\text{time}$ and is called the diffusion coefficient, or molecular diffusivity. For diffusion in three dimensions Fick's law can be written in vector notation as

$$\mathbf{q} = -D\, \nabla C, \tag{2.2}$$

where $\mathbf{q}$ is the mass flux vector with components (q_x, q_y, q_z) in a Cartesian coordinate system.

Fick's law is a statement relating the mass flux to the concentration gradient. We will now show how conservation of mass leads to a second relationship which is true irrespective of the type of transport process. Combination of these two results leads to a partial differential equation that used to describe diffusion processes.

Figure 2.1 illustrates a one-dimensional transport process in which mass is being transferred in the x direction. Two parallel surfaces of unit area are drawn perpendicular to the x axis and separated by a distance Δx. Let $C(x, t)$ be the mass

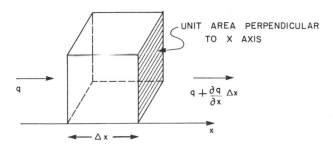

Figure 2.1   The control volume used to derive Eq. (2.3).

per unit volume at the point x at time t. Then there is a mass $C(x, t)\Delta x$ in the line segment bounded by the parallel planes. Since molecules are passing in and out of the "volume" defined by each bounding surface, there is a time rate of change of mass in the volume equal to

$$(\partial C/\partial t)\Delta x.$$

This time rate of change must be equal to the difference in the flux, or rate of passage of molecules, through each surface. Suppose the mass rate of flow across the unit surface located at x is $q(x, t)$ then the mass rate of flow per unit area across the surface at $x + \Delta x$ is simply

$$q(x, t) + \partial q(x, t)/\partial x \, \Delta x,$$

and the difference between the two is $\partial q/\partial x \, \Delta x$. This difference must be equal to the rate of change of mass in the volume in order to satisfy conservation of mass, and equating the two gives

$$(\partial q/\partial x) + (\partial C/\partial t) = 0. \tag{2.3}$$

We have therefore deduced a relationship between the flux $q(x, t)$ and concentration $C(x, t)$ that is true regardless of the mechanism of molecule transport.

However, for molecular diffusion processes we also have Fick's law, Eq. (2.1), which can be substituted into Eq. (2.3) to give

$$(\partial C/\partial t) = D(\partial^2 C/\partial x^2). \tag{2.4}$$

Alternately, differentiation of Eq. (2.3) with respect to x and substitution of $-q/D$ for $\partial C/\partial x$ gives

$$(\partial q/\partial t) = D(\partial^2 q/\partial x^2). \tag{2.5}$$

Equations (2.4) and (2.5) are known as diffusion equations and describe how mass is transferred by Fickian diffusion processes. A large number of problem solutions are available for these equations, primarily because if $q(x, t)$ is regarded as heat flux and $C(x, t)$ as heat concentration, i.e., temperature, Eq. (2.3) becomes the heat equation. There is a direct and complete analogy between heat flow and molecular diffusion and, so far as the mathematical description is concerned, the processes are identical. Two classic texts addressing the mathematics of the solution of the diffusion or heat equation are Carslaw and Jaeger (1959) and Crank (1956), and the solution of most common diffusion problems can be found in either volume.

The previous results can be extended to more than one dimension, most succinctly by using vector notation.† Consider a fixed volume V with surface

---

† We will sometimes use vector notation for completeness or simplicity. The reader unfamiliar with vector calculus may skip these parts, as they are not essential to the understanding of the material.

area S. The concentration of tracer mass is now a function of position $\mathbf{x}$ and time t, so that the total mass in the volume is

$$\int_V C(\mathbf{x}, t) \, dV.$$

If the mass flux is $\mathbf{q}(\mathbf{x}, t)$ then conservation of mass requires that

$$\frac{\partial}{\partial t} \int_V C(\mathbf{x}, t) \, dV + \int_S (\mathbf{q}(\mathbf{x}, t) \cdot \mathbf{n}) \, dS = 0,$$

where $\mathbf{n}$ is the unit vector normal to surface element dS. Using Green's theorem, and noting that V is a fixed volume, we have that

$$\int_V \left(\frac{\partial C}{\partial t} + \nabla \cdot \mathbf{q} \right) dV = 0. \tag{2.6}$$

Since the volume V is arbitrary

$$\partial C / \partial t = -\nabla \cdot \mathbf{q}. \tag{2.7}$$

For molecular processes the flux is specified by Fick's law, Eq. (2.2), so that Eq. (2.7) becomes the diffusion equation

$$\partial C / \partial t = D \nabla^2 C \tag{2.8}$$

or written out fully in Cartesian coordinates,

$$\frac{\partial C}{\partial t} = D \left(\frac{\partial^2 C}{\partial x^2} + \frac{\partial^2 C}{\partial y^2} + \frac{\partial^2 C}{\partial z^2} \right). \tag{2.9}$$

Equation (2.9) describes the spreading of mass in a fluid with no mean velocity. Diffusion in a moving fluid is discussed in Section 2.4.

The most fundamental solution to Eq. (2.4) is that which describes the spreading, by diffusion, of an initial slug of mass M introduced at time zero at the x origin. Since the equation is linear, this solution may be used as a building block to construct solutions to problems with more complex initial or boundary conditions.

The fundamental solution can be obtained by any of several mathematical techniques; the one we prefer is an application of dimensional analysis. The concentration $C(x, t)$ can only be a function of M, x, t, and D. Since the process is linear, C must be proportional to the mass introduced M. In one dimension the units of concentration are mass per unit length; therefore C must be proportional to M divided by some characteristic length.† The units of the diffusion

---

† Alternatively, we may think of C as having units of mass per unit volume as long as M is interpreted as having units of mass per unit area in the y–z plane.

coefficient D are (length)^2/(time) so $\sqrt{Dt}$ is a suitable characteristic length. Thus dimensional analysis gives us a relationship,

$$C = \frac{M}{\sqrt{4\pi Dt}} f\left(\frac{x}{\sqrt{4Dt}}\right). \tag{2.10}$$

The factors 4π and 4 have been added for convenience arbitrarily since we will find that they appear in the final solution. Dimensional analysis only tells us the form of the relationship.

We can now transform Eq. (2.4) into an ordinary differential equation by defining $\eta = x/\sqrt{4Dt}$ and substituting Eq. (2.10) into Eq. (2.4) to obtain:

$$\frac{df}{d\eta} + 2\eta f = 0 \tag{2.11}$$

which has the solution

$$f = C_0 e^{-\eta^2}. \tag{2.12}$$

The total mass contained in the system can be found by integrating the concentration along the whole x axis, and since we postulate that the total quantity of mass is constant at all times we have for any time

$$\int_{-\infty}^{\infty} C \, dx = M. \tag{2.13}$$

Now C_0 in Eq. (2.12) can be found by substituting Eq. (2.12) and (2.10) into (2.13) and performing the integration; the result is that $C_0 = 1$ for all times, and therefore

$$C(x, t) = (M/\sqrt{4\pi Dt}) \exp -(x^2/4Dt) \tag{2.14}$$

is the required fundamental solution.

If preferred, the reader may now skip directly to Section 2.3, which explores the properties of the fundamental solution and how it is used to construct a variety of solutions to the diffusion equation. Section 2.2 presents two rationalizations for the diffusion equation and Fick's law based on the idea that the transport is a consequence of the random motion of individual fluid particles. These descriptions are not offered as proofs of Fick's law, but we think they will aid in understanding the basic nature of diffusion processes. The reader wishing to study in more detail the statistical analysis of random motion may also wish to see the fundamental paper by Einstein (1927) on Brownian motion, and the papers by Uhlenbeck and Ornstein (1930) and Chandrasekhar (1943).

2.2   THE RANDOM WALK AND MOLECULAR DIFFUSION

The molecules of a fluid are constantly in motion and forever colliding with one another and with any small particles held in suspension in the fluid. Any small particle†, or molecule, experiences a number of collisions per second, numbering in the billions of billions (there are $\sim 10^{26}$ molecules in a cubic centimeter of water). The collision rate depends on the fluid, the size of the particles, and the density and temperature of the fluid. As a consequence of being hit so frequently a particle very quickly loses the memory of its previous velocity and its motion describes a random path, much like that of a single billiard ball surrounded by thousands of other constantly stimulated billiard balls on a huge billiard table. It is apparent that the description of the motion of such a particle must be a statistical one, and further, if we are seeking to describe the motion of a large number of such particles, this must also be done statistically.

We can approach the problem in two ways: either by studying the statistics of motion of a single molecule (or particle) and generalizing, or by studying the integrated effect of random motion of a large number of particles simultaneously. It is educational to study both methods, as each has its analogues in models used to describe dispersal of tracer materials in the environment.

2.2.1   The Random Walk

Suppose that the motion of a tracer molecule or particle consists of a series of random steps. Although in reality the motion is in three dimensions, nothing important is lost if for the moment we assume that the motion is one dimensional. Assume further that each step is of equal length Δx and takes an interval of time Δt, but that whether the step is forward or backward is entirely random with equal probability. We ask how far the particle is likely to get; on the average the answer is nowhere, but after a number of steps sometimes the particle will have moved forward and sometimes backward. By means of the central limit theorem (see books on probability theory such as Feller (1950, Vol. 1, p. 323)) it can be shown that in the limit of many steps the probability of the particle being between $m\Delta x$ and $(m + 1)\Delta x$ approaches the normal distribution with zero mean and a variance $\sigma^2 = t(\Delta x)^2/\Delta t$, provided the quotient $(\Delta x)^2/\Delta t$ approaches a constant value as $\Delta t \to 0$. Designating this quotient by $2D$ then the probability that the particle is between the point x and $x + dx$ is given by

$$p(x, t)\, dx = (1/\sigma\sqrt{2\pi}) \exp(-x^2/2\sigma^2)\, dx = (1/\sqrt{4\pi Dt}) \exp(-x^2/4Dt)\, dx.$$
(2.15)

---

† By particle we mean any sufficiently small, but identifiable, neutrally buoyant entity whose size is such that its dynamical behavior is essentially indistinguishable from that of the fluid. The word is used interchangeably with molecule in this context.

Now if a whole group of particles begins its wandering at the origin at time zero, the concentration at station x at any later time t will be proportional to the likelihood of any one being in the neighborhood of the station, or in other words

$$C(x, t) = (M/\sqrt{4\pi Dt}) \exp(-x^2/4Dt).$$

Comparing the result with Eq. (2.14) we see that the constant D takes the role of the diffusion coefficient, and that the random walk process postulated here leads to the same result as postulating that an initial slug of tracer diffuses according to the diffusion equation.

2.2.2 The Gradient-Flux Relationship

Now let us rederive the results just given, but by taking an integrated view of the result of the random motion of a large number of particles at the same time. Suppose that in a host medium we have a dilute solution of molecules with some distribution of concentration $C(x, t)$, and with a gradient of concentration along a line as sketched in Fig. 2.2a. Of course, the "line" has to have some thickness to include some molecules; what we really mean is that in the second and third dimensions the concentration is constant. We take it as given that each of the molecules is executing a random motion, just as in the previous derivation. If we consider again any surface perpendicular to our line, the probability of a molecule passing through the surface should be proportional to the average number of molecules near it. Furthermore, the probability of a molecule passing from right to left should be proportional to the average number of molecules on the right, and the probability of a particle passing from left to right should be proportional to the average number of molecules on the left. Figure 2.2b shows an example. Suppose we consider one realization of an experiment in which we start with 10 molecules on the left and 20 on the right. If the probability of any particle crossing the line during time Δt is 0.2, then at the end of Δt we expect there to be, on the average, 8 of the original molecules on the left, plus 4 new ones from the right, a total 12 on the left, and similarly a total of 18 molecules on the right. To keep things simple we have neglected any transfer through the other surfaces. Note that if concentration is defined to be the average number of molecules per box, the difference in concentration between the boxes has changed from ten to six during the time step. This is the most fundamental aspect of diffusion; differences in mean concentration are, on the average, always reduced, never increased.

We can now define the flux of material across the bounding surface to be the net rate at which tracer mass is exchanged per unit time and per unit area of the surface. For simplicity take a unit area perpendicular to the line. The flux of material from left to right is equal to the number of particles in the box on the left, times the mass of each (all assumed equal), times the probability of transfer.

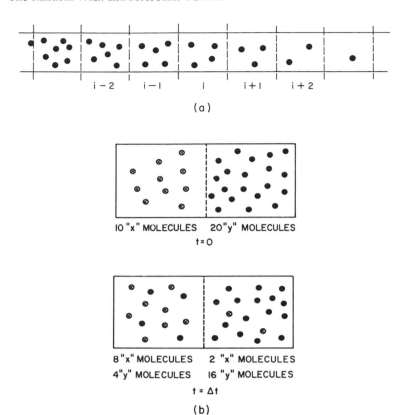

Figure 2.2 (a) A longitudinal concentration gradient. (b) An illustration of molecular diffusion. During the time interval $\triangle t$ the probability is that two-tenths of the molecules in each box will cross into the adjacent box.

If we let M_1 be the mass of tracer in the left-hand box, the flux from left to right is kM_1, where k is the transfer probability. Similarly, if M_r is the mass of tracer in the right-hand box the flux from right to left is kM_r. The net flux is

$$q = k(M_1 - M_r). \tag{2.16}$$

If we now define $C_1 = \overline{M}_1/\Delta x$ and $C_r = \overline{M}_r/\Delta x$, where $\overline{M}_1$ and $\overline{M}_r$ are the average masses in the boxes on the left and right after many repetitions of the experiment, and we note that as Δx becomes small we can write $\partial C/\partial x = (C_r - C_1)/\Delta x$ we have

$$q = -k(\Delta x)^2 \, \partial C/\partial x. \tag{2.17}$$

Equation (2.17) shows that the net transport is always down the gradient, that is from an area of high concentration to one of lower concentration.

The transfer probability is a function of the molecular motion, and also a function of the size of the box since the larger the box the fewer molecules that are close to the boundary. The mass-transfer rate should not of course, depend on an arbitrarily defined box size. The only way to avoid having q depend on Δx is for $k(\Delta x)^2$ to be a constant, the diffusion coefficient. Thus we have obtained Fick's law as stated in Section 2.1.

We will see later that although this derivation was based on molecules in random motion we could equally well apply the same argument to larger elements of fluid in random turbulent motion, thereby leading to a rationalization of the concept of Fick's law for turbulent diffusion.

2.3   SOME MATHEMATICS OF THE DIFFUSION EQUATION

In Section 2.1 we derived the fundamental solution to the diffusion equation, Eq. (2.14), which is the solution given an initial slug of mass M introduced at time zero at the x origin, if there are no boundaries to prevent the mass from

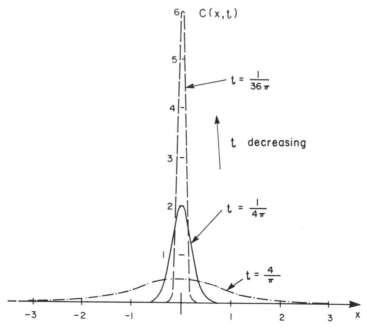

Figure 2.3   The reduction of the Gaussian distribution [Eq. (2.14)] to a "spike" as t decreases. The illustration uses the values $M = 1$, $D = \frac{1}{4}$.

diffusing to infinity in both directions. Mathematically this initial condition can be written as

$$C(x, 0) = M\delta(x), \tag{2.18}$$

where $\delta(x)$ is known as the Dirac delta function. Physically a delta function is a way of representing a unit mass of tracer concentrated into an infinitely small space with an infinitely large concentration, and $M\delta(x)$ represents a mass M concentrated into a very small space. For example, if a bucket of concentrated dye is dumped into a large river we can represent the initial concentration distribution by a delta function. The delta function can be thought of as a "spike" distribution; the way that the fundamental solution coalesces to a "spike" for very small time is illustrated in Fig. 2.3.

Equation (2.14) is known as the Gaussian distribution; if the mass is unity it is also known as the normal distribution. In this section we discuss some of the properties of the Gaussian distribution and give some of the most commonly used solutions of the diffusion equation. A much more comprehensive discussion of solutions and solution techniques is available in the book by Carslaw and Jaeger (1959).

2.3.1 Some Properties of Concentration Distributions

In what follows we will often be concerned with various moments of a concentration distribution, which are defined as follows:

$$\text{zeroth moment} = M_0 = \int_{-\infty}^{\infty} C(x, t) \, dx,$$

$$\text{first moment} = M_1 = \int_{-\infty}^{\infty} xC(x, t) \, dx,$$

$$\text{second moment} = M_2 = \int_{-\infty}^{\infty} x^2 C(x, t) \, dx,$$

$$\vdots$$

$$p\text{th moment} = M_p = \int_{-\infty}^{\infty} x^p C(x, t) \, dx,$$

in general the moments are functions of time, although as we have just seen the zeroth moment is simply the mass and is constant. The mean μ and the variance σ^2 of a distribution are found from the moments by the equations

$$\mu = M_1/M_0, \tag{2.19}$$

$$\sigma^2 = \int_{-\infty}^{\infty} (x - \mu)^2 C(x, t) \, dx/M_0 = (M_2/M_0) - \mu^2. \tag{2.20}$$

The various quantities for the normal distribution can be found by carrying out the indicated integrations. The results are

$$M_0 = 1, \qquad \mu = 0, \qquad \sigma^2 = 2Dt, \qquad (2.21)$$

where μ is the location of the centroid of the concentration distribution. Equation (2.21) states that the mean of a normal distribution is independent of time. The variance σ^2 is a measure of the spread of the distribution, and the third moment is a measure of the skew; it is easy to show that the third moment (and all odd moments) of the normal distribution are zero, and that all even moments can be written in terms of the second moment.

Often the standard deviation σ (the square root of the variance) is used as a measure of the spread. As can be seen from Table 2.1, for the Gaussian distribution a spread of 4σ includes approximately 95% of the total mass, or area under the concentration distribution. Figure 2.4 shows the relationship between

TABLE 2.1

Values of the Error Function and the Integral
of the Normal Distribution C^a

$\dfrac{x}{\sigma}$	$\operatorname{erf}\dfrac{x}{\sigma}$	$\dfrac{1}{\sigma\sqrt{2\pi}}\displaystyle\int_0^{x/\sigma}\exp(-x^2/2\sigma^2)\,dx$
0.0	0.0	0.0
0.1	0.1129	0.0398
0.2	0.2227	0.0793
0.3	0.3286	0.1197
0.4	0.4284	0.1554
0.5	0.5205	0.1915
0.6	0.6309	0.2257
0.7	0.6778	0.2580
0.8	0.7421	0.2881
0.9	0.7969	0.3159
1.0	0.8427	0.3413
1.2	0.9103	0.3849
1.4	0.9523	0.4192
1.6	0.9763	0.4452
1.8	0.9891	0.4641
2.0	0.9953	0.4773
2.5	0.9996	0.4938
3.0	0.99998	0.4987
4.0		0.49996
∞	1.0000	0.5000

$$^a\ C = \frac{1}{\sqrt{4\pi Dt}}\exp(-x^2/4Dt) = \frac{1}{\sigma\sqrt{2\pi}}\exp(-x^2/2\sigma^2),$$

where $\sigma = \sqrt{2Dt}$.

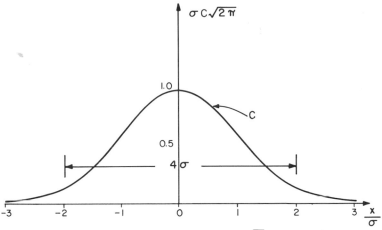

Figure 2.4 The normal distribution $C = (\sigma\sqrt{2\pi})^{-1}\exp(-x^2/2\sigma^2)$.

4σ and the normal distribution. In many practical problems a simple and adequate estimate of the width of a dispersing cloud is 4σ or $4\sqrt{2Dt}$.

An additional property often used to compute values of the diffusion coefficient in practical studies is that

$$d\sigma^2/dt = 2D. \tag{2.22}$$

For the normal distribution this is obvious, but it is also true for any concentration distribution, provided that it is dispersing in accord with the diffusion equation in a one-dimensional system of infinite extent, and that the concentration is always zero at $x = \pm\infty$.† To see this begin with the diffusion equation [Eq. (2.4)], multiply each side by x^2, and integrate from $x = -\infty$ to $x = +\infty$ to obtain

$$\int_{-\infty}^{\infty} \frac{\partial C}{\partial t} x^2 \, dx = \int_{-\infty}^{\infty} Dx^2 \frac{\partial^2 C}{\partial x^2} \, dx. \tag{2.23}$$

On the left-hand side the time derivative can be taken outside the integral, while the right-hand side can be integrated by parts to give

$$\frac{\partial}{\partial t} \int_{-\infty}^{\infty} Cx^2 \, dx = 2D \int_{-\infty}^{\infty} C \, dx. \tag{2.24}$$

Similarly it can be shown that $(\partial/\partial t)\int Cx\,dx = 0$, so the position of the mean is constant and can be taken without loss of generality to be at $x = 0$. Recalling the definition of the variance [Eq. (2.20)], we have the desired result

$$2D = \frac{\partial}{\partial t} \int_{-\infty}^{\infty} Cx^2 \, dx \bigg/ \int_{-\infty}^{\infty} C \, dx, \qquad = \frac{\partial}{\partial t} \sigma^2. \tag{2.25}$$

† More precisely, it is necessary that $\lim_{x \to \pm\infty} [x^2 \, \partial C/\partial X] = 0$.

Equation (2.25) states that the variance of a finite distribution increases at the rate $2D$ *no matter what its shape*. It is also a property of the diffusion equation that any finite initial distribution eventually decays into a Gaussian distribution; we will not demonstrate that here. A consequence of Eq. (2.25), which we will use later, is that if the variance $\sigma_1{}^2$ of a tracer distribution is known at time t_1 and D is constant, then the variance at any later time t_2 is given by

$$\sigma_2{}^2 = \sigma_1{}^2 + 2D(t_2 - t_1). \tag{2.26}$$

2.3.2 Solutions of the Diffusion Equation for Various Initial and Boundary Conditions

2.3.2.1 An Initial Spatial Distribution $C(x, 0)$

Consider first the solution corresponding to a mass M released at time $t = 0$ at the point $x = \xi$. The initial condition is

$$C(x, 0) = M\delta(x - \xi), \tag{2.27}$$

and the solution corresponding to an infinite domain, for which $C(\pm\infty, t) = 0$, is

$$C(x, t) = \frac{M}{\sqrt{4\pi Dt}} \exp\left[\frac{-(x - \xi)^2}{4Dt}\right], \tag{2.28}$$

as can be seen from the previous discussion. Suppose now that the initial conditions were

$$C(x, 0) = f(x), \qquad -\infty < x < \infty, \tag{2.29}$$

where $f(x)$ is some arbitrary function. Then we can imagine that this initial distribution is composed from a distributed series of separate slugs, which all diffuse independently because of the fundamental premise that the motion of individual particles is independent of the concentration of other particles. Figure 2.5 shows how a distribution is idealized by a set of slugs, each distributed over a distance $d\xi$ and with mass specified by the local value of $f(\xi)$. Each slug contains a mass $M = f(\xi)\,d\xi$. The concentration at point x and time t resulting from the spike centered at ξ and of width $d\xi$ and height $f(\xi)$ is therefore

$$\frac{f(\xi)\,d\xi}{\sqrt{4\pi Dt}} \exp\left[\frac{-(x - \xi)^2}{4Dt}\right].$$

The total contribution at x and t from all such slugs is simply the integral sum of all the individual contributions, which is

$$C(x, t) = \int_{-\infty}^{\infty} \frac{f(\xi)}{\sqrt{4\pi Dt}} \exp\left[\frac{-(x - \xi)^2}{4Dt}\right] d\xi. \tag{2.30}$$

This is known as a superposition integral.

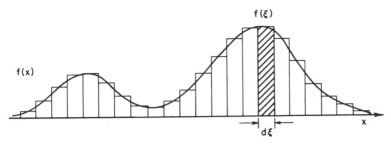

Figure 2.5 Approximation of $f(x)$ by a series of slugs each containing mass $f(\xi)\,d\xi$.

A particular example of a superposition integral is the case where $f(x)$ is given by a step function, viz.,

$$C(x, 0) = \begin{cases} 0, & x < 0, \\ C_0, & x > 0. \end{cases} \qquad (2.31)$$

as illustrated in Fig. 2.6. In this case

$$C(x, t) = \int_0^\infty \frac{C_0}{\sqrt{4\pi Dt}} \exp\left[\frac{-(x - \xi)^2}{4Dt}\right] d\xi, \qquad (2.32)$$

which can be transformed by setting $u = (x - \xi)/\sqrt{4Dt}$ to

$$
\begin{aligned}
C(x, t) &= \frac{C_0}{\sqrt{\pi}} \int_{-\infty}^{x/(4Dt)^{1/2}} e^{-u^2}\, du \\
&= \frac{C_0}{\sqrt{\pi}} \left[\frac{\sqrt{\pi}}{2} + \int_0^{x/(4Dt)^{1/2}} e^{-u^2}\, du\right] \\
&= \frac{C_0}{2}\left[1 + \mathrm{erf}\left(\frac{x}{\sqrt{4Dt}}\right)\right],
\end{aligned}
\qquad (2.33)
$$

where erf means the "error function," which is defined as

$$\mathrm{erf}\, z = \frac{2}{\sqrt{\pi}} \int_0^z \exp(-\xi^2)\, d\xi. \qquad (2.34)$$

The solution is depicted graphically in Fig. 2.6 and erf z is tabulated in Table 2.1.

2.3.2.2 *Concentration Specified as a Function of Time, $C(0, t)$*

The next problem we wish to solve is when the concentration is specified as a function of time at some fixed point. If the domain is infinite we can take the point to be $x = 0$ without loss of generality. As a first step, suppose that at initial time $t = 0$ the concentration is zero everywhere along the x axis. The

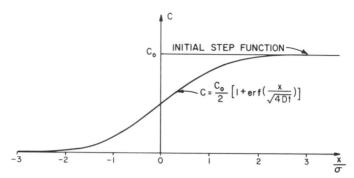

Figure 2.6 Spread of a step distribution as given by Eq. (2.33).

concentration is suddenly raised to C_0 at the point $x = 0$ and is held at that value. Find the concentration distribution $C(x, t)$.

The solution is most easily obtained by a dimensional analysis. The concentration at any point must depend on C_0 and on the physical descriptors of the problem, x, D, and t. The only dimensionally correct statement of this relationship is

$$C = C_0 f(x/\sqrt{Dt}), \tag{2.35}$$

where f indicates an as yet undetermined relationship. Now set $\eta = x/\sqrt{Dt}$ and note that

$$\frac{dC}{\partial t} = \frac{dC}{d\eta}\frac{\partial \eta}{\partial t} = -\frac{1}{2t}\eta\frac{dC}{d\eta}$$

and

$$\frac{\partial^2 C}{\partial x^2} = \frac{1}{tD}\frac{d^2 C}{d\eta^2}.$$

When these relationships are substituted into the diffusion equation (2.4) we obtain an ordinary differential equation,

$$-\tfrac{1}{2}\eta \, df/d\eta = d^2 f/d\eta^2 \tag{2.36}$$

with boundary conditions $f(0) = 1$ and $f(\infty) = 0$. Since $C(-x, t) = C(x, t)$, the solution can be found only along the positive x axis and is

$$C = C_0\left(1 - \operatorname{erf}\left(\frac{x}{\sqrt{4Dt}}\right)\right)$$

$$= C_0 \operatorname{erfc}\left(\frac{x}{\sqrt{4Dt}}\right) \qquad (x > 0). \tag{2.37}$$

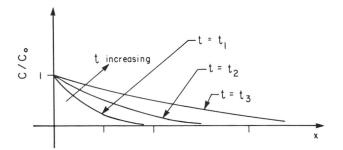

Figure 2.7   The concentration distribution given by Eq. (2.37).

"Erfc" stands for the "complimentary error function," which is defined by

$$\text{erfc}(Z) = 1 - \text{erf}(Z). \tag{2.38}$$

The solution is sketched in Fig. 2.7, which shows an advancing front with the same error function shape seen in Fig. 2.6. Note that the distance to a point having any particular value of C/C_0 increases as $\sqrt{Dt}$.

Now consider the same problem as before, except that $C_0(\tau)$ is a time-variable concentration specified at $x = 0$.† The solution is obtained by a super-position of the solutions just obtained, according to the scheme diagramed in Fig. 2.8. In each time increment $\delta\tau$ the concentration at $x = 0$ changes by an amount $\partial C/\partial\tau \, \delta\tau$. For a change occurring at time τ the result for all future times, due to the incremental change, is by Eq. (2.37)

$$\delta C = \frac{\partial C_0}{\partial\tau} \, \delta\tau \, \text{erfc}\left(\frac{x}{\sqrt{4D(t - \tau)}}\right) \qquad (t > \tau). \tag{2.39}$$

The total concentration at time t is the sum of the contributions at all prior times

$$C = \int_{-\infty}^{t} \frac{\partial C_0}{\partial\tau} \, \text{erfc}\left(\frac{x}{\sqrt{4D(t - \tau)}}\right) d\tau. \tag{2.40}$$

2.3.2.3   Input of Mass Specified as a Function of Time

Instead of specifying concentrations, suppose that we specify the rate at which mass is being added. The concentration distribution resulting from input of a single slug of mass M at $x = 0$ is the Gaussian distribution given as Eq. (2.14). A continuous injection of mass at the rate $\dot{M}$ is equivalent to injecting a slug of amount $\dot{M}\delta t$ after each time increment δt, where δt is infinitesimally small. The concentration resulting from the continuous injection is the sum of

---

† Here we use τ as a time variable for the source function to distinguish from the later time of observation t.

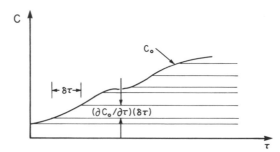

Figure 2.8 The superposition used to obtain the solution for a variable concentration $C_0(\tau)$ [Eq. (2.40)].

the concentrations resulting from the individual slugs injected at all times prior to the time of observation:

$$C = \int_{-\infty}^{t} \frac{\dot{M}(\tau)}{\sqrt{4\pi D(t-\tau)}} \exp\left[-\frac{x^2}{4D(t-\tau)}\right] d\tau, \qquad (2.41)$$

where $\dot{M}(\tau)$ is the rate of input of mass at time τ and may vary with time.

If the concentration is initially zero everywhere and a source of mass of constant strength $\dot{M}$ (mass units per unit time) is switched on at $t = 0$ at $x = 0$, Eq. (2.41) gives

$$C(x, t) = \frac{\dot{M}}{\sqrt{4\pi D}} \int_{0}^{t} \frac{1}{\sqrt{t-\tau}} \exp\left[-\frac{x^2}{4D(t-\tau)}\right] d\tau$$

$$= \frac{\dot{M}x}{4D\sqrt{\pi}} \int_{0}^{4Dt/x^2} u^{-1/2} e^{-1/u} du. \qquad (2.42)$$

This solution is sketched in Fig. 2.9. Finally, if we have a distributed source of mass $m(x, t)$, where m has units of mass per unit length per unit time, we can

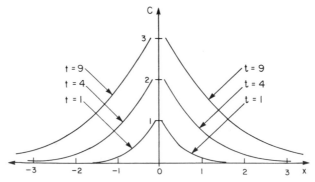

Figure 2.9 The distribution resulting from a source of mass of constant strength $\dot{M}$ as given by Eq. (2.42). Values used for the figure are $\dot{M} = 1$ and $D = \frac{1}{4}$.

superpose in space as in Eq. (2.30) and then in time to get the general solution to Eq. (2.4) as

$$C(x, t) = \int_{-\infty}^{t} \int_{-\infty}^{\infty} \frac{m(\xi, \tau)}{\sqrt{4\pi D(t - \tau)}} \exp\left[- \frac{(x - \xi)^2}{4D(t - \tau)} \right] d\xi \, d\tau. \qquad (2.43)$$

2.3.2.4   Solutions Accounting for Boundaries

Another class of problems occurs when spreading is restricted by the presence of boundaries. Here again, it is often useful to use the principle of superposition, which states that if the equation and boundary conditions are linear it is possible to superpose any number of individual solutions of the equation to obtain a new solution.

To illustrate the method, suppose that a unit mass of solute is concentrated at the origin of a one-dimensional system at $t = 0$ and that there is a wall through which concentration cannot diffuse located at $x = -L$ (a distance L from the source). According to Fick's law the boundary condition of no transport through the wall states that

$$q = -D \, \partial C / \partial x = 0 \qquad \text{at} \quad x = -L \qquad (2.44)$$

or in other words that the concentration gradient must be zero at the wall. As illustrated in Fig. 2.10a, this condition would be met if an additional unit mass of solute was concentrated at the point $x = -2L$ at $t = 0$ and if the wall was removed so that both slugs could diffuse to infinity in both directions. The solution to the real problem with the real boundary is the same as the sum of the solutions for the real plus the image source without the boundary, or

$$C = \frac{1}{\sqrt{4\pi Dt}} \left[\exp\left[- \frac{x^2}{4Dt} \right] + \exp\left[- \frac{(x + 2L)^2}{4Dt} \right] \right]. \qquad (2.45)$$

Now suppose we need the solution for the case where the source is surrounded by boundaries, one at $x = -L$ and one at $x = +L$. To satisfy the boundary condition at $-L$ we add an image source at $x = -2L$, as before. Similarly, to satisfy the boundary condition at $x = +L$ we add an image source at $x = +2L$ (see Fig. 2.10b). However, the slug at $x = -2L$ causes a positive gradient at the boundary at $+L$, which must be counteracted by another slug located at $x = +4L$. The same reasoning requires slugs at $-6L, +8L$, *ad infinitum*. Similarly, the slug at $x = +2L$ requires slugs at $-4L, +6L$, etc. The solution to the problem is the sum of the effects of all the slugs, namely,

$$C(x, t) = \sum_{n = -\infty}^{\infty} \frac{1}{\sqrt{4\pi Dt}} \exp\left[\frac{-(x + 2nL)^2}{4Dt} \right]. \qquad (2.46)$$

If the boundary condition is zero concentration at $x = \pm L$, negative slugs must

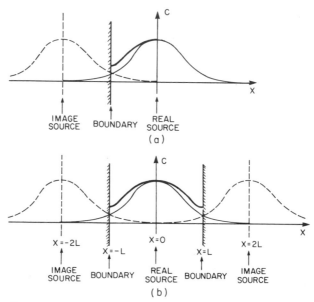

Figure 2.10   The method of superposition for matching the boundary condition of zero transport through the walls. (a) Single boundary; one image source needed. (b) Two boundaries; an
infinite number of image sources needed. ——— Distribution matching boundary conditions.
——— Distribution from real source if no boundaries. ----- Distribution from image sources.

be placed at $x = \pm 2L$, positive slugs at $x = \pm 4L$, etc., and the complete solution is

$$C(x, t) = \frac{1}{\sqrt{4\pi Dt}} \sum_{n=-\infty}^{\infty} \exp\left[\frac{-(x + 4nL)^2}{4Dt}\right] - \exp\left[\frac{-[x + (4n - 2)L]^2}{4Dt}\right].$$

$$(2.47)$$

In both (2.46) and (2.47) it usually suffices to use only a few terms, for instance
$n = 0, \pm 1$, and ± 2.

It is left as an exercise for the reader to see how other combinations of boundary
conditions lead to different requirements for distributions of image sources. A
useful starter is to take the case of nondiffusive boundaries at $x = \pm L$ and a
unit source at $x = L/2$.

2.3.3   Solutions in Two and Three Dimensions

Suppose that a mass M of tracer is deposited at $t = 0$ at the origin of an x–y
coordinate system in a two-dimensional fluid. The initial condition can be
written

$$C(x, y, 0) = M\delta(x)\delta(y).$$

$$(2.48)$$

Let us write the diffusion equation in two dimensions as

$$\frac{\partial C}{\partial t} = D_x \frac{\partial^2 C}{\partial x^2} + D_y \frac{\partial^2 C}{\partial y^2}, \tag{2.49}$$

where D_x is the diffusion coefficient in the x direction and D_y is the coefficient in the y direction. In molecular diffusion $D_x = D_y = D$, but we will also use the diffusion equation to solve many environmental problems in which the diffusivities are different in different directions so we may as well begin introducing the difference here.

The solution can be obtained by use of the "product rule." Set

$$C(x, y, t) = C_1(x, t)C_2(y, t), \tag{2.50}$$

where C_1 is not a function of y and C_2 is not a function of x. Then

$$\frac{\partial}{\partial t}(C_1 C_2) = C_1 \frac{\partial C_2}{\partial t} + C_2 \frac{\partial C_1}{\partial t} = D_x C_2 \frac{\partial^2 C_1}{\partial x^2} + D_y C_1 \frac{\partial^2 C_2}{\partial y^2}. \tag{2.51}$$

Rewriting,

$$C_2 \left[\frac{\partial C_1}{\partial t} - D_x \frac{\partial^2 C_1}{\partial x^2} \right] + C_1 \left[\frac{\partial C_2}{\partial t} - D_y \frac{\partial^2 C_2}{\partial y^2} \right] = 0. \tag{2.52}$$

This equation will be satisfied if the quantities within the brackets are zero separately, i.e., if C_1 and C_2 satisfy one-dimensional diffusion equations for which the solution is of the form given by Eq. (2.14). Multiplying the two results together, and noting that $\iint C \, dx \, dy = M$, gives the complete result

$$C = C_1 C_2 = \frac{M}{4\pi t \sqrt{D_x D_y}} \exp\left(-\frac{x^2}{4D_x t} - \frac{y^2}{4D_y t} \right). \tag{2.53}$$

Note that in one dimension the units of concentration were mass per unit length; in two dimensions they are mass per unit area, and in three dimensions mass per unit volume.† Equation (2.53) gives lines of constant concentration that are a set of concentric ellipses, the lengths of whose major and minor axes are in the ratio of $[D_x/D_y]^{1/2}$.

The product rule can easily be extended to three dimensions. It is left as an exercise to show that if a mass M is deposited at the origin of coordinates x, y, z in a three-dimensional fluid at time $t = 0$ the resulting concentration distribution is given by

$$C(x, y, z, t) = \frac{M}{(4\pi t)^{3/2}(D_x D_y D_z)^{1/2}} \exp\left(-\frac{x^2}{4D_x t} - \frac{y^2}{4D_y t} - \frac{z^2}{4D_z t} \right). \tag{2.54}$$

† Alternatively, C may be considered to have units of mass per unit volume, but then M has, respectively, units of mass per unit area in one dimension (a plane source) and mass per unit length in two dimensions (a line source).

Equations (2.53) and (2.54) are the fundamental solutions for two and three dimensions. Solutions corresponding to other initial and boundary conditions, analogous to the one-dimensional solutions given in Section 2.3.2, can be obtained by superposition methods similar to the ones we have used for one dimension.

2.4 ADVECTIVE DIFFUSION

Up to now we have assumed that the fluid was stationary and that mass transport was by diffusion alone. Now suppose that the fluid itself is moving with velocity $\mathbf{u}$, whose components in the x, y, and z directions are u, v, and w. We call the transport by the mean motion of the fluid "advection," and assume that transport by advection and by diffusion are separate, additive processes. This is equivalent to assuming that diffusion takes place within the moving fluid just as though the fluid were stationary. For the time being let us also assume that we are dealing with molecular diffusion in laminar flow, so that the diffusion coefficient has a constant value D in all directions. We will take up the complexities of turbulent flow in the next chapter.

The rate of mass transport through a unit area in the yz plane by the component of velocity in the x direction is the quantity (uC), because this is the rate at which fluid volume passes through the unit area ($u \times$ unit area = volume/unit time) multiplied by the concentration of mass in that volume. The total rate of mass transport is the advective plus the diffusive flux

$$q = \underset{\substack{\text{advective} \\ \text{flux}}}{uC} + \underset{\substack{\text{diffusive} \\ \text{flux}}}{(-D\,\partial C/\partial x)}. \tag{2.55}$$

when this is substituted into the equation for conservation of mass in one dimension [Eq. (2.3)] we obtain the diffusion equation as before plus the additional advective term:

$$\frac{\partial C}{\partial t} + \frac{\partial}{\partial x}(uC) = D\frac{\partial^2 C}{\partial x^2}. \tag{2.56}$$

The equation in three dimensions can be obtained by following the same steps as led to Eq. (2.8). It is left as an exercise to the reader to show that Eq. (2.6) becomes

$$(\partial C/\partial t) + \nabla \cdot (C\mathbf{u}) = D\,\nabla^2 C \tag{2.57}$$

or, making use of the equation for conservation of fluid volume $\nabla \cdot \mathbf{u} = 0$,

$$(\partial C/\partial t) + \mathbf{u} \cdot \nabla C = D\,\nabla^2 C. \tag{2.58}$$

Written out fully in Cartesian coordinates, the equation is

$$\frac{\partial C}{\partial t} + u\frac{\partial C}{\partial x} + v\frac{\partial C}{\partial y} + w\frac{\partial C}{\partial z} = D\left[\frac{\partial^2 C}{\partial x^2} + \frac{\partial^2 C}{\partial y^2} + \frac{\partial^2 C}{\partial z^2}\right]. \tag{2.59}$$

This equation is often referred to as the "advective diffusion" equation, but since advection is such a common feature of environmental problems we will simply call it the "diffusion equation."

We now turn to two simple limiting solutions of the diffusion equation (2.59) in a fluid moving with a constant velocity u in the x direction only. In the first the gradients in the y direction are small,

$$\frac{\partial C}{\partial t} + u\frac{\partial C}{\partial x} = D\frac{\partial^2 C}{\partial x^2}, \tag{2.60}$$

and in the second the diffusive transport in the x direction is smaller than the advective transport, so that

$$\frac{\partial C}{\partial t} + u\frac{\partial C}{\partial x} = D\frac{\partial^2 C}{\partial y^2}, \tag{2.61}$$

where y is a transverse direction. In the system described by Eq. (2.60) there is an advective transport in the same direction as the diffusion. For example, one could consider the problem of a pipe filled with one fluid and being displaced at a mean flow velocity u by another fluid with a tracer in concentration C_0. At time $t = 0$ there is a sharp front so that

$$C(x, 0) = \begin{cases} 0, & x > 0, \\ C_0, & x < 0. \end{cases} \tag{2.62}$$

If we let $x' = x - ut$ then Eq. (2.60) becomes

$$(\partial C/\partial t) = D(\partial^2 C/\partial x'^2).$$

In other words, the problem is the same as diffusion in a stagnant fluid when viewed in a coordinate system moving at speed u. We have already solved this problem in a mirror image form and the solution is given by Eq. (2.33). Adjusted for the moving coordinates and the mirror image effect we have

$$C(x, t) = \frac{C_0}{2}\left[1 - \text{erf}\left(\frac{x - ut}{\sqrt{4Dt}}\right)\right]. \tag{2.63}$$

An example of a lateral diffusion problem is illustrated in Fig. 2.11. This represents the simplest case of transverse mixing of two streams of different uniform

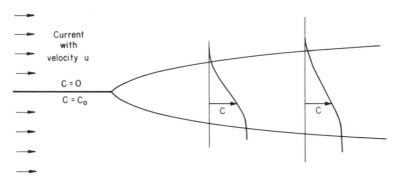

Figure 2.11   Growth of a lateral mixing zone.

concentration flowing side by side. Since the input is constant the solution must not depend on time, so Eq. (2.61) simplifies to

$$u(\partial C/\partial x) = D(\partial^2 C/\partial y^2),$$

with boundary conditions

$$C(0, y) = \begin{cases} 0, & y > 0, \\ C_0, & y < 0, \end{cases}$$

and

$$C(x, \infty) \to 0, \qquad C(x, -\infty) \to C_0.$$

The solution is immediately evident from the previous example if we recognize the equivalence of x' and y, and t and x/u. We have

$$C = \frac{C_0}{2}\left[1 - \text{erf}\left(\frac{y}{\sqrt{4Dx/u}}\right)\right]. \tag{2.64}$$

A somewhat more complicated example is the boundary value problem specified by

$$C(0, t) = C_0, \qquad 0 < t < \infty,$$

$$C(x, 0) = 0, \qquad 0 < x < \infty,$$

$$\frac{\partial C}{\partial t} + u\frac{\partial C}{\partial x} = D\frac{\partial^2 C}{\partial x^2}, \qquad 0 < x < \infty.$$

Physically, this problem could represent a river for which a steady concentration C_0 is introduced at the origin of the coordinate system at time $t = 0$ and continued. Obviously the final solution will correspond to a concentration C_0

down the entire river. It is left as an exercise for the reader to show that for time $t < \infty$,

$$C(x, t) = \frac{C_0}{2}\left[\operatorname{erfc}\left(\frac{x - ut}{\sqrt{4Dt}}\right) + \operatorname{erfc}\left(\frac{x + ut}{\sqrt{4Dt}}\right)\exp\left(\frac{ux}{D}\right)\right]. \qquad (2.65)$$

The case of a maintained point discharge in a two- or three-dimensional flow is interesting because it is usually possible to simplify the problem by one dimension. Suppose that a point source discharges mass at the rate $\dot{M}$ at the origin of an (x, y, z) coordinate system in a three-dimensional flow, and let the mean velocity be u in the x direction. For simplicity assume that the diffusion coefficient is D in all directions, so that the diffusion equation is

$$\frac{\partial C}{\partial t} + u\frac{\partial C}{\partial x} = D\left(\frac{\partial^2 C}{\partial x^2} + \frac{\partial^2 C}{\partial y^2} + \frac{\partial^2 C}{\partial z^2}\right). \qquad (2.66)$$

A general solution can be obtained by superposing point sources in space and time, using the fundamental solution given by Eq. (2.14) and the procedures described in Section 2.3.2. In most practical cases, however, it is possible to reduce the three-dimensional problem to that of the spread of a instantaneous point source in two dimensions, for which we already have the solution in Eq. (2.53). To see this, visualize the flow as consisting of a series of parallel slices of thickness δx bounded by infinite parallel y–z planes, as illustrated in Fig. 2.12. The slices are being advected past the source, and during the passage each one receives a slug of mass of amount $\dot{M}\,\delta t$; δt is the time taken for the slice to pass the source and is equal to $\delta x/u$. Subsequently the mass per unit area in the slice, by Eq. (2.53), is

$$\frac{\dot{M}\delta x}{4\pi Dtu}\exp\left[-\frac{(y^2 + z^2)}{4Dt}\right].$$

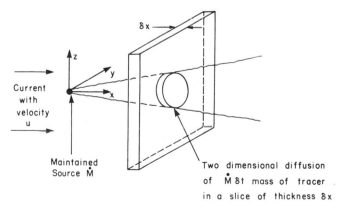

Figure 2.12 Reduction of a three-dimensional problem to two dimensions by considering diffusion in a moving slice.

Now if we recognize that the location of the slice is given by $x = ut$ and that the three-dimensional concentration is the mass per unit area in the slice divided by the thickness of the slice, we have

$$C(x, y, z) = \frac{\dot{M}}{4\pi Dx} \exp\left[-\frac{(y^2 + z^2)u}{4Dx} \right]. \tag{2.67}$$

It should be noted, however, that we have obtained this solution by neglecting diffusion in the direction of flow. Diffusion in the flow direction produces spreading characterized by a length proportional to $(2Dt)^{1/2}$ (the standard deviation of a diffusing cloud), and distance from the source to the slice is $x = ut$ so x-direction diffusion can be neglected if approximately $ut \gg (2Dt)^{1/2}$, or $t \gg 2D/u^2$. In practical problems the value of t required to meet this condition is often very small, so that Eq. (2.67) can be used without difficulty. The reader is left to show that the approximate solution for spreading from a maintained point source in two dimensions is

$$C(x, y) = \frac{\dot{M}}{u\sqrt{4\pi Dx/u}} \exp\left(-\frac{y^2 u}{4Dx} \right) \tag{2.68}$$

where $\dot{M}$ is now the strength of a line source in units of mass per unit length per unit time. This solution will be used in Section 5.1.3 to analyze transverse mixing of a pollutant discharge from a pipe into a river. As for Eq. (2.67), Eq. (2.68) should be used only when $t \gg 2D/u^2$.

Chapter 3

Turbulent Diffusion

Although Chapter 2 dealt with molecular diffusion in laminar flow, fluid motions in the environment are almost always turbulent. This chapter begins with a description of turbulence and some simple aspects of its statistics. Then we show how and under what conditions Fickian "turbulent mixing" coefficients, analogous to the molecular diffusion coefficient, can be used to describe mixing in turbulent flow. Finally, we treat the case of the spread of single clouds of particles, where the coefficient describing the rate of spread increases with the size of the cloud.

3.1 INTRODUCTION

It is difficult to define a turbulent fluid motion, but as a disease may be recognized by its symptoms, turbulence may be detected by the following occurrences:

(a) Mass introduced at a point will spread much faster in turbulent flow than in laminar flow. A classic demonstration is Reynold's experiment in pipe flow, illustrated in Fig. 3.1a, in which a filament of dye is introduced on the centerline at the upstream end of a pipe. In laminar flow the filament makes a straight streak along the centerline, but in turbulent flow the streak is quickly broken up and spreads across the pipe.

   (b)   Velocities and pressures measured at a point in the fluid are unsteady and possess an appreciable random component. Figure 3.1b shows the longitudinal velocity observed at the center of a pipe in laminar and turbulent flow. In steady laminar flow the velocity is constant; in steady turbulent flow we see random excursions above and below the constant mean. Turbulent flow occurs when the pipe Reynolds number is greater than approximately 2000.

   A consequence of the random motions is that we can think of turbulent flow occurring in a range of sizes or "scales" of motion. To illustrate this concept consider what happens if a straight plane of dye is painted on fluid particles across the cross section of pipe. Figure 3.1c shows four cases. In laminar flow the plane is distorted by the velocity gradient with two possible scales. If the plane is painted near the inlet, the velocity is uniform near the center of the pipe and reduces to zero at the wall through a boundary layer. The only scale of distortion of the plane is the thickness of the boundary layer. If the plane is painted further downstream in a region where the pipe flow is fully developed, the dye is distorted into a parabolic surface extending over the diameter of the pipe. In this case the pipe size itself provides the only measure, or "scale," of the extent of the distortion. Now consider the case of turbulent flow. The dye plane can be distorted by the shape of the boundary layer near the inlet or by the fully developed velocity profile further downstream, but it is also distorted by the random turbulent excursions; we may be able to observe radii of curvature of the dye plane varying all the way from the diameter of the pipe down to very small. For want of a better physical description we often refer to these observed curvatures as resulting from "eddies", and speak of the range of "eddy sizes" or, equivalently, "scales of turbulence."
   The range of scales over which the velocity varies in both space and time is a result of the importance of the strong nonlinearity of the equations of fluid motion at large Reynolds numbers. The effect of these nonlinear terms is a spreading of the kinetic energy of the fluid motion over a range of eddy sizes through the interaction of the large and smaller scales of motion. At the very smallest scales viscosity has a strong effect and turns the kinetic energy of the fluid motion into heat. Because of this energy dissipation there must be a process continually feeding energy to the smallest scales of motion. If the flow is in equilibrium the transfer process must be in equilibrium with the energy dissipation rate. At the larger scales of motion the viscosity is relatively unimportant and the nonlinear terms in the equations of motion are responsible for redistributing the kinetic energy from the scales at which it is generated by gravity or pressure forces. Again, the redistribution process must be in equilibrium in steady flows, such as in a pipe or channel, where there is an exact matching between the production of kinetic energy by gravity or pressure forces and the dissipation of energy by viscosity.
   Since small scale motions tend to have small time scales, one may assume that

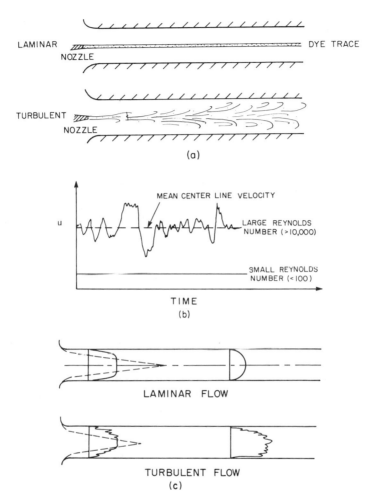

Figure 3.1   (a) Dye introduced at the upstream end of a pipe in laminar and turbulent flow. (b) Record of longitudinal velocity at the center of a pipe at a large and a small Reynolds number. (c) Deformation of a dye surface in laminar and turbulent flow.

these motions are statistically independent of the relatively slow, large-scale turbulence and of the mean flow. If this assumption is correct, then the small scale motion should depend only on the rate of energy transfer from the larger scales and on the viscosity of the fluid. Furthermore, since the kinetic energy of the small and intermediate scale motions varies only at the rate at which the mean flow varies, it may be postulated that the behavior of the intermediate scales is governed only by the transfer of energy which, in turn, is exactly balanced by dissipation at the very small scales. This is the basis of Kolmogorov's universal equilibrium theory of turbulence.

Since the units for rate of energy dissipation per unit mass ϵ are length squared divided by time cubed, the length, time, and velocity scales of the very smallest scale motions must be (by dimensional analysis): $(\nu^3/\epsilon)^{1/4}$, $(\nu/\epsilon)^{1/2}$, and $(\nu\epsilon)^{1/4}$. These are generally referred to as the Kolmogorov scales. For the open ocean the energy dissipation rate varies depending on the current velocity and wind speed, but an average value of 0.01 cm^2/sec^3 (10^{-7} W/kg). This implies dissipation length, time, and velocity scales of the order of 0.1 cm, 1 sec, and 0.1 cm/sec, respectively.

Consider now the spreading of a slug of tracer, or a group of marked particles in a steady high Reynolds number flow. Suppose a mass M of tracer (or a total number M of marked particles) is released at a fixed coordinate point in the flow; the subsequent spread of the tracer is to be viewed by an observer moving with the mean (i.e., time-averaged) velocity of the fluid. We assume also that the character of the turbulence remains steady, or in other words, the variance of the velocity is steady and does not change with time or position. Fluid dynamicists call this stationary, homogeneous turbulence. Now suppose, for the sake of visualization, that the flow is essentially two dimensional so that we can take a series of photographs looking down on the tracer cloud as it spreads. We use a camera traveling with the mean speed of the fluid. Figures 3.2a and b depict the result of two identical experiments with the photos taken at equal times after the release. The experiment could be repeated many times. The results of the two tests shown are obviously quite different, and the differences are of two sorts. First, the small scale fluctuations, which are different for each cloud, distort the shape of the cloud and produce steep concentration differences over short distances. These local differences will eventually be smoothed out by molecular diffusion. Second, the large scale fluctuations, especially those substantially larger than the cloud itself, transport the entire cloud. Each cloud of particles encounters a different set of large scale motions, so the motion of the center of mass of each cloud is different.

Now suppose that we release a large number of clouds of particles, one after the other, and watch the spread of each cloud over a long period of time. Since we have subtracted out the mean motion, the average position of the center of mass will be at the origin, but the center of mass of each cloud may diverge from the origin because of the large scale eddies. If we wait long enough each separate cloud will grow to be bigger than the largest eddies and will average out their effects; the center of mass of each cloud will tend to return to the origin through the process of averaging the random motions. This may take a long time, however, and one of the central problems of turbulent diffusion is what to do until that time occurs. What ought to be done will depend on the problem at hand, but there are basically two strategies. One is to average over all of the releases to obtain an "ensemble average," and the other is try to follow each release separately. The difficulty is that the ensemble average concentration at a point in space and time is likely to be an average of a large number of zeros (times

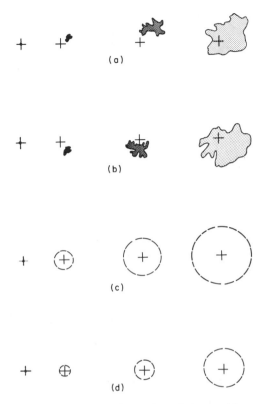

Figure 3.2 Diffusion in homogeneous, isotropic turbulence with zero mean velocity. The largest scale of motion is slightly larger than the largest cloud shown. (a) Spread of a single cloud. (b) Spread of a second cloud. (c) Spread of the ensemble mean. (d) Spread of the ensemble mean of clouds after superposition of centers of mass.

when the tracer cloud does not cover the point) plus a few large values. This type of average may be meaningless to an organism subject to release of a pollutant, because the few high concentrations may kill the organism and the large numbers of zeros cannot bring it back to life. On the other hand, not enough is known about turbulent flow to permit the computation of the spread of each individual cloud or the peak concentration at any instant. The best we can do, as explained in Section 3.4, is a statistical estimate of the size of an individual cloud. The difference is illustrated by Figs. 3.2c and d. Figure 3.2c shows the growth of the ensemble average; Fig. 3.2d shows an average obtained by superposing the centers of mass of each of the individual clouds and then averaging over the ensemble of releases. The average extent of each individual cloud is, of course, smaller than the extent of the ensemble average because the ensemble average includes the distribution of the centers of mass.

Before going further we need to be more specific about defining the various kinds of averages, and to introduce some of the concepts of statistical analysis of turbulence. We will do that in Section 3.2, and will then return to a detailed analysis of the spread of the ensemble average (Section 3.3) and of single clouds (Section 3.4).

3.2   SOME STATISTICAL CONCEPTS

It is not the purpose of this section to present an exposition on the analysis of random signals, but rather to introduce the concepts used in the next sections and already alluded to in Section 3.1. We will thus confine our attention to the various ways of computing averages, variances, and correlation coefficients of a random time series. For a much fuller treatment the reader is referred to the books by Cramér and Leadbetter (1967) and Bendat and Piersol (1971).

Suppose that a particle in a turbulent fluid is, at time t_0, located at the point whose position vector is ξ and which has Cartesian coordinates (ξ, η, ζ). At a subsequent time t the particle moves to a new point whose position vector is $\mathbf{X}$ with Cartesian coordinates (X, Y, Z). The trajectory of the particular particle that happened to be at ξ at time t_0 is therefore specified by a vector functional relationship of the form

$$\mathbf{X} = \mathbf{X}(\xi, t, t_0) \qquad (3.1)$$

where

$$X = X(\xi, \eta, \zeta, t, t_0), \qquad Y = Y(\xi, \eta, \zeta, t, t_0), \qquad Z = Z(\xi, \eta, \zeta, t, t_0). \quad (3.2)$$

For each release of a particle there will be a different functional form of $\mathbf{X}$ reflecting the random nature of the trajectories particles take in turbulent fluids. Alternatively, $\mathbf{X}$ may be interpreted as a random variable in time for any one release since the position a particle takes at some later time $t + \tau$ may bear little correlation to the position at time t.

These two interpretations of Eq. (3.1) are important because the statistics of the particle motion may be different if the average is across the totality of experiments performed or is an average over a period of time in any one experiment. Figure 3.3a illustrates the first type of average; a number of particles are released at different times, and the displacement $\mathbf{X}$ of each particle a time T after its release is observed. $\mathbf{X}$ is a random variable whose statistical properties can be determined if a large number of experiments are carried out; an average obtained in this way is called an ensemble average. Figure 3.3b illustrates the second type of average. A single particle is released and followed through a large number of time increments each of duration T. The displacement $\mathbf{X}$ during each time increment is observed. This $\mathbf{X}$ is also a random variable, but its statistical properties may differ from those obtained by an ensemble of experi-

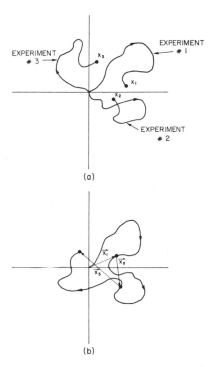

Figure 3.3 (a) Trajectories of three particles, each released at the origin but at different times, and each wandering for a time T. (b) Trajectory of a single particle wandering for a time $3T$, divided into three separate wanderings of duration T each.

ments. If the time series average and the ensemble average are the same we say that the process satisfies the ergodic property.

Equation (3.1) may also be interpreted in a somewhat different sense. In any one realization, instead of taking the position ξ as fixed, and letting $\mathbf{X}$ trace out a distorted curving line in space as t is varied in that realization, the inverse may be considered. That is, we can consider the point $\mathbf{X}$ as fixed and the point ξ as a function of time. This view is that which an observer would see if he fixed his attention at a particular $\mathbf{X}$ in a particular experiment and at each time interpreted from where the particle originated which is at $\mathbf{X}$ at the time of observation. Thus by fixing ξ, Eq. (3.1) traces out a path of a particular particle, and by fixing the observation station $\mathbf{X}$, the inverse of Eq. (3.1) reveals the initial coordinates of the particle presently at $\mathbf{X}$. Equation (3.1) is central to the turbulent diffusion problem since molecular diffusion merely smooths out sharp small scale irregularities and it is the turbulent motion of the fluid particles that really spreads the tracer.

Let us now focus on the x coordinate of the random variable $\mathbf{X}$, and consider the problem of describing the statistical properties of $X = X(\xi, \eta, \zeta, t, t_0)$. Because of the correlation of turbulent motions, the probability that the value is between X and $X + dX$ at time t will almost certainly not be independent of where the particle was at some previous instant, say $t - \Delta t$. It is therefore insufficient to specify merely the probability of finding the value between X and $X + dX$, we also have to specify the *joint* probability that a value can be found between X_1 and $X_1 + dX_1$ at time t_1, and between X_2 and $X_2 + dX_2$ at time t_2. And, in general, to specify the random nature of the variable fully it is necessary to give the joint probability that the value will be between X_1 and $X_1 + dX_1$ at t_1, between X_2 and $X_2 + dX_2$ at t_2, between X_3 and $X_3 + dX_3$ at t_3, and so on, to between X_n and $X_n + dX_n$ at t_n, for *arbitrary n*.

Two turbulent mixing processes are therefore identical if *all* the joint probabilities are the same. (This does not mean of course that any two trials will produce the same position time history!)

We can define an ensemble mean† as the average taken over a large number of trials. For the variable X it is

$$\langle X \rangle = E[X(\xi, t, t_0)] = \int_{-\infty}^{\infty} X p(X|\xi, t, t_0)\, dX, \tag{3.3}$$

where the angle brackets are shorthand for ensemble average, E stands for "expected value of," and

$$p(X|\xi, t, t_0)\, dX = \text{probability that } X \text{ has a value between } X \text{ and}$$
$$X + dX \text{ at time } t \text{ given that it was } \xi \text{ at time } t_0,$$

where $p(X|\xi, t, t_0)$ is defined as the probability density function of the process X. Similarly, the kth moment of the probability distribution is defined as

$$\langle X^k \rangle = E[X^k(\xi, t, t_0)] = \int_{-\infty}^{\infty} X^k p(X|\xi, t, t_0)\, dX. \tag{3.4}$$

The autocovariance of the particle displacement X is defined as the expectation of X having the value X_1 at time t_1 and X_2 at time t_2:

$$B_{XX}(\xi, t_1, t_2, t_0) = E[X_1 X_2]$$
$$= \int_{-\infty}^{\infty} \int_{-\infty}^{\infty} X_1 X_2 p(X_1, X_2|\xi, t_1, t_2, t_0)\, dX_1\, dX_2, \tag{3.5}$$

where

$$p(X_1, X_2|\xi, t_1, t_2, t_0)\, dX_1\, dX_2 = \text{probability that } X \text{ has a value between } X_1$$
$$\text{and } X_1 + dX_1 \text{ at time } t_1 \text{ and between } X_2$$
$$\text{and } X_2 + dX_2 \text{ at time } t_2 \text{ given that the}$$
$$\text{value was } \xi \text{ at time } t_0.$$

Finally, the cross-covariance between the components of particle displacement, say X and Y or X and Z, is defined as the expectation of XY or XZ and so forth. For instance,

$$B_{XY}(\xi, t, t_0) = E[XY]$$
$$= \int_{-\infty}^{\infty} \int_{-\infty}^{\infty} XY p(X, Y|\xi, t, t_0)\, dX\, dY, \tag{3.6}$$

---

† In this chapter we will always use $\langle\ \rangle$ to denote the mean over many trials, the ensemble mean. Note that in Chapter 7 the same brackets are used to denote an average over a tidal cycle.

where $p(X, Y \mid \xi, t, t_0)$ is the joint probability density function for the X and Y components of particle displacement.

We shall refer to the particle displacement $\mathbf{X}$ as stationary if all the moments of the distribution of the displacement are independent of the time origin and depend only on the time difference $t - t_0$, and in addition the covariances depend only on the time difference. Similarly, the particle displacement is called homogeneous if these same quantities depend only on the relative displacement $|\mathbf{X} - \xi|$ and not the initial position ξ.

We are also concerned with the statistical distribution of concentration at a point. Let $n(\mathbf{x}, t)$ be the concentration observed at point $\mathbf{x}$ at time t, and let

$p(n \mid \mathbf{x}, t) \, dn$ = probability that the concentration of tracer material has a value between n and $n + dn$ at the point $\mathbf{x}$ at time t.

With this density function we can write

$$C(\mathbf{x}, t) = \langle n(\mathbf{x}, t) \rangle = \int_0^\infty n p(n \mid \mathbf{x}, t) \, dn. \tag{3.7}$$

$C(\mathbf{x}, t)$ is the ensemble average of the concentration $n(\mathbf{x}, t)$ measured at point $\mathbf{x}$ at time t after many repeated trials in which identical clouds of particles are released under the same statistical conditions. Note that $C(\mathbf{x}, t)$ implicitly depends on the position at which the cloud was initially released, and the time of release t_0. For stationary turbulence t_0 can arbitrarily be chosen to be zero.

In the next section we will see that there is a simple relationship between the probability density function $p(\mathbf{x} \mid \xi, t, t_0)$ and the ensemble mean concentration $C(\mathbf{x}, t)$. However, we first define the x coordinate of the center of mass of a cloud resulting from a single trial†. It is given by

$$\bar{X} = \frac{1}{M} \int_{-\infty}^\infty \int_{-\infty}^\infty \int_{-\infty}^\infty x n(x, y, z, t) \, dx \, dy \, dz, \tag{3.8}$$

where M is the total mass (or number of particles) in the cloud and is given by

$$M = \int_{-\infty}^\infty \int_{-\infty}^\infty \int_{-\infty}^\infty n(x, y, z, t) \, dx \, dy \, dz. \tag{3.9}$$

We can similarly define $\bar{Y}$ and $\bar{Z}$, the other coordinates of the center of mass.

The variance of the cloud, that is the mean square x displacement about the center of mass of particles in a single cloud is given by

$$\sigma_x^2 = \frac{1}{M} \int_{-\infty}^\infty \int_{-\infty}^\infty \int_{-\infty}^\infty (x - \bar{X})^2 n(x, y, z, t) \, dx \, dy \, dz. \tag{3.10}$$

† We use an overbar to define the spatial mean in *one* trial, as opposed to the ensemble mean $\langle \ \rangle$.

Now, considering the ensemble of clouds, we can define the expected position of the center of mass for the ensemble. This is given by

$$\langle \overline{X} \rangle = \frac{1}{M} \int_{-\infty}^{\infty} \int_{-\infty}^{\infty} \int_{-\infty}^{\infty} xC(x, y, z, t) \, dx \, dy \, dz. \tag{3.11}$$

The overall variance of the concentration in the ensemble of clouds with respect to the expected position of the overall center of mass is given by

$$\Sigma_x^{\ 2} = \frac{1}{M} \int_{-\infty}^{\infty} \int_{-\infty}^{\infty} \int_{-\infty}^{\infty} (x - \langle \overline{X} \rangle)^2 C(x, y, z, t) \, dx \, dy \, dz. \tag{3.12}$$

If now we expand Eq. (3.10) and then take the ensemble average it can be seen that

$$\Sigma_x^{\ 2} = \langle \sigma_x^{\ 2} \rangle + \langle (\overline{X} + \langle \overline{X} \rangle)^2 \rangle, \tag{3.13}$$

where, as before, the angle brackets indicate an ensemble average. In other words, the variance of the ensemble distribution about its expected position is equal to the ensemble average of the variance of each cloud about its center of mass plus the ensemble mean square displacement of an individual cloud's center of mass from its expected position. We can define the size of an individual cloud as

$$l(t) = [\tfrac{1}{3}(\sigma_x^{\ 2} + \sigma_y^{\ 2} + \sigma_z^{\ 2})]^{1/2}, \tag{3.14}$$

and the size of an average cloud as

$$L(t) = [\tfrac{1}{3}(\Sigma_x^{\ 2} + \Sigma_y^{\ 2} + \Sigma_z^{\ 2})]^{1/2}. \tag{3.15}$$

Equation (3.13) plus the similar relations in the y and z coordinate directions indicate that the average size of a cloud as formed from the ensemble is

$$L^2(t) = \langle l^2(t) \rangle + \tfrac{1}{3}[\langle (\overline{X} - \langle \overline{X} \rangle)^2 \rangle + \langle (\overline{Y} - \langle \overline{Y} \rangle)^2 \rangle + \langle (\overline{Z} - \langle \overline{Z} \rangle)^2 \rangle]. \tag{3.16}$$

In other words, the width of the ensemble mean concentration profile is larger than the average width of a cloud, as already indicated in Figs. 3.2c and 3.2d.

We will see subsequently that there are quite reasonable theories for predicting the *rates* of change of both $\langle l^2(t) \rangle$ and $L^2(t)$. We are therefore able to predict how clouds of tracer material grow either when considered relative to a coordinate system located at the center of mass, or when considered relative to an absolute system.

3.3 DIFFUSION OF THE ENSEMBLE MEAN CONCENTRATION

Let us first turn to the problem depicted in Fig. 3.2c, that of finding the properties of the ensemble mean concentration and the length L. We will use a classic analysis by G. I. Taylor published in 1921 under the title "Diffusion by Continuous Movements"; this remarkable analysis has withstood the subsequent 57 years without significant change or improvement. The analysis does require, however, that we specify a particular type of turbulent field, namely a field of stationary homogeneous turbulence. An example of such a field is turbulence in flow in an infinitely long, straight pipe, if diffusion only in the axial direction is considered. This is because the turbulence has the same properties at every point along the axis and is not decaying in time. In addition, Taylor's analysis requires that the mean flow velocity be zero; in the case of the pipe we can satisfy this requirement by moving the coordinate system at the mean velocity of the flow.

Now consider a series of experiments in a stationary homogeneous turbulent fluid with zero mean velocity where, in each experiment, mass m of matter is added to a single fluid particle at time t_0 at point ξ. For simplicity let us consider diffusion only in the x direction; the extension to three dimensions is simple but complicates the equations.† Since the mean flow is zero the mean position of the marked particles $\langle X \rangle$ is zero. Therefore Eqs. (3.15) and (3.12) simplify to

$$L^2(t) = \Sigma_x^2 = \frac{1}{M} \int_{-\infty}^{\infty} \int_{-\infty}^{\infty} \int_{-\infty}^{\infty} x^2 C(x, y, z, t) \, dx \, dy \, dz, \tag{3.17}$$

where M/m is the number of particles released. If the effects of molecular diffusion are neglected the mass m added to each particle remains with that particle at all times so the average concentration $C(\mathbf{x}, t)$ measured at a fixed point in space will just be proportional to the probability that the particle is at $\mathbf{x}$, that is

$$C(\mathbf{x}, t) = M p(\mathbf{x} | \xi, t, t_0). \tag{3.18}$$

Substituting Eq. (3.18) into (3.17) and using Eq. (3.4) gives

$$L^2(t) = \int_{-\infty}^{\infty} \int_{-\infty}^{\infty} \int_{-\infty}^{\infty} x^2 p(\mathbf{x} | \xi, t, t_0) \, dx \, dy \, dz$$

$$= \int_{-\infty}^{\infty} x^2 p(\mathbf{x} | \xi, t, t_0) \, dx = \langle x^2 \rangle. \tag{3.19}$$

This result holds if all the particles begin their motion at the same point ξ, but it can also be generalized to the case in which we begin with a cloud of particles

---

† Note that in one dimension we can set $L^2 = \Sigma_x^2$, omitting the factor $\frac{1}{3}$ needed in Eq. (3.15).

having a spatial distribution $C(\xi, t_0)$ at time t_0. By superposition of the point source result Eq. (3.18), as discussed in Section 2.3.2.1, we have

$$C(\mathbf{x}, t) = \int_V C(\xi, t_0) p(\mathbf{x} \mid \xi, t, t_0) \, d\xi. \tag{3.18a}$$

By substituting Eq. (3.18a) into (3.17), changing the order of integration and origin of the integration domain, the more general result follows:

$$L^2(t) = \langle x^2 \rangle + L^2(t_0), \tag{3.19a}$$

where $L^2(t_0)$ is the ensemble mean size of the cloud at the initial time t_0. Thus the problem of finding the ensemble mean size of the cloud is equivalent to finding the ensemble mean square displacement of the fluid particles. The latter is the problem solved by Taylor in 1921.

We proceed as follows. Let U be the velocity of the particle, with zero mean (in the case of pipe flow we move the coordinate system at the mean velocity, so that with respect to the moving coordinate system $\langle U \rangle = 0$). Without loss of generality take $\xi = 0$ and $t_0 = 0$. Then the location of the particle is

$$X(t) = \int_0^t U \, dt$$

and

$$X^2(t) = \left(\int_0^t U \, d\tau_1 \right) \left(\int_0^t U \, d\tau_2 \right) = \int_0^t \int_0^t U(\tau_1) U(\tau_2) \, d\tau_1 \, d\tau_2. \tag{3.20}$$

The ensemble mean is

$$\langle X^2 \rangle = \int_0^t \int_0^t \langle U(\tau_1) U(\tau_2) \rangle \, d\tau_1 \, d\tau_2. \tag{3.21}$$

The ensemble average $\langle U(\tau_1) U(\tau_2) \rangle$ means the average over a large number of trials of the product of the velocity of a single particle at τ_1 multiplied by the velocity of the same particle at time τ_2; since the turbulence is stationary this can only be a function of the difference between τ_1 and τ_2, and we can define a correlation coefficient as

$$R_x(\tau_2 - \tau_1) = \langle U(\tau_1) U(\tau_2) \rangle / \langle U^2 \rangle, \tag{3.22}$$

where $\langle U^2 \rangle = \langle U(0) U(0) \rangle$ (the square root of this quantity is known as the "intensity" of the turbulence). R_x is called the Lagrangian autocorrelation function, the word "Lagrangian" being used to indicate that U is the velocity observed following the motion of a particular particle, not the velocity at a fixed point in space. Equation (3.22) can be used to rewrite Eq. (3.21) as

$$\langle X^2(t) \rangle = \langle U^2 \rangle \int_0^t \int_0^t R_x(\tau_2 - \tau_1) \, d\tau_2 \, d\tau_1. \tag{3.23}$$

The remainder of Taylor's analysis requires nothing but manipulations of the double integral; there are no further hypotheses or appeals to physical intuition, and that is why the analysis is such a classic. Changing the variables of integration to

$$s = \tau_2 - \tau_1 \quad \text{and} \quad \tau = (\tau_1 + \tau_2)/2,$$

and performing the integration with respect to τ gives

$$\langle X^2(t) \rangle = 2\langle U^2 \rangle \int_0^t (t - s) R_x(s) \, ds. \tag{3.24}$$

In a similar way we can obtain the mean square particle displacements in the y and z coordinate directions for the case of diffusion in three dimensions as

$$\langle Y^2(t) \rangle = 2\langle V^2 \rangle \int_0^t (t - s) R_y(s) \, ds, \tag{3.25}$$

$$\langle Z^2(t) \rangle = 2\langle W^2 \rangle \int_0^t (t - s) R_z(s) \, ds, \tag{3.26}$$

and also the cross-variances $\langle XY \rangle$, $\langle YZ \rangle$, and $\langle XZ \rangle$. In most cases, however, we can choose the coordinate axes to coincide with the principal axes of the turbulent fluctuations so that the cross-variances are zero. Therefore we will ignore cross-variances from this point; a more general treatment is given by Batchelor (1949).

Let us continue to focus on diffusion in the x direction and see what useful results can be obtained from Eq. (3.24). Two limiting cases can be derived. For very short times after release of the particles the particle velocity is very nearly constant and R_x is very nearly unity. Setting $R_x = 1$ gives in this limit

$$\langle X^2 \rangle = \langle U^2 \rangle t^2. \tag{3.27}$$

On the other hand, for very long times it seems likely, although not strictly proven, that $R_x \to 0$ because the motions become less and less correlated at longer and longer times. If we assume that the integrals $\int_0^t R_x(s) \, ds$ and $\int_0^t s R_x(s) \, ds$ are bounded as t becomes large Eq. (3.24) gives in this limit

$$\langle X^2 \rangle \to 2\langle U^2 \rangle T_x t + \text{const.} \tag{3.28}$$

in which

$$T_x = \int_0^\infty R_x(s) \, ds \tag{3.29}$$

is known as the "Lagrangian time scale." Equation (3.28) can be differentiated to give

$$d\langle X^2 \rangle/dt = 2\langle U^2 \rangle T_x \tag{3.30}$$

which shows that after some long enough time the variance of the ensemble averaged concentration distribution for clouds dispersing in a stationary homogeneous field of turbulence grows linearly with time. The time required before this will occur will be proportional to the Lagrangian time scale. Thus a more exact statement of the meaning of "short" and "long" times in Eqs. (3.27) and (3.28) is that Eq. (3.27) is true for $t \ll T_x$ and (3.28) is true for $t \gg T_x$. If $R_x(t)$ is monotonically decreasing, as is often assumed, the Lagrangian time scale is a measure of how long the particle takes to lose memory of its initial velocity.

This is as far as Taylor went, and as far as it seems possible to go without introducing some degree of conjecture.

It is now useful to recall what we learned about molecular diffusion in Chapter 2. We found that Fick's law led to the derivation of the diffusion equation [Eq. (2.7)], that the fundamental solution to the diffusion equation is the Gaussian distribution [Eq. (2.23)], and that a property of the diffusion equation is that the variance of a concentration distribution always grows linearly with time [Eq. (2.35)]. We also found that after some start-up time the random walk generates the Gaussian distribution, that the Gaussian distribution implies that the diffusion equation describes the process, and that the variance of a spreading cloud of molecules undergoing a random walk grows linearly with time. Taylor's analysis has shown that after some start-up time the variance of a spreading cloud of particles in stationary homogeneous turbulent motion grows linearly with time, and suggests that we can define a turbulent mixing coefficient, analogous to the molecular diffusion coefficient, by the relationship

$$\varepsilon_x = \frac{1}{2} \frac{d\langle X^2 \rangle}{dt} = \langle U^2 \rangle T_x. \tag{3.30a}$$

However, this suggestion needs to be investigated further to see when it applies and whether it implies the existence of a "turbulent diffusion equation." Linear growth of the variance is a necessary condition for the diffusion equation to apply, but it is not a sufficient one. On the other hand, the velocity $U(t)$ is a random variable for any fixed time t, so that $X(t) = \int_0^t U(t)\,dt$ is the sum or integral of random variables. The central limit theorem from probability theory tells us that such sums approach normality as $t \to \infty$ provided that the variable $U(t)$ satisfies certain weak independence requirements. It is most likely, although not proven, that a stationary homogeneous turbulent velocity field satisfies these requirements and we may expect that $X(t)$ becomes a normal or Gaussian random variable for large time. If so, the discussion in Section 2.2 shows that the spread of the ensemble mean concentration may be described by a diffusion equation whose simplest three-dimensional form is

$$\frac{\partial C}{\partial t} = \varepsilon_x \frac{\partial^2 C}{\partial x^2} + \varepsilon_y \frac{\partial^2 C}{\partial y^2} + \varepsilon_z \frac{\partial^2 C}{\partial z^2}. \tag{3.31}$$

This equation is written for zero mean flow velocity, as in Taylor's analysis; if the fluid has a mean velocity it is necessary to add the terms $u\,\partial C/\partial x$, etc., as was done for molecular diffusion in Eqs. (2.15) and (2.49).

The analogy with molecular diffusion can be taken further as follows. The equation for conservation of matter in turbulent flow, neglecting molecular diffusion and letting C be, for the moment, the time-varying point concentration, is

$$\frac{\partial C}{\partial t} + U\frac{\partial C}{\partial x} + V\frac{\partial C}{\partial y} + W\frac{\partial C}{\partial z} = 0, \tag{3.32}$$

where U, V, and W are randomly varying velocities. If this equation is averaged over a time long enough to average the turbulent velocity and concentration fluctuations, and if the time averages of U, V, and W are zero, we obtain

$$\frac{\partial \bar{C}}{\partial t} = -\frac{\partial}{\partial x}\overline{UC} - \frac{\partial}{\partial y}\overline{VC} - \frac{\partial}{\partial z}\overline{WC}, \tag{3.33}$$

where, in this equation and the next one only, the overbar indicates the time average. $\overline{UC}$, $\overline{VC}$, and $\overline{WC}$ are the time-averaged turbulent fluxes in the x, y, and z directions, and it follows by comparison with Eq. (3.31) that

$$\overline{UC} = -\varepsilon_x\frac{\partial C}{\partial x}, \qquad \overline{VC} = -\varepsilon_y\frac{\partial C}{\partial y}, \qquad \overline{WC} = -\varepsilon_z\frac{\partial C}{\partial z}. \tag{3.34}$$

Comparison with Fick's law for molecular diffusion, $q = -D\,\partial C/\partial x$ [Eq. (2.1)] shows that ε_x, ε_y, and ε_z are the turbulent equivalents to the molecular diffusion coefficient, i.e., the constants in the relationship that the flux is proportional to the concentration gradient. For this reason ε_x, etc., are often referred to as "Fickian turbulent diffusion coefficients"; since they result from a process involving larger scale random motions they are also often called "eddy diffusivities."

We must now resolve the question of under what conditions a Fickian turbulent diffusion equation like Eq. (3.31) ought to be used. If we are dealing with a cloud of particles originating at a point the requirement given by Taylor's analysis is that more time has elapsed than the Lagrangian time scale. In three-dimensional diffusion the definition of the time scale can be generalized to $T_L = \frac{1}{3}(T_x + T_y + T_z)$, where T_y and T_z are defined in a similar way to T_x, and the requirement for use of constant diffusion coefficients is that $t > T_L$. In many cases, however, the time of origin of the cloud as a point source may not be known, and it is useful to define the validity of the equation in terms of the size of the dispersing cloud. Let us define a "Lagrangian length scale," l_L, by the relation

$$l_L{}^2 = \langle U^2\rangle T_L{}^2. \tag{3.35}$$

The scale l_L gives the order of magnitude of the distance a fluid particle will travel before losing memory of its initial velocity; we will see in later chapters that this scale can sometimes be estimated in bounded problems, such as turbulent mixing in rivers (Chapter 5) in which we will use an estimate that l_L is the depth of flow. By Eq. (3.28) $L^2 \cong 2\langle U^2 \rangle T_L t$, neglecting the constant. Therefore the requirement that $t > T_L$ is equivalent, in terms of the size of the cloud L, to setting a requirement for use of the diffusion equation that

$$L^2 > 2l_L{}^2 \tag{3.36}$$

or, in other words, that the size of the dispersing cloud should substantially exceed the distance over which turbulent motions are correlated.

Equation (3.31) is not the most general form of the turbulent mixing equation, because up to now we have used the simpler forms to stress the concepts. In practical problems the turbulence is often not homogeneous, and it is common to find the diffusion equation written with spatially variable coefficients in the form

$$\frac{\partial C}{\partial t} + u \frac{\partial C}{\partial x} + v \frac{\partial C}{\partial y} + w \frac{\partial C}{\partial z} = \frac{\partial}{\partial x}\left(\varepsilon_x \frac{\partial C}{\partial x}\right) + \frac{\partial}{\partial y}\left(\varepsilon_y \frac{\partial C}{\partial y}\right) + \frac{\partial}{\partial z}\left(\varepsilon_z \frac{\partial C}{\partial z}\right).$$

$$\tag{3.37}$$

We will use this form in Chapter 5 to describe turbulent mixing in rivers, for example. The theoretical foundation for the use of the equation with spatially varying coefficients was given by Kolmogorov (1931, 1933). In order for a concentration at point x at time t to depend only on the initial conditions and time from release as implied by Eq. (3.37), it is necessary that the probability of a particle released at point ξ at time t_0 being at point $\mathbf{x}$ at time t be equal to the probability that the particle reaches some intermediate point $\mathbf{x}_1$ at an intermediate time t_1 times the probability that the particle goes from point $\mathbf{x}_1$ to point $\mathbf{x}$ in time $t - t_1$, integrated over all possible intermediate points or, in other words,

$$p(\mathbf{x}|\xi, t, t_0) = \int_{-\infty}^{\infty} \int_{-\infty}^{\infty} \int_{-\infty}^{\infty} p(\mathbf{x}|\mathbf{x}_1, t, t_1)p(\mathbf{x}_1|\xi, t_1, t_0)\, dx_1\, dy_1\, dz_1. \tag{3.38}$$

This can only be true if the motion of the particle consists of a series of independent random steps, which happens only if $t > T_L$. Kolmogorov showed that it is this independence property which is critical for the validity of the diffusion equation and not the homogeneity or stationarity of the velocity field. He was able to show quite generally that the diffusion of matter in a turbulent velocity field with independent properties satisfies the diffusion equation

$$\frac{\partial C}{\partial t} + \mathbf{U} \cdot \nabla C = \frac{\partial}{\partial x}\left(\varepsilon_{xx}\frac{\partial C}{\partial x}\right) + \frac{\partial}{\partial y}\left(\varepsilon_{yy}\frac{\partial C}{\partial y}\right) + \frac{\partial}{\partial z}\left(\varepsilon_{zz}\frac{\partial C}{\partial z}\right)$$

$$+ \frac{\partial}{\partial x}\left(\varepsilon_{xy}\frac{\partial C}{\partial y}\right) + \frac{\partial}{\partial x}\left(\varepsilon_{xz}\frac{\partial C}{\partial z}\right) + \text{similar terms}, \quad (3.39)$$

where

$$\varepsilon_{xx}(\xi, t) = \frac{1}{2}\frac{d}{dt}\langle X^2(\xi, t)\rangle, \qquad \varepsilon_{xy}(\xi, t) = \frac{1}{2}\frac{d}{dt}\langle X(\xi, t)Y(\xi, t)\rangle,$$

and similar terms are in general functions of position and time. We mentioned before that if the coordinate axes are chosen to coincide with the principal axes of the flow the cross-variances are zero and Eq. (3.39) simplifies to Eq. (3.37). In practical work it is difficult enough to measure or predict the values of the scalar mixing coefficients ε_x, etc., in Eq. (3.37), let alone the values of the tensor components in Eq. (3.39); therefore Eq. (3.37) is the most complex form we are likely to use in practice, and the limits on its use are those given before, namely that $t > T_L$, which allows sufficient time for any sharp gradients to be smoothed over a distance greater than $\sqrt{2}l_L$.

In summary, the analysis given in this section has shown that there is such a thing as a turbulent mixing coefficient that is analogous to the molecular diffusion coefficient; that we can use such a coefficient in a turbulent diffusion equation, of which the most usual practical form is Eq. (3.37); and that the equation can be applied only after the diffusing particles have been in the flow longer than the Lagrangian time scale [Eq. (3.29)] and have spread to cover a distance larger than the Lagrangian length scale [Eq. (3.35)]. In addition, Taylor's analysis shows, by combination of Eqs. (3.30a) and (3.35) that

$$\varepsilon_x = l_L[\langle U^2\rangle]^{1/2}, \quad (3.40)$$

which says that the turbulent mixing coefficient is a product of the Lagrangian length scale (a measure of how far a particle travels before it forgets its initial velocity) and the intensity of the turbulence. In Section 5.1 we will show how Eq. (3.40) gives a way of estimating the magnitude of turbulent mixing coefficients in open channel flow.

3.4   RELATIVE DIFFUSION OF CLOUDS

Let us now turn to the problem of describing the growth of clouds of particles when the time from release is less than the Lagrangian time scale, $t < T_L$.

As we noted in Section 3.1 there are two ways of forming ensemble average concentrations when a sequence of identical clouds are released from a specified point in a turbulent fluid. In the first method (considered in the previous section) an ensemble mean is formed by averaging the concentration at each point in space at identical times. If $n(\mathbf{x}, t)$ is the concentration in one trial at point $\mathbf{x}$ and time t then we defined

$$C(\mathbf{x}, t) = \langle n(\mathbf{x}, t) \rangle. \tag{3.41}$$

For the measure of the spread of the concentration distribution we used

$$\Sigma_x^2 = \frac{1}{M} \int_{-\infty}^{\infty} \int_{-\infty}^{\infty} \int_{-\infty}^{\infty} (x - \langle \overline{X} \rangle)^2 C(\mathbf{x}, t) \, dx, \tag{3.42}$$

where

$$\langle \overline{X} \rangle = \frac{1}{M} \int_{-\infty}^{\infty} \int_{-\infty}^{\infty} \int_{-\infty}^{\infty} x C(\mathbf{x}, t) \, dx, \tag{3.43}$$

and, as the Taylor diffusion theory considered in the previous section enabled us to define how

$$L^2(t) \tag{3.44}$$

changes with time, we defined a size of the cloud as

$$L^2(t) = \tfrac{1}{3}(\Sigma_x^2 + \Sigma_y^2 + \Sigma_z^2). \tag{3.45}$$

In the second method an ensemble mean concentration is formed by averaging the concentration at points equidistant from the center of mass of each cloud in the trial. We described the rate of growth of the ensemble of such clouds by

$$\langle \sigma_x^2 \rangle = \left\langle \frac{1}{M} \int_{-\infty}^{\infty} \int_{-\infty}^{\infty} \int_{-\infty}^{\infty} (x - \overline{X})^2 n(\mathbf{x}, t) \, dx \, dy \, dz \right\rangle, \tag{3.46}$$

where

$$\overline{X} = \frac{1}{M} \int_{-\infty}^{\infty} \int_{-\infty}^{\infty} \int_{-\infty}^{\infty} x n(\mathbf{x}, t) \, dx \, dy \, dz. \tag{3.47}$$

The size of an ensemble cloud in this context is defined as

$$\langle l^2(t) \rangle = \tfrac{1}{3} \langle \sigma_x^2 + \sigma_y^2 + \sigma_z^2 \rangle. \tag{3.48}$$

We can define the ensemble mean concentration formed by aligning the centers of mass by

$$\psi(\chi, t) = \langle n(\mathbf{x} - \overline{\mathbf{X}}, t) \rangle, \tag{3.49}$$

where

$$\chi = \mathbf{x} - \overline{\mathbf{X}}. \tag{3.50}$$

By translating the origin to $\chi = 0$, Eq. (3.46) becomes

$$\langle \sigma_x^2 \rangle = \frac{1}{M} \int_{-\infty}^{\infty} \int_{-\infty}^{\infty} \int_{-\infty}^{\infty} \chi_x^2 \psi(\chi, t) \, d\chi_x \, d\chi_y \, d\chi_z. \tag{3.51}$$

The importance of this description of the turbulent diffusion of clouds is that the mean concentrations as described are actually higher than those described by the previously considered method (although still not as high as the peak concentrations in single clouds). This is apparent from the result

$$L^2(t) = \langle l^2(t) \rangle + \tfrac{1}{3}\langle [(\overline{X} - \langle \overline{X} \rangle)^2 + (\overline{Y} - \langle \overline{Y} \rangle)^2 + (\overline{Z} - \langle \overline{Z} \rangle)^2] \rangle, \tag{3.52}$$

as shown in Section 3.2.

Now, whereas Taylor's theory enabled us to predict the rate of growth of $L^2(t)$, we consider here a theory that enables us to predict $\langle \sigma_x^2 \rangle$ and therefore $\langle l^2(t) \rangle$. The theory is based on arguments first presented by Richardson (1926) and later refined by Batchelor (1952).

As can perhaps be appreciated, there is a significant degree of confusion in the literature regarding the concentration actually used in an analysis, i.e., whether it is $n(\mathbf{x}, t)$, $C(\mathbf{x}, t)$, or $\psi(\chi, t)$. It is not clear exactly which concentration definition Richardson actually had in mind and for this reason we prefer Batchelor's approach as it appears better defined.

We begin by considering the statistics of the separation of a particular pair of particles in a cloud and to keep the arguments simple we discuss only a one-dimensional cloud. The extension to higher dimensions follows from similar arguments.

We define the probability density function

$Q(s, t; s_0, t_0) \, ds$ = probability that a pair of particles separated
  by a distance between s_0 and $s_0 + ds_0$ at time t_0
  will be separated by a distance between s and
  $s + ds$ at time t.

$$\tag{3.53}$$

The thrust of Batchelor's argument is as follows. If the pair of particles are not widely separated in relation to the scale of the turbulent eddying motion then only two length scales are important to the statistics of their *relative* (or separat-ing) motion. These two scales must be their initial separation s_0 and the Kolmogorov scale $(\nu^3/\epsilon)^{1/4}$. Similarly, only two time scales are important to the motion: the scale $t - t_0$ and the Kolmogorov scale $(\nu/\epsilon)^{1/2}$. Simple dimensional analysis then implies that $\langle s^2 \rangle$, as defined by

$$\langle s^2 \rangle = \int_{-\infty}^{\infty} s^2 Q(s, t; s_0, t_0) \, ds, \tag{3.54}$$

must be given by

$$\frac{d\langle s^2 \rangle}{dt} = \epsilon \tau^2 f\left(\frac{s_0}{\epsilon^{1/2}\tau^{3/2}}, \frac{\bar{v}\epsilon^{1/2}}{v^{1/2}}\right),$$ (3.55)

where $\tau = t - t_0$.

For very long times from release the two particles will wander independently and

$$\langle s^2 \rangle \to 2\langle X^2 \rangle$$ (3.56)

because the variance of the sum is equal to the sum of the variances. Since the mean motion is zero the position of the center of mass of the cloud will tend towards zero, which implies that

$$\psi(\chi, t) \to C(\mathbf{x}, t) \qquad \text{as} \quad t \to \infty.$$ (3.57)

For very short times, the rate of increase is a function of viscosity and the initial separation; however viscosity is not important if s is much larger than the Kolmogorov scale $v^{3/4}\epsilon^{-1/4}$, which we have already mentioned is of the order of 1 mm in the open ocean. If enough time has also passed that the initial separation of the particles s_0 has been forgotten, Eq. (3.55) becomes

$$\frac{d}{dt}\langle s^2 \rangle \sim C_2 \epsilon(\tau - t_1)^2,$$ (3.58)

where t_1 is proportional to $s_0^{2/3}\epsilon^{-1/3}$ and C_2 is a universal constant. Integrating this result gives $\langle s^2 \rangle \sim t^3$ so that we also have

$$\frac{d\langle s^2 \rangle}{dt} \sim \epsilon^{1/3}[\langle s^2 \rangle]^{2/3},$$ (3.59)

or, in other words, the rate of increase of the mean square separation of particles is proportional to the mean square separation to the power $\frac{2}{3}$.

At this point it is worthwhile to review the experimental results collated by Richardson (1926). Figure 3.4 is a reproduction of Richardson's original curve

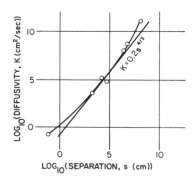

Figure 3.4   Richardson's original plot of $\frac{1}{2}d\langle s^2 \rangle/dt = K$ versus particle separation s.

relating $d\langle s^2\rangle/dt$ to the separation of the particles in a large number of experiments. It will be noted that

$$\frac{d\langle s^2\rangle}{dt} \propto s^{4/3}, \tag{3.60}$$

and the similarity to the result given in Eq. (3.59) cannot be overlooked.

Now, it might well be asked how is the probability density function for the separation of a pair of particles related to the function of interest $\psi(\chi, t)$ which is the ensemble mean concentration of clouds whose centers of mass are made to coincide. The answer lies in the recognition that the probability density function for the separation of a pair of particles is formally equivalent to the probability density function describing the distance of a particle from the center of mass of an identical pair. (The center of mass of a pair always lies midway between such a pair.) Then, *provided that the initial separation* of pairs of particles in a cloud has been forgotten, and that *the turbulence is homogeneous*, the description of the mean square separation for all pairs of points in an ensemble will be identical and will be exactly the same as the description of the mean square displacement from the center of mass.

If we let $\chi = |\boldsymbol{\chi}|$ then we have that

$$\langle\chi^2\rangle = \frac{1}{M}\int_{-\infty}^{\infty}\int_{-\infty}^{\infty}\int_{-\infty}^{\infty}\chi^2\psi(\boldsymbol{\chi}, t)\,d\chi_x\,d\chi_y\,d\chi_z = 3(l^2(t)), \tag{3.61}$$

and so

$$\frac{d\langle l^2(t)\rangle}{dt} = k\epsilon^{1/3}\langle l^2(t)\rangle^{2/3}, \tag{3.62}$$

for some universal constant k. From this result we postulate the existence of a differential equation describing $\psi(\boldsymbol{\chi}, t)$ of the form

$$\frac{\partial\psi}{\partial t} = \frac{\partial}{\partial\chi_x}\left(K\frac{\partial\psi}{\partial\chi_x}\right) + \frac{\partial}{\partial\chi_y}\left(K\frac{\partial\psi}{\partial\chi_y}\right) + \frac{\partial}{\partial\chi_z}\left(K\frac{\partial\psi}{\partial\chi_z}\right), \tag{3.63}$$

with

$$K = \alpha\langle\chi^2\rangle^{2/3}, \tag{3.64}$$

for some constant α which satisfies the condition

$$K = \frac{1}{2}\frac{d\langle\chi^2\rangle}{dt} = \alpha\langle\chi^2\rangle^{2/3}. \tag{3.65}$$

The solution† of Eq. (3.63) corresponding to a point source at the origin $|\chi| = 0$ is

$$\psi(\chi, t) = \frac{1}{[2\pi\langle\chi^2\rangle]^{n/2}} \exp\left(\frac{-\chi^2}{2\langle\chi^2\rangle}\right) \tag{3.66}$$

with $n = 3$ and

$$\langle\chi^2\rangle = (\tfrac{2}{3}\alpha t)^3. \tag{3.67}$$

It should be recognized that this solution should only apply where

$$t \gg \langle\chi_0^2\rangle^{2/3}\epsilon^{-1/3},$$

which implies that the original cloud size has been forgotten. However, the remarkable thing about the result given in Eq. (3.65) is that it seems to apply in the ocean to length scales ranging from 10 m to 1000 km. Figure 3.5 is a plot of data obtained in numerous two-dimensional field diffusion studies and collected together by Okubo (1974). It is clear that the "$\frac{4}{3}$ laws" is apparently valid over a much larger scale than the Batchelor–Kolmogorov theory would indicate as appropriate. The reasons for this are not clear but other turbulent diffusion theories, not so limited in length scale, also give rise to the same diffusion law (Onsager, 1945; von Weisacker, 1948; Heisenberg 1948; Obukhov, 1959; Lin, 1960; Bowden, 1970) so that it appears that other processes leading to the $\frac{4}{3}$ law are likely ‡

One of the difficulties with the acceptance of the $\frac{4}{3}$ law has been the apparent decrease in the constant α with an increase in the scale of the diffusion. According to Eq. (3.62) α is proportional to the one-third power of the energy dissipation rate and the coefficient of proportionality should be a universal constant.

The difficulty has been resolved by Ozmidov (1965). He reasoned that since the parameter ϵ is actually the energy pass-through rate in the energy cascade from large energy containing eddies to small dissipative eddies it will decrease with an increase in scale. The reason for this is that energy is actually fed to ocean turbulence at roughly three scales; by wind waves at about 10 m, by inertial and tidal motions at about 10 km, and by atmospheric pressure systems influencing the general circulation patterns at about 1000 km. Okubo and Ozmidov (1970) have shown that allowing for the increase in the energy passed through the spectrum as the scale decreases does result in a better data fit. However, unless the scale of the problem contemplated is very large a reasonable estimate for α for engineering purposes is given by taking the universal constant to have a value in the range of 0.002–0.01 $cm^{2/3}/sec$, as shown by Fig. 3.5.

---

† Note that the solution given is for three-dimensional diffusion. Most field work where this theory is appropriate involves two-dimensional diffusion so that $n = 2$ in Eq. (3.66).

‡ See, for example, the discussion of dispersion in unbounded shear flow in Section 4.5.

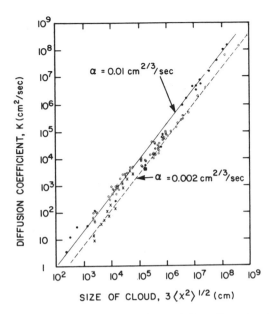

Figure 3.5 A unified diagram of patch diffusion and float dispersion (after Okubo, 1974). ● surface float dispersion. ○ upper mixed layer, × coastal thermocline, ■ 300 m depth patch diffusion. ——— $K = 0.01 \langle \chi^2 \rangle^{2/3}$; --- $K = 0.002 \langle \chi^2 \rangle^{2/3}$.

One further consequence of the description above should be noted. It is that Fick's law also applies to the concentration gradient and flux as defined by

$$\mathbf{q} = -\frac{1}{2} \frac{d\langle \chi^2 \rangle}{dt} \cdot \nabla \psi,$$

provided that the coordinate system (χ_x, χ_y, χ_z) is based on the center of mass of the ensemble of clouds, and the concentration is an ensemble mean formed in the appropriate way.

3.5 SUMMARY

This chapter has given an introduction to the kinematics of turbulence, particularly as it affects the mixing of pollutants. We have discussed the range of scales of motion in turbulent flow, and how they interact with the size of a dispersing cloud of tracer. We have defined three concentrations, as follows:

$n(\mathbf{x}, t)$ is the concentration observed at point $\mathbf{x}$ at time t after the release of a single cloud of particles from point ξ.

$\psi(\chi, t)$ is the concentration computed at point χ at time t by releasing a large number of clouds of particles at point ξ, superposing the centers of mass of each cloud, and then averaging the values of n over all the clouds. χ is measured relative to the center of mass of the cloud and t is measured relative to the time of release.

$C(\mathbf{x}, t)$ is the ensemble average obtained by releasing a large number of clouds of particles at point ξ at various times and averaging the values of n observed at point $\mathbf{x}$ for all clouds at time t after their release.

From the point of view of pollution control n is the most relevant concentration because it is what is actually seen by an organism in the water. Unfortunately we have not been able to give a general method for predicting values of n, and none seems to exist. We have seen, however, that spreading of both ψ and C can be modeled by Fickian diffusion equations similar to the equation introduced in Chapter 2. Their uses and limitations are as follows:

$\psi(\chi, t)$.   The average concentration we have called ψ is not an observable concentration because it requires an averaging by superposition of centers of mass that never actually occurs in the environment. Nevertheless, the concept is important because it leads to the celebrated "$\frac{4}{3}$ law" which says that the diffusion coefficient is proportional to the $\frac{4}{3}$ power of the size of the cloud. We have shown that this "law" only applies in reality to the average over a large number of clouds, and then only in homogeneous turbulence far from any boundaries and only when the size of the cloud is large enough to have forgotten its original size but still smaller than the largest scales of turbulence. As has been observed in many field experiments, the growth of any single cloud may differ greatly from the ensemble average, and may bear little relation to the dictates of the $\frac{4}{3}$rds law. The $\frac{4}{3}$ law is useful for studies in the ocean, but a great deal of care must be used in applying it in lakes and estuaries where the turbulence is usually not homogeneous and boundaries are usually nearby.

$C(x, t)$.   The ensemble average concentration C is obtained by averaging at a fixed point over a large number of releases of tracer. We have also shown that it is likely that C is what will be observed after even just one release of tracer if the tracer has been in the flow longer than the Lagrangian time scale of the turbulence [as defined in Eq. (3.29)]. After the tracer has been in the flow longer than the Lagrangian time scale, further changes in C are governed by the Fickian diffusion equation with constant coefficients. The magnitude of the coefficients results from the product of a length scale of the turbulence and the turbulence intensity [Eq. (3.40)]. The magnitude is of course much greater than for molecular diffusion, but otherwise the mathematical description is the same as for molecular diffusion and the results given in Chapter 2 apply. This is the basis for most practical applications of the theory of turbulent diffusion.

To summarize again in brief, there can be a great difference between the spread of individual small clouds of tracer and the spread of a widely distributed tracer. Individual clouds grow at a rate which increases with their size, and which is different for each cloud, until they reach a size larger than the largest scales of turbulent motion. After that, but only after that, the further spread is described by a diffusion equation with constant coefficients. There is an intermediate stage in which the average growth of the cloud can be described by the diffusion equation with a diffusion coefficient proportional to the $\frac{4}{3}$ power of the size of the cloud, but only if the turbulence is homogeneous and no boundaries affect the growth. The diffusion equation with constant coefficients is used for practical studies of rivers and estuaries (Chapters 5 and 7) because the size of the pollutant cloud is usually greater than the largest eddy sizes. In lakes and coastal areas (Chapters 6 and 10) the pollutant cloud is often smaller than the large eddies and special techniques, sometimes based on the $\frac{4}{3}$ law and sometimes not, must be devised.

Chapter 4

Shear Flow Dispersion

In 1953 the great English fluid mechanician, Sir Geoffrey Taylor, published a paper describing the spread of dissolved contaminants in laminar flow through a pipe. A year later (Taylor, 1954) he extended his analysis to turbulent flow. Taylor's analysis has since been extended to a variety of environmental flows; in later chapters we will see how it can be used to give a reasonably accurate estimate of the rate of longitudinal dispersion in rivers, and a partial estimate of longitudinal dispersion in estuaries. Common to all these flows is that spreading in the direction of flow is caused primarily by the velocity profile in the cross section; flows with velocity gradients are often referred to as "shear flows," and the mechanism Taylor analysed is often known as the "shear effect."

This chapter explains the fundamentals of shear flow dispersion. Some readers may find the mathematics difficult, but understanding them is important, for he who understands the fundamental mechanics of shear flow dispersion will have little trouble understanding the remaining material in this book, or, for that matter, knowing how to approach almost any problem of dispersion in the environment.

4.1 DISPERSION IN LAMINAR SHEAR FLOW

4.1.1 Introductory Remarks

Taylor began with the velocity profile for laminar flow in a pipe, as sketched in Fig. 4.1; he realized that if two molecules are being carried in the flow, say one in the center and one near the wall, the rate of separation caused by the difference in advective velocity will greatly exceed that caused by molecular motion. He also realized that, given enough time, any single molecule would wander randomly throughout the cross section of the pipe because of molecular diffusion, and would sample at random all the advective velocities. Therefore if a long enough averaging time was available, a single molecule's time-averaged velocity would be equal to the instantaneous cross-sectional average of all the molecules' velocities. The separation between any two molecules with different advective velocities would, however, increase much more rapidly than if their advective velocities were the same and they could separate only because of their thermal molecular motion.

The velocity of any single molecule is essentially that of the stream line on which it is located, a function of cross-sectional position. Because of molecular diffusion each molecule moves at random back and forth across the cross section and after some long enough "forgetting time" we can expect that its location is independent of the location at which it started. Therefore its velocity is independent of its initial velocity. Thus we can imagine that the motion of a single molecule is the sum of a series of independent steps of random length. If we adopt a coordinate system moving at the mean velocity the random steps are, with respect to the moving coordinate system, equally likely to be backward as forward, since the mean motion is zero. The motion is similar to the random walk discussed in Section 2.2, provided that the flow continues unchanged for a time much longer than the "forgetting time." Hence we should expect to find that Eq. (2.4) describes the spread of particles along the axis of the pipe, except that since the step length and time increment are very much different from those of molecular diffusion we expect to find a different value of the "diffusion coefficient" (which we will rename the "dispersion coefficient," since it results from the process of shear flow dispersion).

Let us try to make these ideas more specific by comparing the process with the analysis of diffusion by continuous movements (Taylor, 1921). In the previous chapter we showed that the rate of spreading in a field of turbulence is

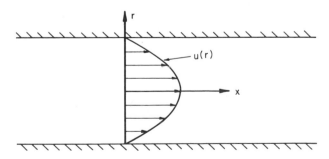

Figure 4.1 The parabolic velocity distribution in laminar pipe flow.

described by a turbulent diffusion coefficient, $\varepsilon = \langle U^2 \rangle T_L$, where T_L is the Lagrangian time scale [see Eq. (3.30a)]. The motion of a single molecule in laminar pipe flow is similar to the motion of a fluid particle in turbulent flow in that the velocity of the molecule is a stationary random function of time. The Lagrangian time scale will be proportional to the time required to sample the whole field of velocities, which is proportional to the time scale for cross-sectional mixing a^2/D [see Eq. (1.1)], where a is the pipe radius and D the molecular diffusion coefficient. The mean square velocity deviation of the molecule $\langle U^2 \rangle$ results primarily from the wandering of the molecule across the cross section, during which it samples velocities ranging from zero at the wall to the peak velocity u_0 at the centerline. Therefore we should expect that $\langle U^2 \rangle$ will be proportional to u_0^2, and that spreading of the molecules in the direction of the pipe axis should be described, at least in the limit $t \gg T_L$, by a relation of the form

$$K = \langle U^2 \rangle T_L \propto u_0^2 a^2/D, \qquad (4.1)$$

where K is the longitudinal dispersion coefficient mentioned in the previous paragraph.

The proportionality given in Eq. (4.1) can be predicted from the results of the paper published by Taylor in 1921, but it remained until 1953 for Taylor himself to publish a wholly different analysis to supply the constant of proportionality. Rather than continue directly to study Taylor's 1953 analysis, however, it is easier to present the concept and method of analysis in a two-dimensional x-y plane. In so doing we also see that Taylor's methodology is applicable to a much wider range of flows than laminar flow in a pipe. After a general introduction, we will return to the details of what Taylor did.

4.1.2   A Generalized Introduction

Consider the two-dimensional flow sketched in Fig. 4.2a. The flow is guided between parallel walls separated by a distance h so that all the flow lines are parallel to the walls. The velocity variation between the walls is given by $u(y)$, and the mean velocity is $\bar{u}$. Note that whatever the distribution, the mean velocity can be found by integration

$$\bar{u} = \frac{1}{h} \int_0^h u \, dy \qquad (4.2)$$

and the deviation of the velocity from the cross-sectional mean is defined as

$$u'(y) = u(y) - \bar{u}. \qquad (4.3)$$

In this chapter an overbar will always denote a cross-sectional average. Let the flow carry a solute with concentration $C(x, y)$ and molecular diffusion

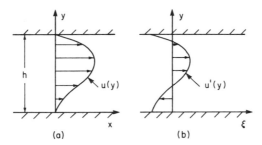

Figure 4.2 (a) An example velocity distribution. (b) The same velocity distribution as in (a) but transformed to a coordinate system moving at the mean velocity.

coefficient D, and define the mean concentration at any cross section of the flow to be

$$\bar{C} = \frac{1}{h} \int_0^h C \, dy. \tag{4.4}$$

As before, the deviation from the mean is defined by $C'(y) = C(y) - \bar{C}$. Since the only flow is in the x direction, the diffusion equation is

$$\frac{\partial}{\partial t}(\bar{C} + C') + (\bar{u} + u')\frac{\partial}{\partial x}(\bar{C} + C') = D\left[\frac{\partial^2}{\partial x^2}(\bar{C} + C') + \frac{\partial^2 C'}{\partial y^2}\right]. \tag{4.5}$$

For the present let us discuss only a laminar flow, so that we need not worry about turbulent fluctuations or the effect of turbulence on mass transport. We confine our view thus for the first part of the present chapter so that we can concentrate on the analytical method. Later (Section 4.2) we will show that the analysis applies equally well to turbulent flows; turbulence will be treated simply by allowing u and u' in Eq. (4.5) to represent the means of the turbulent fluctuations, and replacing D by the "eddy coefficient" of the turbulent flow.

Equation (4.5) can be simplified by a transformation to a coordinate system whose origin moves at the mean flow velocity. Let

$$\xi = x - \bar{u}t, \qquad \tau = t \tag{4.6}$$

and note that by the "chain rule" of calculus

$$\frac{\partial}{\partial x} = \frac{\partial \xi}{\partial x}\frac{\partial}{\partial \xi} + \frac{\partial \tau}{\partial x}\frac{\partial}{\partial \tau} = \frac{\partial}{\partial \xi}$$

$$\frac{\partial}{\partial t} = \frac{\partial \xi}{\partial t}\frac{\partial}{\partial \xi} + \frac{\partial \tau}{\partial t}\frac{\partial}{\partial \tau} = -\bar{u}\frac{\partial}{\partial \xi} + \frac{\partial}{\partial \tau} \tag{4.7}$$

so that Eq. (4.5) becomes

$$\frac{\partial}{\partial \tau}(\bar{C} + C') + u'\frac{\partial}{\partial \xi}(\bar{C} + C') = D\left[\frac{\partial^2}{\partial \xi^2}(\bar{C} + C') + \frac{\partial^2 C'}{\partial y^2}\right]. \qquad (4.8)$$

The transformation to the ξ, τ system allows us to view the flow as an observer moving at the mean velocity. In the moving system the only observable velocity is u', as sketched in Fig. 4.2b; hence, the transformed equation does not contain $\bar{u}$.

We argued in our introductory comments that the rate of spreading along the flow direction due to the velocity profile should greatly exceed that due to molecular diffusion. We will see later that this is true. If so we can neglect the longitudinal diffusion term in Eq. (4.8), leaving

$$\frac{\partial}{\partial \tau}\bar{C} + \frac{\partial}{\partial \tau}C' + u'\frac{\partial \bar{C}}{\partial \xi} + u'\frac{\partial C'}{\partial \xi} = D\frac{\partial^2 C'}{\partial y^2}. \qquad (4.9)$$

Unfortunately, we are still left with an intractable equation, because u' varies with y. A general procedure for dealing with differential equations with variable coefficients is not available, and no general solution of (4.9) can be found. Taylor obtained his solution by a display of the sort of brilliance we can only admire; he discarded three of the first four terms, including the term $\partial \bar{C}/\partial \tau$ which expresses precisely what we are trying to find, the rate of decay of concentration, to leave the easily solvable equation for $C'(y)$

$$u'\frac{\partial \bar{C}}{\partial \xi} = D\frac{\partial^2 C'}{\partial y^2}, \quad \text{with} \quad \frac{\partial C'}{\partial y} = 0 \quad \text{at} \quad y = 0, h. \qquad (4.10)$$

The reader may wish to pass immediately to the next paragraph to see how Taylor's analysis continues and, in the end, gives the value of the discarded term $\partial \bar{C}/\partial \tau$, or he may wish to pause and wonder, as many researchers have, what is the justification for the major surgery performed on (4.9). Essentially the justification depends on the orders of magnitude of the terms in the equation. If we apply the operator $1/h \int_0^h (\ ) \, dy$ to each term of (4.9), we have

$$\frac{\partial \bar{C}}{\partial \tau} + \overline{u'\frac{\partial C'}{\partial \xi}} = 0, \qquad (4.11)$$

the overbar indicating a cross-sectional average of the whole term. Subtracting (4.11) from (4.10) gives

$$\frac{\partial C'}{\partial \tau} + u'\frac{\partial \bar{C}}{\partial \xi} + u'\frac{\partial C'}{\partial \xi} - \overline{u'\frac{\partial C'}{\partial \xi}} = D\frac{\partial^2 C'}{\partial y^2}. \qquad (4.12)$$

We can now imagine that in some circumstances, which may have to be investigated in more detail later, $\bar{C}$ and C' will both be well behaved, slowly varying functions, and C' will be much smaller than $\bar{C}$. If so, the third and fourth terms will nearly balance each other and will be much smaller than the second term. Dropping terms three and four leaves the diffusion equation for a cross sectional slice

$$\frac{\partial C'}{\partial \tau} - D\frac{\partial^2 C'}{\partial y^2} = -u'\frac{\partial \overline{C}}{\partial \xi},\tag{4.13}$$

in which the term $-u'\,\partial\overline{C}/\partial\xi$ plays the role of a source term of variable strength. The net addition by the source term is zero because the average of u' is zero. If, in addition, circumstances are such that $\partial\overline{C}/\partial\xi$ remains constant for a long time, so that the source is constant, the solution to the reduced equation (4.13) is the steady-state solution obtained by solving (4.10), or in other words, Taylor's discard of the first three terms in (4.9) is justified. The solution to (4.10) implies that the cross-sectional concentration profile, $C'(y)$ is established by a simple balance between longitudinal advective transport and cross-sectional diffusive transport as illustrated in Fig. 4.3. Taylor assumed that the balance would be reached; let us now complete the analysis and then return to the question of when it actually is.

Equation (4.10) has the solution

$$C'(y) = \frac{1}{D}\frac{\partial \overline{C}}{\partial x}\int_0^y\int_0^y u'\,dy\,dy + C'(0).\tag{4.14}$$

Now consider the rate of mass transport in the streamwise direction. The mass transport, relative to the moving coordinate axis, is given by

$$\dot{M} = \int_0^h u'C'\,dy = \frac{1}{D}\frac{\partial \overline{C}}{\partial x}\int_0^h u'\int_0^y\int_0^y u'\,dy\,dy\,dy.\tag{4.15}$$

The extra term $\int_0^h u'\{C'(0)\}\,dy = 0$ because $\int_0^h u'\,dy = 0$.

We now note an essential and perhaps surprising result: *The total mass transport in the streamwise direction is proportional to the concentration gradient in the streamwise direction.* This is exactly the result we found for molecular diffusion, but now we find it in an integrated sense for diffusion in the flow direction due to the whole field of flow.

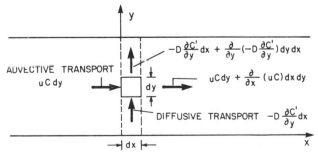

Figure 4.3   The balance of advective flux versus diffusive flux expressed by Eq. (4.10).

Because of this remarkable result, we are now able to define a bulk transport coefficient, or "dispersion" coefficient, in analogy to the molecular diffusion coefficient, by the equation

$$\dot{M} = -hK \frac{\partial \bar{C}}{\partial x} \qquad (4.16)$$

in which h, the depth, is the area per unit width of flow. The dispersion coefficient K expresses the diffusive property of the velocity distribution and is generally known as the "longitudinal dispersion coefficient." Comparing (4.15) and (4.16) we see that

$$K = \frac{-1}{hD} \int_0^h u' \int_0^y \int_0^y u' \, dy \, dy \, dy. \qquad (4.17)$$

K plays the same role for the whole cross section as does D, the molecular diffusion coefficient, on a microscopic scale. Thus we can write a one-dimensional diffusion equation for cross-sectional averages, which in the moving coordinate system is

$$\frac{\partial \bar{C}}{\partial \tau} = K \frac{\partial^2 \bar{C}}{\partial \xi^2}. \qquad (4.18)$$

To return to the fixed coordinate system we must reintroduce the term containing the mean advective velocity, to give

$$\frac{\partial \bar{C}}{\partial t} + \bar{u} \frac{\partial \bar{C}}{\partial x} = K \frac{\partial^2 \bar{C}}{\partial x^2}. \qquad (4.19)$$

This equation is known as the "one-dimensional dispersion equation"; we will see in later chapters that it is used widely in the analysis of dispersion in environmental flows such as rivers and estuaries.

Finally, we must return to the question of when the balance of advection and diffusion assumed by Taylor and expressed by Eq. (4.10) actually occurs. Let us suppose, for example, that at some initial time $t = 0$ a line source of tracer is deposited in the flow shown in Fig. 4.4a. The actual initial distribution of the tracer doesn't much matter, but it is easiest to visualize a line source. Initially the line source is advected and distorted by the velocity profile. At the same time the distorted line source begins to diffuse across the cross section; shortly we see a smeared cloud, with trailing stringers along the boundaries, as in Fig. 4.4b. During this period advection and diffusion are by no means in balance, and Taylor's analysis does not apply. Instead, what we find if we plot the cross-sectional average concentration is the appearance of a skewed longitudinal distribution as shown in Fig. 4.4c.

If we wait a much longer time than that shown in the figures, it is experimentally observed and intuitively apparent that the cloud of tracer extends over a long

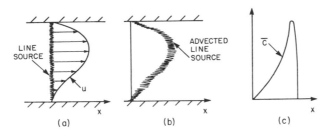

Figure 4.4 The production of a skewed longitudinal distribution by advection of a line source.
(a) The line source and the velocity profile. (b) The advected line source. (c) The longitudinal
distribution of the cross-sectional mean concentration corresponding to the distribution shown
in (b).

distance in the x direction, $\bar{C}$ varies slowly along the channel, and $\partial \bar{C}/\partial x$ is
essentially constant over a long period of time. C' becomes small because
cross-sectional diffusion evens out cross-sectional concentration gradients.
These are the circumstances which we said were required to establish the balance
shown in Eq. (4.10). Once the balance is established further longitudinal
spreading follows (4.19), whose solution after sufficiently long time is a normally
distributed cloud moving at the mean speed $\bar{u}$, and continuing to spread according to Eq. (2.25) with $d\sigma^2/dt = 2K$.

As for exactly how long we must wait, Chatwin (1970) has shown that the
dispersing cloud will become normally distributed approximately a time
h^2/D after its injection. The skewed longitudinal distribution shown in Fig.
4.4c is produced in the period $t < 0.4h^2/D$. When $t > 0.4h^2/D$, the variance of
the dispersing cloud grows linearly with time, and the initial skew degenerates
into the normal distribution in the period $0.4 < tD/h^2 < 1$. Equation (4.19)
applies with reasonable accuracy for $t > 0.4h^2/D$.

4.1.3 A Simple Example

In this section we carry out the computations described in the previous section,
using the simplest possible physically realisable geometry. Consider two parallel
plates of infinite extent, separated by a distance h. The top plate is moving at
velocity U relative to the bottom one and the space between the plates is filled
with fluid. For simplicity assume that the top plate is moving to the right with
velocity $U/2$, and the bottom plate to the left with velocity $U/2$. In laminar flow
the velocity distribution between the plates is given by

$$u(y) = Uy/h, \tag{4.20}$$

where y is the vertical coordinate as defined in Fig. 4.5.

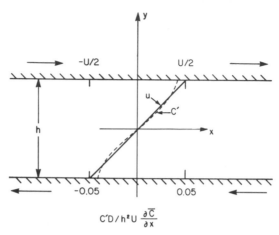

Figure 4.5 The velocity profile used in the simple example, and the resulting concentration profile. The concentration profile corresponds to $\bar{C}$ decreasing in the positive x direction, which leads to positive transport.

Suppose that a slug of a tracer material is injected between the plates, and that a time greater than h^2/D has elapsed since the injection so that the tracer is well distributed. Then the concentration profile is given by (4.14),

$$C'(y) = \frac{1}{D}\frac{\partial \bar{C}}{\partial x}\int_{-h/2}^{y}\int_{-h/2}^{y}\frac{Uy}{h}\,dy\,dy + C'\left(-\frac{h}{2}\right)$$

$$= \frac{1}{D}\frac{\partial \bar{C}}{\partial x}\frac{U}{2h}\left(\frac{y^3}{3} - \frac{h^2 y}{4} - \frac{h^3}{12}\right) + C'\left(-\frac{h}{2}\right). \tag{4.21}$$

C' at $y = -h/2$ can be found from the condition that the average value of C' must be zero, or more easily from the condition that by symmetry $C' = 0$ at $y = 0$. Figure 4.5 shows a plot of the concentration profile. It is not necessary to compute the value of $C'\,(-h/2)$ since the term containing it will integrate to zero in the next step. We now determine the dispersion coefficient from (4.17),

$$K = -\frac{1}{h\,\partial \bar{C}/\partial x}\int_{-h/2}^{h/2} u'C'\,dy$$

$$= -\frac{U}{2h^2 D}\int_{-h/2}^{h/2}\frac{Uy}{h}\left[\frac{y^3}{3} - \frac{h^2 y}{4} - \frac{h^3}{12} + C'\left(-\frac{h}{2}\right)\right]dy$$

$$= \frac{U^2 h^2}{120D}. \tag{4.22}$$

4.1.4 Taylor's Analysis of Laminar Flow in a Tube

Taylor, who presented the concepts used in the previous sections in his classic paper in 1953, analyzed the dispersion of a solute in laminar flow in a tube. It is assumed that the solute has been in the tube long enough to become well distributed over the cross section, so that axial symmetry can be used. The velocity distribution (see Fig. 4.1) is

$$u(r) = u_0(1 - r^2/a^2), \tag{4.23}$$

where a is the radius of the tube and u_0 the maximum velocity on the centerline. By integration it can be shown that the mean velocity is $u_0/2$. In cylindrical coordinates the advective diffusion equation becomes

$$\frac{\partial C}{\partial t} + u_0\left(1 - \frac{r^2}{a^2}\right)\frac{\partial C}{\partial x} = D\left(\frac{\partial^2 C}{\partial r^2} + \frac{1}{r}\frac{\partial C}{\partial r} + \frac{\partial^2 C}{\partial x^2}\right). \tag{4.24}$$

Shifting to a coordinate system moving at velocity $u_0/2$, neglecting $\partial^2 C/\partial x^2$ and $\partial C/\partial t$ as before, and setting $Z = r/a$ gives

$$\frac{u_0 a^2}{D}(\tfrac{1}{2} - Z^2)\frac{\partial \bar{C}}{\partial x} = \frac{\partial^2 C'}{\partial Z^2} + \frac{1}{Z}\frac{\partial C'}{\partial Z}. \tag{4.25}$$

Integrating twice and using the boundary condition that $\partial C'/\partial Z = 0$ at $Z = 1$, we have

$$C' = \frac{u_0 a^2}{8D}(Z^2 - \tfrac{1}{2}Z^4)\frac{\partial \bar{C}}{\partial x} + \text{const.} \tag{4.26}$$

As before we can set

$$K = -\frac{\dot{M}}{A\,\partial\bar{C}/\partial x} = \frac{1}{A\,\partial\bar{C}/\partial x}\int_A u'C'\,dA, \tag{4.27}$$

where A is the cross-sectional area πa^2, and obtain by another integration

$$K = a^2 u_0^2/192D. \tag{4.28}$$

Note the surprising result that the longitudinal dispersion coefficient is inversely proportional to the molecular diffusion coefficient. For salt in water $D \approx 10^{-5}$ cm²/scc. As an example of orders of magnitude, a flow at a peak velocity of 1 cm/sec, in a tube of radius 2 mm has a value of $K = 20$ cm²/sec, which is more than a million times the magnitude of D. The initialization time $0.4a^2/D = 1600$ sec, during which a slug of tracer would flow 800 cm, or 4,000 tube radii. Hence, the dispersion during the first 4,000 radii of flow is not described by the one-dimensional dispersion equation (4.19), an important limitation on this equation's use.

4.1.5   Aris's Analysis

Aris (1956) showed that it is possible to obtain Taylor's main results without stipulating the features of the concentration distribution. Aris also proved that the spreading by longitudinal molecular diffusion is directly additive to that caused by the velocity profile. The approach taken by Aris is often referred to as the "concentration moment" method; it is worth study, as many environmental problems studied more recently can be approached either from Taylor's or from Aris's point of view, and often with complimentary results.

Aris presented his analysis using a cross section of arbitrary shape and a variable cross-sectional diffusion coefficient, but the method is more easily studied without important loss of generality if we return to the problem analysed in Section 4.1.2, a laminar flow with arbitrary velocity profile $u(y)$ between parallel plates. We begin with the diffusion equation written in the moving coordinate system

$$\frac{\partial C}{\partial t} + u'\frac{\partial C}{\partial \xi} = D\left(\frac{\partial^2 C}{\partial x^2} + \frac{\partial^2 C}{\partial y^2}\right). \tag{4.29}$$

We now define the moments of the concentration distribution,

$$C_p(y) = \int_{-\infty}^{\infty} \xi^p C(\xi, y)\, d\xi, \tag{4.30}$$

$$M_p = \overline{C_p}, \tag{4.31}$$

where M_p is the cross sectional average of C_p, and p can take any positive integer value. We can also take the moment of (4.29) by applying the operator $\int_{-\infty}^{\infty} \xi^p(\ )\, d\xi$ to each term to obtain

$$\frac{\partial}{\partial t}C_p + \int_{-\infty}^{\infty} \xi^p u'\, dC = \int_{-\infty}^{\infty} \xi^p d\left(D\frac{\partial C}{\partial \xi}\right) + D\frac{\partial^2 C_p}{\partial y^2}. \tag{4.32}$$

We postulate an infinitely long channel so that $C = 0$ at $\xi = \pm\infty$; hence the two central terms can be integrated by parts [see a similar integration of Eq. (2.23)] to yield

$$\frac{\partial}{\partial t}C_p - pu'C_{p-1} = D\left(p(p-1)C_{p-2} + \frac{\partial^2 C_p}{\partial y^2}\right) \tag{4.33}$$

with the boundary conditions

$$D\frac{\partial C_p}{\partial y} = 0 \qquad \text{at} \quad y = 0, h.$$

Averaging over the cross section gives

$$\frac{d}{dt}M_p - p\overline{u'C_{p-1}} = p(p-1)D\overline{C_{p-2}}. \tag{4.34}$$

Equation (4.33) can, in principle, be solved sequentially for $p = 0, 1, 2, \ldots$ to any desired value of p, and the concentration distribution constructed to any desired degree of accuracy for any value of time. Thus the analysis is more general than Taylor's in that it applies for low values of time. Taylor's results can be recovered by writing the equations for $p = 0, 1,$ and 2 and observing the solution as $t \to \infty$, as follows:

Equation	Consequence as $t \to \infty$
$dM_0/dt = 0$	Mass is conserved
$\dfrac{\partial}{\partial t} C_0 = \dfrac{\partial}{\partial y} D \dfrac{\partial C_0}{\partial y}$	$C_0 \to M_0$ (uniform distribution over the cross section)
$\dfrac{d}{dt} M_1 = \overline{u'C_0}$	$M_1 \to$ constant
$\dfrac{\partial C_1}{\partial t} - u'C_0 = \dfrac{\partial}{\partial y} D \dfrac{\partial C_1}{\partial y}$	C_1 satisfies the same equation as C' in Section 4.1.2, except that $-C_0$ replaces $\partial \bar{C}/\partial \xi$
$\dfrac{dM_2}{dt} = 2\overline{u'C_1} + 2D\overline{C_0}$	$d\sigma^2/dt = 2K + 2D$, where K is defined by (4.17), i.e., molecular diffusion and shear flow dispersion are additive

4.2 DISPERSION IN TURBULENT SHEAR FLOW

The extension of Taylor's or Aris's analysis to turbulent flow is straightforward. In turbulent flow the velocity profile will be somewhat different than in a laminar flow in the same channel, and the cross-sectional turbulent mixing coefficient will play the role of molecular diffusion in laminar flow. Otherwise there are no differences, and the conclusions we have reached about the use of the one-dimensional dispersion equation apply unchanged. The only significant difference in the mathematics is that the cross-sectional mixing coefficient $\varepsilon(y)$ may be a function of cross-sectional position y. In the analysis of unidirectional turbulent flow between parallel plates, for example, Eqs. (4.10) and (4.17) become

$$u' \frac{\partial \bar{C}}{\partial \xi} = \frac{\partial}{\partial y} \varepsilon(y) \frac{\partial C'}{\partial y} \tag{4.35}$$

and

$$K = -\frac{1}{h} \int_0^h u' \int_0^y \frac{1}{\varepsilon} \int_0^y u' \, dy \, dy \, dy. \tag{4.36}$$

The extension of Taylor's analysis to turbulent flow was first done by Taylor himself in 1954 for the case of a long straight pipe. Taylor noted the previously established experimental result that all turbulent pipe flows have similar velocity profiles,

$$u = u_0 - \sqrt{\frac{\tau_0}{\rho}}\, f(Z) = u_0 - u^* f(Z), \tag{4.37}$$

in which τ_0 is the shear stress at the pipe wall and the other notation is as for the laminar analysis. The quantity, $\sqrt{\tau_0/\rho} = u^*$, is usually known as the "shear velocity," and will be used frequently in analysis of turbulent flows. $f(Z)$ is an empirical function tabulated in Taylor's (1954) paper [approximately given by Eq. (1.27)].

The cross-sectional mixing coefficient was obtained from the "Reynolds analogy" that the mixing coefficients for momentum and mass are the same. The flux of momentum through a surface, equivalent to the mass flux q, is the shear stress at the surface divided by the fluid density, τ/ρ. Also it is easy to show by a force balance that at any distance r from the center of a pipe of radius a, the local shear stress is given by $\tau(r) = (r/a)\tau_0$. Hence

$$\varepsilon = \frac{q}{-\partial C/\partial r} = \frac{\tau}{-\rho\, \partial u/\partial r} = \frac{aZ}{df/dZ}\, u^*. \tag{4.38}$$

It is now possible to tabulate $u'(r) = u(r) - \bar{u}$ and $\varepsilon(r)$, numerically integrate (4.35) expressed in radial coordinates,

$$u'\frac{\partial \bar{C}}{\partial \xi} = \varepsilon\left[\frac{\partial^2 C'}{\partial r^2} + \frac{1}{r}\frac{\partial C'}{\partial r}\right] \tag{4.39}$$

to obtain $C'(r)$, and again numerically integrate to find K. Taylor's result, probably the best known as well as the simplest of all equations describing turbulent dispersion, was

$$K = 10.1\, au^*. \tag{4.40}$$

An even more straightforward application of Taylor's method was that by Elder (1959). Elder postulated a flow down an infinitely wide inclined plane and took the von Karman logarithmic velocity profile,

$$u' = (u^*/\kappa)(1 + \ln y'), \tag{4.41}$$

where κ is the von Karman constant, usually taken to be 0.4, and $y' = y/d$. The flow is sketched in Fig. 1.3. A force balance similar to that used for the pipe gives

$$\tau = \rho\varepsilon\frac{du}{dy} = \tau_0(1 - y') \tag{4.42}$$

from which

$$\varepsilon = \kappa y'(1 - y')du^*. \tag{4.43}$$

In (4.42) τ_0 is the shear stress on the bottom, $u^* = \sqrt{\tau_0/\rho}$, and ε is assumed to describe mass and momentum transport equally, as in (4.38).

TABLE 4.1

Values of the Dispersion Coefficient and the Integral Defined by Eq. (4.48) in Various Parallel Shear Flows

Flow	Velocity profile	Dispersion coefficient, K	Velocity deviation intensity, $\overline{u'^2}$	Characteristic length, h	Average cross-sectional mixing coefficient, E	$I = KE/\overline{u'^2}h^2$
Laminar flow in a tube of radius a (see Section 4.1.4)	$u_0(1 - r^2/a^2)$	$\dfrac{a^2 u_0^2}{192D}$	$\dfrac{1}{12}u_0^2$	a	D	0.0625
Laminar flow at depth d down an inclined plane	$u_0[2(y/d) - y^2/d^2]$	$\dfrac{8}{945}\dfrac{d^2 u_0^2}{D}$	$\dfrac{4}{45}u_0^2$	d	D	0.0952
Laminar flow with a linear velocity profile across a spacing h (see Section 4.1.3)	$U(y/h)$	$\dfrac{U^2 h^2}{120D}$	$\dfrac{1}{12}U^2$	h	D	0.10
Step velocity profile (see Fig. 4.9) with constant cross-stream mixing coefficient D	u_1 in upper half of flow $h/2 < y < h$; u_2 in lower half, $0 < y < h/2$	$\dfrac{1}{48}\dfrac{(u_2 - u_1)^2 h^2}{D}$	$\dfrac{(u_2 - u_1)^2}{4}$	h	D	0.083
Turbulent flow in a pipe of radius a	Empirical distribution (see Taylor, 1954)	$10.1au^*$	$10.1u^{*2}$	a	$0.054au^*$	0.054
Turbulent flow at depth d down an inclined plane	$\bar{u} + u^*/\kappa(1 + \ln y/d)$	$\dfrac{0.404}{\kappa^3}du^*$	$\dfrac{u^{*2}}{\kappa^2}$	d	$\dfrac{1}{6}\kappa du^*$	0.067

K is found by substituting (4.41) and (4.43) into (4.36), although the integrations are not simple. The first two integrals yield

$$C' = \frac{\partial \bar{C}}{\partial x} \frac{d}{\kappa^2} \left(\sum_{n=1}^{\infty} \frac{1}{n^2} \left(\frac{d-y}{d} \right)^n - 0.648 \right) \qquad (4.44)$$

and the third gives

$$K = \frac{0.404}{\kappa^3} du^*. \qquad (4.45)$$

Elder took $\kappa = 0.41$ to give the well known result

$$K = 5.93 du^*. \qquad (4.46)$$

The general form for the longitudinal dispersion coefficient in unidirectional shear flow can be seen by introducing the dimensionless quantities $y' = y/h$, $u'' = u'/\sqrt{\overline{u'^2}}$, and $\varepsilon' = \varepsilon/E$, into (4.36). E can be taken as the cross-sectional average of ε. $\sqrt{\overline{u'^2}}$ is the "intensity" of the velocity deviation; note that it is not the turbulent intensity but rather a measure of how much the turbulent averaged velocity deviates throughout the cross section from its cross-sectional mean. We can now write

$$K = \frac{h^2 \overline{u'^2}}{E} I, \qquad (4.47)$$

in which I is a dimensionless integral

$$I = -\int_0^1 u'' \int_0^{y'} \frac{1}{\varepsilon'} \int_0^{y'} u'' \, dy' \, dy' \, dy'. \qquad (4.48)$$

The range of values of I for flows of practical interest is small, as can be seen in Table 4.1. In most practical cases it may suffice to take $I = 0.1$, and not worry about the details of the triple integration.

4.3   DISPERSION IN UNSTEADY SHEAR FLOW

Real environmental flows are often unsteady, as for example the reversing flow in a tidal estuary or the wind-driven flow in a lake caused by a passing storm. Since an unsteady flow can often be obtained by adding an oscillatory component to a steady component, it is of interest to study the application of Taylor's analysis to an oscillatory shear flow. As an example, let us analyze the linear velocity profile of Section 4.1.3 with a sinusoidal oscillation. The velocity profile is given by

$$u = U(y/h) \sin(2\pi t/T), \qquad (4.49)$$

in which T is the period of oscillation.

We will see that it is possible to obtain an analytical solution for this simple case, but the computations are lengthy. We will not attempt to deal with more complicated velocity profiles because numerical solutions have shown that the results obtained for the simple case can be applied to the more complicated profiles with adequate accuracy.

Before beginning the analysis, let us consider qualitatively what we should expect to find. The steady flow dealt with in Section 4.1.3 gives a concentration profile given by Eq. (4.21). If the flow were exactly reversed ($u = -Uy/h$) the concentration profile would be exactly reversed (substitute $-y$ for y in (4.21)) and the dispersion coefficient, as computed by (4.22), would be the same. Now consider a "flip-flop" sort of flow that reverses instantaneously between $u = Uy/h$ and $u = -Uy/h$ after every time interval $T/2$. After each reversal the concentration profile has to change, but a time of approximately $T_c = h^2/D$ is required before the concentration profile is completely adopted to a new velocity profile. We can discuss two limiting cases, that in which the period of reversal T is much longer than T_c, and that in which T is much shorter than T_c. Okubo (1967) showed by use of the Aris moment method what would happen in the two limits, and we will show the same result by an extension of Taylor's method, but the limiting results are apparent without doing the analysis. First suppose that the period of reversal is very long ($T \gg T_c$). The concentration profile will have sufficient time to adopt itself to the velocity profile in each direction, i.e., the time required for C' to reach the profile given by (4.21) will be short compared to the time during which C' has that profile. Therefore the dispersion coefficient will be the same as that in a steady flow, except during a short period after the flow reversal which, in the limit, is negligible. Now consider the opposite limit in which the period of reversal is very short compared to the cross-sectional mixing time ($T \ll T_c$). In this limit the concentration profile does not have time to respond to the velocity profile; we can expect C' to oscillate around the mean of the symmetric limiting profiles, which is $C' = 0$. Therefore in this limit the dispersion coefficient tends toward zero.

Therefore we have the two limits:

$T \gg T_c$ dispersion as if flow were steady in either direction

$T \ll T_c$ no dispersion due to the velocity profile

The second case can also be visualized as sketched in Fig. 4.6, which shows the fate of an instantaneous line source when $T \ll T_c$. During the time that the flow is in the one direction the line source is stretched out, but after the flow reversal the line is returned exactly to its original position. This result occurs, of course, only when there is essentially no cross-sectional mixing before the flow reverses. G. I. Taylor once made a movie of essentially this limit.† He

---

† A portion of *Low Reynolds Number Flows*, a film produced by the Education Development Center for the National Committee for Fluid Mechanics Films (available from Encyclopedia Britannica Educational Corporation, 425 N. Michigan Avenue, Chicago, Illinois, Film No. 21617.)

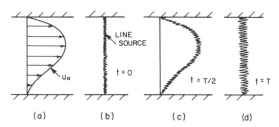

Figure 4.6   The shear effect in oscillating flow for the case $T \ll T_c$. (a) A hypothetical velocity distribution $u = u_0 \sin(2\pi t/T)$. (b) A line source introduced at $t = 0$. (c) The distribution at $t = T/2$. (d) The distribution at $t = T$.

mounted a cylinder in the center of a round container so that the cylinder and the container were coaxial and the cylinder could be rotated inside the container. He filled the annulus separating the cylinder from the container with glycerine and drew a straight line on the glycerine surface with dye. When the cylinder was rotated, producing a parallel shear flow, the dye line was distorted until after several turns it was unrecognizable. Then the cylinder was rotated the opposite direction the same number of turns, and when it returned to its original position, the dye line reappeared on the glycerine surface as if by magic.

To quantify these arguments we need a solution of Eq. (4.13), which is the diffusion equation simplified by Taylor's method but retaining the unsteady term for C', with the velocity profile given by (4.49). The boundary conditions are $\partial C'/\partial y = 0$ at $y = \pm h/2$, and without loss of generality we can assume an initial condition $C'(y, 0) = 0$. The advective term in (4.13) is an unsteady source term; therefore the solution can be found by the method described by Carslaw and Jaeger (1959, p. 32). In brief the method is to replace the unsteady source by a source of constant strength by setting $t = t_0$ where t_0 is a constant. Equation (4.13) becomes

$$\frac{\partial C^*}{\partial t} - D \frac{\partial^2 C^*}{\partial y^2} = -\frac{U y}{h} \frac{\partial \bar{C}}{\partial x} \sin\left(\frac{2\pi t_0}{T}\right)$$

$$\frac{\partial C^*}{\partial y} = 0 \quad \text{at} \quad y = \pm h/2, \qquad C^*(y, 0) = 0.$$

(4.50)

C^* will be the distribution resulting from a suddenly imposed source distribution of constant strength; therefore, as diagrammed in Fig. 2.8 $\partial C^*/\partial t$ will be the result of an instantaneously applied source and the result we seek, the solution for a series of sources of variable strength, can be obtained by

$$C'(y, t) = \int_0^t \frac{\partial}{\partial t} C^*(y, t - t_0; t_0) \, dt_0.$$

(4.51)

Also as $t - t_0$ becomes large $\partial C^*/\partial t \to 0$ so the solution depends only on recent value of $C^*(y, t; t_0)$. Hence for large t we can write

$$C'(y, t) = \int_{-\infty}^{t} \frac{\partial}{\partial t} C^*(y, t - t_0; t_0) \, dt_0. \tag{4.52}$$

Solutions for C^* and C' can be found by straightforward if tedious analysis. C^* can be expressed by the sum $C^*(y, t) = u(y) + w(y, t)$, with $w(y, 0) = -u(y)$, yielding an integrable equation for u and an equation for w that can be solved by separation of variables and a Fourier expansion. Further integration of the result leads to

$$C' = \frac{2Uh^2}{\pi^3 D} \frac{T}{T_c} \frac{\partial \bar{C}}{\partial x} \sum_{n=1}^{\infty} \frac{(-1)^n}{(2n-1)^2} \sin(2n - 1)\pi \frac{y}{h}$$

$$\times \left[\left(\frac{\pi}{2}(2n-1)^2 \frac{T}{T_c} \right)^2 + 1 \right]^{-1/2} \sin\left(\frac{2\pi t}{T} + \theta_{2n-1} \right), \tag{4.53}$$

where

$$\theta_{2n-1} = \sin^{-1}(-\{[\tfrac{1}{2}\pi(2n-1)^2 T/T_c]^2 + 1\}^{-1/2}). \tag{4.54}$$

The average over the period of oscillation of the one-dimensional dispersion coefficient is

$$K = \frac{1}{T} \int_0^T \left(-\int_{-h/2}^{h/2} u'C' \, dy/h \frac{\partial \bar{C}}{\partial x} \right) dt$$

$$= \frac{u^2 h^2}{\pi^4 D} \left(\frac{T}{T_c} \right)^2 \sum_{n=1}^{\infty} (2n-1)^{-2} \left\{ \left[\frac{\pi}{2}(2n-1)^2 \left(\frac{T}{T_c} \right)^2 \right]^2 + 1 \right\}^{-1}. \tag{4.55}$$

For $T \ll T_c$, $K \to 0$. For $T \gg T_c$ we have

$$K_0 = (1/240)U^2 h^2/D. \tag{4.56}$$

Now recall that for a linear steady profile $u = U(y/h) \sin \alpha$, where α is a constant, the dispersion coefficient $K = (1/120)U^2 h^2 \sin^2 \alpha/D$, and the ensemble average of this result over all values of α gives $\bar{K} = (1/240)u^2 h^2/D$. Thus (4.56) is consistent with the qualitative argument that for very short period oscillations the dispersion coefficient is zero, and for very long period oscillations dispersion proceeds as though the flow were steady at a succession of flows which make up the cycle. The intermediate behavior given by (4.55) is plotted in Fig. 4.7.

Now suppose that the flow includes an oscillating and a steady component, let us say

$$u(y) = u_1(y) \sin 2\pi t/T + u_2(y). \tag{4.57}$$

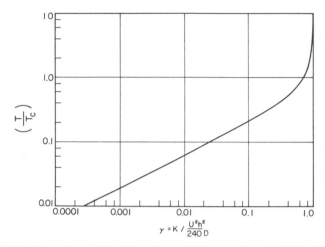

Figure 4.7   The dependence of the dispersion coefficient on the period of oscillation, as given by Eq. (4.55). γ is the ratio of K in a flow oscillating with period T to K in the same flow as $T \rightarrow \infty$.

For example, if $u_1 = u_2 = Uy/h$, $u(y)$ is a pulsating flow such as might be found in a blood vessel. It is easy to show that the results of the separate velocity profiles are additive. Let $C_1{'}$ be the solution to the equation

$$\partial C_1{'}/\partial t + u_1 \sin(2\pi t/T) \, \partial \bar{C}/\partial x = \varepsilon \, \partial^2 C_1{'}/\partial y^2$$

and $C_2{'}$ be the solution to

$$\partial C_2{'}/\partial t + u_2 \, \partial \bar{C}/\partial x = \varepsilon \, \partial^2 C_2{'}/\partial y^2$$

Then $C_1{'} + C_2{'} = C'$ is the solution to $\partial C'/\partial t + u(t) \, \partial C/\partial x = \varepsilon \, \partial^2 C'/\partial y^2$. The cycle-averaged dispersion coefficient is

$$\bar{K} = \frac{1}{T} \int_0^T - \frac{1}{h \, \partial\bar{C}/\partial x} \int_{-h/2}^{h/2} \left(u_1 \sin\frac{2\pi t}{T} + u_2 \right)(C_1{'} + C_2{'}) \, dy \, dt. \quad (4.58)$$

We have already seen that $C_1{'}$ is sinusoidal so integration over the period of oscillation will zero the cross product terms in the integrand. Thus

$$\bar{K} = - \frac{1}{h \, \partial\bar{C}/\partial x} \left(\frac{1}{T} \int_0^T \int_{-h/2}^{h/2} u_1 C_1{'} \sin\frac{2\pi t}{T} \, dy \, dt + \int_{-h/2}^{h/2} u_2 C_2{'} \, dy \right)$$

$$= K_1 + K_2, \quad (4.59)$$

where K_1 is the result of the oscillatory profile and K_2 the result of the steady profile. K_1 is, of course, affected by the ratio T/T_c, but K_2 is not.

It is possible to apply these results to a wide variety of flows, and to analyse more complicated velocity profiles both numerically or by more sophisticated methods of analysis. The reader interested in further details may consult the papers by Chatwin (1975) and Fukuoka (1974). In this book we will only discuss the application to tidal rivers and estuaries, as given in Section 7.2.2.1.

4.4 DISPERSION IN TWO DIMENSIONS

In the problems we have discussed so far the velocity vector lay entirely in the x direction, and its magnitude varied only in the y direction. In many environmental flows the velocity vector rotates with depth

$$\mathbf{u} = u(z) + v(z),\tag{4.60}$$

where u and v are the components of the velocity vector $\mathbf{u}$ in the x and y directions and both may vary arbitrarily with depth. For example, Fig. 4.8 shows the skewed shear flow in the surface layer of Lake Huron, as reported by Csanady (1966).

Taylor's method of analysis is easily applied to a skewed shear flow with velocity profiles in two directions. The two-dimensional form of Eq. (4.10) is

$$u'\frac{\partial \overline{C}}{\partial x} + v'\frac{\partial \overline{C}}{\partial y} = \frac{\partial}{\partial z}\varepsilon\frac{\partial C'}{\partial z}\tag{4.61}$$

$$\frac{\partial C'}{\partial z} = 0 \quad \text{at} \quad z = 0, h.$$

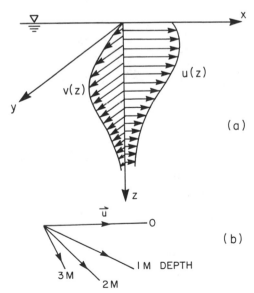

Figure 4.8 A typical skewed shear flow near the surface of a lake. (a) Velocity profile. (b) Hodograph. [After Csanady (1966).]

The same assumptions are required to get (4.61) as (4.10), of course, in particular that the flow is uniform in the x and y directions in a two-dimensional space of constant width h. We now follow the one-dimensional analysis with little change; (4.61) is integrated to give

$$C'(z) = \int_0^{-z} \frac{1}{\varepsilon} \int_0^{-z} \left(u' \frac{\partial \bar{C}}{\partial x} + v' \frac{\partial \bar{C}}{\partial y} \right) dz \, dz. \tag{4.62}$$

A bulk dispersion tensor can be defined by

$$\dot{M}_x = \int u'C' \, dy = -K_{xx} \frac{\partial \bar{C}}{\partial x} - K_{xy} \frac{\partial \bar{C}}{\partial y},$$

$$\dot{M}_y = \int v'C' \, dy = -K_{yx} \frac{\partial \bar{C}}{\partial x} - K_{yy} \frac{\partial \bar{C}}{dy}, \tag{4.63}$$

and the comparison of (4.62) and (4.63) gives

$$K_{xx} = -\int u' \int \frac{1}{\varepsilon} \int u' \, dz \, dz \, dz,$$

$$K_{xy} = -\int u' \int \frac{1}{\varepsilon} \int v' \, dz \, dz \, dz,$$

$$K_{yx} = -\int v' \int \frac{1}{\varepsilon} \int u' \, dz \, dz \, dz, \tag{4.64}$$

$$K_{yy} = -\int v' \int \frac{1}{\varepsilon} \int v' \, dz \, dz \, dz.$$

The components K_{xx} and K_{yy} are the same as if there were only velocity profiles in those directions, but we also have terms K_{xy} and K_{yx} which depend on the interaction of the x and y velocity profiles. These terms mean that a gradient in the x direction can produce mass transport in the y direction and vice versa.

An interesting example is the idealization of the mean flow on a continental shelf discussed by Fischer (1978). The profiles are sketched in Fig. 4.9. The direction of net motion is parallel to the coast (the x direction) and the flow is assumed to have a linear profile with variation of U_0 from top to bottom. The cross-shelf flow is assumed to be a step function with velocity V_0 away from the coast in the top half and $-V_0$ toward the coast in the bottom half. When these profiles are introduced into Eq. (4.64) we obtain

$$K = \frac{d^2}{\varepsilon} \begin{pmatrix} U_0^2/120 & 5U_0 V_0/192 \\ 5U_0 V_0/192 & V_0^2/12 \end{pmatrix}. \tag{4.65}$$

The results of a numerical example are shown in Fig. 4.10. The values used are a mean down-coast flow (in the $+x$ direction) of 5 cm/sec, a variation $U_0 = 5$ cm/sec, and a cross-shelf velocity $V_0 = 5$ cm/sec. The figure shows the spread of

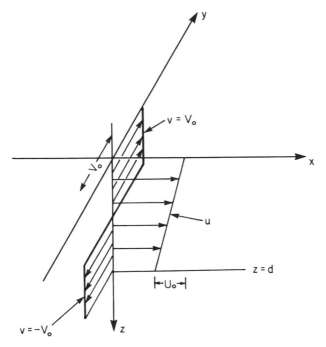

Figure 4.9 Fischer's (1978) representation of the shear flow on the continental shelf of the middle Atlantic bight.

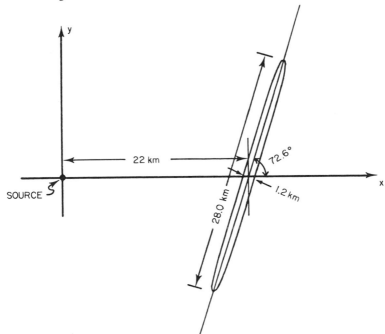

Figure 4.10 The distribution of a concentrated slug of dye after five days in the shear flow illustrated in Fig. 4.9.

a slug of dye after five days. The mean of the cloud moves down coast at the mean flow velocity and the spread is in the form of ellipses about the principal axes of the tensor, which are found to be oriented at $17.5°$ clockwise from the $x-y$ axes.

4.5 DISPERSION IN UNBOUNDED SHEAR FLOW

Dispersion of pollutants from the ground into the atmosphere in the absence of atmospheric inversion layers has been studied as a problem of dispersion in an unbounded shear flow. The problem is the same as that of dispersion in flow between parallel plates, except that the plates are so far apart that the material released at one plate does not feel the effect of the other. Hence if material is released from one side into a bounded shear flow the early stages of the dispersion process, before the material spreads across the flow, will be described by the solution to the unbounded problem.

A solution for the dispersion of an instantaneous release from the boundary into a linear shear flow has been obtained by Saffman (1962) by use of the Aris moment method. As yet water quality engineers seem to have made little use of the unbounded solution, but its main features are worth noting as it may describe dispersion from a point source on an open coast, for example. The form of the solution can be obtained by a very simple argument based on Taylor's analysis of diffusion by continuous movements, Eq. (3.30a). Suppose that a point source of tracer is instantaneously released half way between the plates in the linear shear flow sketched in Fig. 4.5. The width of the cloud, in the y direction, grows by diffusion at the rate $\sigma_y \sim (\varepsilon_y t)^{1/2}$, at least until the cloud begins to "feel" the effects of the walls. The velocity deviation seen by the cloud is that contained within its own width, so $\langle U^2 \rangle \propto (u/h)^2 \sigma_y^2$. The Lagrangian time scale can be expected to be proportional to σ_y^2/ε_y since this gives the time required to sample the velocities within the cloud. Making these substitutions we have from Eq. (3.30)

$$\sigma_x^2 \propto \langle U^2 \rangle Tt \propto [(u/h)^2 \sigma_y^2][\sigma_y^2/\varepsilon_y]t \propto (u/h)^2 \varepsilon_y t^3. \qquad (4.66)$$

A rigorous proof of the arguments leading to Eq. (4.66) is given by Monin and Yaglom. The constant of proportionality can be obtained by Saffman's work. Saffman analysed the spread from a point source concentrated at the ground and found that the contribution from the velocity profile to the longitudinal variance at ground level is given by

$$\sigma_x^2 = \left(\frac{7}{30} - \frac{\pi}{16}\right)\alpha^2 \varepsilon_y t^3, \qquad (4.67)$$

where α is the slope of the linear profile [equivalent to u/h in (4.66)].

Saffman described these results as applying to an "intermediate zone" which may occur between the initial spreading for very short time and the asymptotic result for very long time in a bounded flow. The characteristics of the three zones may be set out as follows:

Initial zone	$\sigma_x{}^2 \sim t^2$	Initial spreading immediately after injection of tracer, before significant cross sectional mixing has occurred.
Intermediate zone	$\sigma_x{}^2 \sim \varepsilon_y t^3$	Tracer has spread over a width of flow which is still much less than the separation of the boundaries.
Final zone	$\sigma_x{}^2 \sim t/\varepsilon_y$	Tracer has spread to fill the space between the boundaries $(t > 0.4h^2/\varepsilon_y)$

Note the great difference between the behavior in the intermediate and final zones. In the intermediate zone the rate of spread is proportional to the rate of cross-sectional mixing because the faster the tracer mixes across the flow the faster it comes into contact with larger velocity deviations. The larger the velocity deviations the more quickly the cloud spreads, so the rate of increase of the variance increases with increasing time. Both these results are essentially opposite to those in the final zone, in which we have already seen that the rate of increase of the variance is constant and inversely proportional to the rate of cross-sectional mixing. It is also interesting to note that if we define a longitudinal dispersion coefficient for the intermediate zone by $K_x = \frac{1}{2} d\sigma_x/dt$ we have

$$K_x \propto \varepsilon_y t^2 \propto \varepsilon_y^{1/3} \sigma_x^{4/3}, \tag{4.68}$$

a result which is similar to the celebrated $\frac{4}{3}$ law for the distance neighbor function discussed in Section 3.4.

Chapter 5

Mixing in Rivers

Consider a stream of effluent discharged into a river, as sketched in Fig. 5.1. What happens can be divided into three stages. In the first stage the initial momentum and buoyancy of the discharge determine the rate of dilution, as discussed in Chapter 9. As the waste is diluted, the effects of the initial momentum and buoyancy are also diluted, leading to a second stage in which the waste is mixed across the receiving channel primarily by turbulence in the receiving stream. The photograph of dye mixing across the Columbia River (see Fig. 5.6 below) illustrates this second stage. Finally, when the waste is fully mixed across the channel the process of longitudinal shear flow dispersion will tend to erase any longitudinal concentration variations. This last stage is like the zone where we could apply Taylor's analysis of longitudinal dispersion in pipes, and we shall find that there is an equivalent analysis for longitudinal dispersion in rivers. Sometimes the first stage may extend over the entire channel, effectively eliminating the second stage. A large discharge of heated water from a once-through cooling system of a riverside power plant might be an example because the discharge contains relatively large amounts of initial momentum and buoyancy; on the other hand, many industrial and municipal effluents contribute a negligible amount of flow, momentum, or buoyancy to the receiving stream, and may be effectively treated as point sources of mass.

In this chapter we restrict our discussion to the effects of turbulence and shear in the receiving stream, that is, we assume that whatever has happened in the first stage can be computed separately by the methods of Chapter 9 so that we

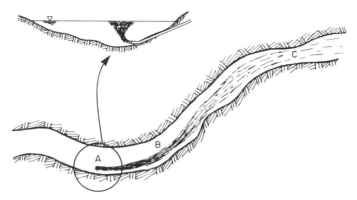

Figure 5.1 Three stages in the mixing of an effluent into a river. (A) Initial momentum and buoyancy determine mixing near the outlet. (B) Turbulence and currents in the receiving stream determine further mixing after initial momentum and buoyancy are dissipated. (C) After cross-sectional mixing is complete, longitudinal dispersion erases longitudinal gradients caused by changes in effluent or river discharge.

can now deal with a source of tracer without its own momentum or buoyancy. We start with a discussion of turbulent mixing in the idealized case of an infinitely wide channel of constant depth, and then show how the results can be applied to compute rates of mixing across real streams. Then, in the major portion of the chapter, we discuss what is known of the process of longitudinal dispersion and how it affects the distributions of conservative and nonconservative effluents.

5.1 TURBULENT MIXING IN RIVERS

5.1.1 The Idealized Case of a Uniform, Straight, Infinitely Wide Channel of Constant Depth

In Chapter 3 we discussed how, in certain circumstances, turbulent mixing can be described by the diffusion equation with a turbulent mixing coefficient in place of the molecular diffusion coefficient. The flow of water in a straight channel of constant depth and great width is one example of a flow which meets all of the requirements for the use of a turbulent mixing coefficient. The turbulence is homogeneous and stationary because the channel is uniform. If the sidewalls are very far apart the width of the flow should play no role, and we should expect that the important length scale is the depth. A cloud of tracer deposited in such a flow will grow until it fills the depth, and then will continue to grow in the directions of the length and breadth. The Lagrangian length scale will be some multiple or fraction of the depth; indeed, experiments have shown that it is approximately equal to the depth. Therefore once the cloud extends over several depths, the requirements for the use of a constant turbulent mixing coefficient will be fulfilled, as explained in Chapter 3.

Equation (3.40) demonstrates that a turbulent mixing coefficient is the product of the Lagrangian length scale and the intensity of the turbulence. Experiments by Laufer (1950) and others have shown that turbulence intensity in any wall shear flow is proportional to the shear stress on the wall. For dimensional reasons the shear stress must be expressed as a velocity, so we use the shear velocity defined in Chapter 4 by $u^* = \sqrt{\tau_0/\rho}$ in which τ_0 is the shear stress on the channel bottom and ρ the density of the fluid. In uniform open channel flow, the bottom shear stress can be evaluated by a force balance (see Henderson, 1966, or any similar book on open channel flow) to yield

$$u^* = \sqrt{gdS}, \tag{5.1}$$

where S is the slope of the channel. Since we have already identified the depth as the relevant length scale, we now see that in our idealized flow any turbulent mixing coefficient must be proportional to the product (du^*). We should expect that there will be one coefficient for vertical mixing and another for transverse and longitudinal mixing because the presence of the horizontal boundaries at surface and bottom means that the turbulence will not be isotropic. On the other hand, there is no reason to expect a nondiagonal tensor coefficient of the form ε_{yz}, as long as the y and z axes are oriented parallel and perpendicular to the bottom, because there is no preferred diagonal direction of motion.

5.1.1.1 Vertical Mixing

As part of the discussion of Elder's analysis in Chapter 4 we showed that a vertical mixing coefficient can be derived from the velocity profile. The logarithmic law velocity profile leads to a vertical mixing coefficient for momentum,[†]

$$\varepsilon_v = \kappa du^*(z/d)[1 - (z/d)]. \tag{5.2}$$

The "Reynolds analogy" states that the same coefficient can be used for transport of mass, and this result has been verified by Jobson and Sayre (1970) by means of an experimental study of the vertical mixing of dye in a flume. Averaging over the depth and taking $\kappa = 0.4$ leads to the useful result,

$$\bar{\varepsilon}_v = 0.067du^*. \tag{5.3}$$

[†] See Eq. (4.43). In Chapter 4 y denoted any cross-stream coordinate direction. In this chapter y is always a transverse direction and z is the vertical coordinate direction (see Fig. 5.4 below).

This result is similar to others found in a wide range of flows; for example, Csanady (1976) gives an average value of $\varepsilon_v = 0.05du^*$ from measurements in an unstratified atmospheric boundary layer, where d is the depth of the boundary layer and u^* the shear velocity at the surface of the earth.

5.1.1.2 Transverse Mixing

There is no transverse velocity profile in an infinitely wide uniform channel, so it is not possible to establish a transverse analogy of Eq. (5.2) to determine the transverse mixing coefficient. Instead, we are forced to rely on experiments. Researchers have conducted a large number of experiments on transverse mixing, both in straight rectangular laboratory channels and in natural or irregular channels. For the moment we restrict our attention to results in straight rectangular channels, which are the closest possible approximation to the idealized infinitely wide stream; irregular and natural channels are treated in Section 5.1.2.

The results of a total of approximately 75 separate experiments in straight, rectangular channels are summarized in Table 5.1 (see Lau and Krishnappan (1977), for a more detailed table of results of separate runs). In addition, the irrigation canal used by Fischer (1967b) approximated a straight, rectangular flume. In almost all cases the nondimensional transverse mixing coefficient, ε_t/du^* has been in the range of 0.1–0.2, the only significant exception being the irrigation canal values of 0.24 and 0.25. An approximate average of the experimental results is

$$\varepsilon_t \cong 0.15du^*. \tag{5.4}$$

More detailed analysis has shown, however, that no matter how wide the channel the width appears to play a role, although it is not clear exactly what that role is. Okoye (1970) gave a plot of ε_t/du^* versus the width to depth ratio W/d based on his own experimental results. Lau and Krishnappan (1977), however, have shown that other experiments do not confirm Okoye's result. Lau and Krishnappan were able to minimize the scatter of a large number of data points by plotting ε_t/Wu^* versus W/d; however, the numerical range of values of ε_t/Wu^* is so large that practical use of this parameter would be difficult. The fundamental difficulty is that if a transverse motion is somehow excited in a wide channel, there is no physical obstacle to its continuation. Movies of the spread of dye in laboratory channels have shown occasional unexplained transverse migrations on a scale larger than the depth; these certainly affect the results of experiments, and since they occasionally run into the walls, there is an effect of width even in a relatively wide channel. The best we can say for practical purposes is that in straight, rectangular channels the result given by Eq. (5.4) is likely to be correct within an error bound of approximately $\pm 50\%$.

TABLE 5.1

Experimental Measurements of Transverse Mixing in Rectangular Open Channels with Smooth Sides

Reference	Type of channel	Type of bottom roughness	Channel width, W (cm)	Mean depth of flow, d (cm)	Mean velocity, $\bar{u}$ (cm/s)	Shear velocity, u^* (cm/s)	Transverse Mixing coefficient, ε_t (cm²/s)	$\dfrac{\varepsilon_t}{du^*}$
Elder (1959)	Laboratory	Smooth	36	1.2	21.6	1.59		0.16
Sayre and Chang (1968)	Laboratory	Wooden cleats	283	14.8–37.1	23.5–37.1	3.81–6.04	9.6–36.9	0.160–0.179
Sullivan (1968)	Laboratory	Smooth	76	7.3–10.2	15.3–22.9	0.83–1.29	0.90–1.18	0.107–0.133
Okoye (1970)	Laboratory	Smooth	85	1.5–17.3	27.1–42.8	1.6–2.2	0.64–2.9	0.09–0.20
Okoye (1970)	Laboratory	Smooth	110	1.7–22.0	30.0–50.4	1.4–2.6	0.79–3.3	0.11–0.24
Okoye (1970)	Laboratory	Stones	110	6.8–17.1	35.3–42.8	3.6–5.2	4.8–7.5	0.11–0.14
Prych (1970)	Laboratory	Smooth	110	4.0–11.1	35.4–46.0	1.9–2.0	1.1–3.6	0.14–0.16
Prych (1970)	Laboratory	Metal lath	110	3.9–6.4	37.3–45.9	3.7–4.0	2.0–3.5	0.14
Miller and Richardson (1974)	Laboratory	Rectangular blocks	59.7	12.5–13.2	30.5–81.4	3.0–16.3	3.7–36.3	0.10–0.18
Lau and Krishnappan (1977)	Laboratory	Smooth	60	3.9–5.0	15.5–33.7	0.9–2.0	0.74–1.4	0.16–0.20
		0.4 mm sand	45–60	1.4–4.0	19.7–20.3	1.6–2.1	0.34–0.88	0.11–0.14
		2.0 mm sand	30	1.6–3.4	20.0–20.4	1.9–2.4	0.74–0.92	0.14–0.20
		2.7 mm sand	45–60	1.3–3.9	19.5–20.4	1.8–2.8	0.59–1.16	0.13–0.26
Fischer (1967b)	Irrigation canal	Sand dunes	1830	66.7–68.3	63–66	6.1–6.3	102	0.24–0.25

5.1.1.3 Longitudinal Mixing

Turbulence causes longitudinal mixing presumably at about the same rate as transverse mixing because there is an equal lack of boundaries to inhibit motion. Sayre and Chang (1968) found that the longitudinal coefficient for spreading of polyethelene particles on the water surface was approximately three times that for transverse spreading; some of the longitudinal spread may have been due to a transverse velocity shear resulting from secondary circulation, however. In any case, longitudinal mixing by turbulent eddies is generally unimportant because the shear flow dispersion coefficient caused by the velocity gradient is much bigger than the mixing coefficients caused by turbulence alone. Recall, for example, that Elder's result for the dispersion coefficient in a logarithmic velocity profile (see Section 4.2) was $K = 5.93du^*$, which is approximately 40 times the expected magnitude of the turbulent mixing coefficient as given by Eq. (5.4). Recall also (Section 4.1.5) that Aris's analysis showed that the coefficients due to turbulent mixing and shear flow are additive, so we can effectively neglect the longitudinal turbulent mixing. Rates of turbulent longitudinal mixing have not been measured by dye spreading experiments, because of the difficulty in separating the effects of longitudinal turbulent fluctuations from the results of the shear flow.

5.1.2 Mixing in Irregular Channels and Natural Streams

Natural channels differ from uniform rectangular ones in three important respects: the depth may vary irregularly, the channel is likely to curve, and there may be large sidewall irregularities such as groins or points of land. None of these factors are thought to have much influence on the rate of vertical mixing, since the scale of vertical motions is limited by the local depth. Hence, it is customary to use Eq. (5.3) to express the rate of vertical mixing, where d is the local depth. We know of no experiments on vertical mixing in a depth-varying flow, but we see no reason why the customary practice should not be adequate. On the other hand, the rate of transverse mixing is strongly affected by the channel irregularities because they are capable of generating a wide variety of transverse motions.

The effect of a cross-sectional depth variation in a straight trapezoidal canal was studied by Holley et al. (1972). These writers showed that the concentration distribution resulting from a side injection in a trapezoidal channel differs from that in a rectangular channel, the details depending on what cross-sectional variations of velocity and mixing coefficient are assumed. We do not show the results here, as the case of a straight trapezoidal channel is unusual and the reader interested therein may see the referenced paper.

Bends and sidewall irregularities, on the other hand, are common to many channels and have a major effect on transverse mixing. A summary of experimental data is given in Table 5.2. Holley and Abraham's (1973a) laboratory

TABLE 5.2

Experimental Measurements of Transverse Mixing in Open Channels with Curves and Irregular Sides

Reference	Channel	Channel geometry	Channel width, W (m)	Mean depth of flow, d (m)	Mean velocity, $\bar{u}$ (m/s)	Shear velocity, u^* (m/s)	Transverse mixing coefficient (m²/s)	$\dfrac{\varepsilon_t}{du^*}$
Yotsukura et al. (1970)	Missouri River near Blair, Nebraska	Meandering river	200	2.7	1.75	0.074	0.12	0.6
Holley and Abraham (1973a)	Laboratory	Smooth sides and bottom; 0.15 m long groins on both sides	2.2	0.097	0.11	—	—	0.36–0.49
Holley and Abraham (1973a)		Smooth sides and bottom; 0.5 m long groins on both sides	2.2	0.097	0.11	—	—	0.3–0.4
Holley and Abraham (1973a)	Laboratory model of the IJssel River	Groins on sides and gentle curvature	1.22	0.9	0.13	0.0078	—	0.45–0.77
Holley and Abraham (1973b)	IJssel River	Groins on sides and gentle curvature	69.5	4.0	0.96	0.075	—	0.51
Mackay (1970)[a]	Mackenzie River from Fort Simpson to Norman Wells	Generally straight alignment or slight curvature; numerous island and sand bars	1240	6.7	1.77	0.152	0.67	0.66
Yotsukura and Sayre (1976) and Sayre and Yeh (1973)	Missouri River downstream of Cooper Nuclear station, Nebraska	Reach includes one 90° and one 180° bend	210–270	4	5.4	0.08	1.1	3.4
Jackman and Yotsukura (1977)	Potomac River; 29 km reach below the Dickerson Power Plant	Gently meandering river with up to 60° bends	350	0.73–1.74	0.29–0.58	0.033–0.051	—	0.52–0.65

[a] The mean width and depth of the Mackenzie River were estimated by the authors from Canadian Geological Survey discharge notes. The transverse mixing coefficient was estimated by comparing Mackay's observations with the curves shown in Fig. 5.9.

studies of groins gave some indication of the effect of sidewall irregularities; values of ε_t/du^* in the range of 0.3–0.7 were observed. However, there have not been enough experiments in flumes, let alone in natural channels, to define how the mixing coefficient varies with the size of the irregularity. The best one can say is that the bigger the irregularity, probably the faster the transverse mixing.

The effects of channel curvature are somewhat better understood and we have enough experimental data to give some general guidelines. The values of ε_t/du^* observed in the model and prototype of the IJssel River by Holley and Abraham (1973), the Potomac by Jackman and Yotsukura (1977), the Missouri near Blair, Nebraska, by Yotsukura et al. (1970), and the Mackenzie in a 300-mile reach downstream from Fort Simpson by Mackay (1970), all fall in the range of 0.4–0.8. Within the test reaches these are nearly straight or slowly meandering rivers; the IJssel has groins on both sides, and the Missouri has groins on the inside of the bends.

Higher values of ε_t/du^* have been observed in sharply curving channels. When a flow rounds a bend the centrifugal forces induce a flow towards the outside bank at the surface, and a compensating reverse flow near the bottom, approximately as sketched in Fig. 5.2. Details of the secondary velocity profile have been given by Rozovskii (1951). Fischer (1969) utilized Rozovskii's velocity profiles and the shear flow analysis described in Chapter 4 to predict a transverse dispersion coefficient based on the transverse shear flow; the result can be written as

$$\frac{\varepsilon_t}{du^*} \propto \left(\frac{\bar{u}}{u^*}\right)^2 \left(\frac{d}{R}\right)^2, \tag{5.5}$$

where R is the radius of the curve. Fischer verified his result in a laboratory flume with constant curvature, and found that the constant of proportionality was ≈ 25; however, Fischer's formula assumes an infinite length of curve, whereas Sayre and Yeh (1973) reported that in a meandering channel ε_t varies from one-half its average value in the upstream portion of a bend to twice the average in the downstream portion. Yotsukura and Sayre (1976) found that Fischer's result agreed better with laboratory and field data if the right-hand

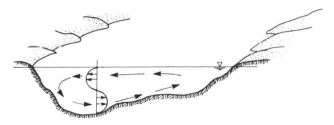

Figure 5.2 An illustration of the cross-sectional component of velocity in a curve, showing the velocity profile used to obtain Eq. (5.5).

side of Eq. (5.5) is multiplied by $(W/d)^2$, but even then the coefficient is different for the laboratory than for the field. Their plot of laboratory and field results is shown in Fig. 5.3.

Clearly our present knowledge of rates of transverse mixing is inadequate, but we may sum up what is known as follows. Values of ε_t/du^* in straight, uniform channels are generally in the range of 0.1–0.2. Curves and sidewall irregularities increase the coefficient such that values of ε_t/du^* in natural streams are hardly ever less than 0.4; if the stream is slowly meandering, and the sidewall irregularities are moderate, ε_t/du^* has been found to be usually in the range of 0.4–0.8, and we can use for practical purposes

$$\varepsilon_t/du^* = 0.6 \pm 50\%. \tag{5.6}$$

Higher values are likely if the channel has sharp curves or rapid changes in geometry, but we do not know enough to specify exactly what we mean by "rapid" and can only advise caution and field experiments if the investigator is in doubt. Similarly, we cannot define exactly the meaning of "sharp" versus "slowly meandering," although Fig. 5.3 suggests the definition that a slowly meandering river is one in which $W\bar{u}/Ru^*$ is generally less than two.

5.1.3 Computation of Concentration Distributions

Now let us consider how to compute the distribution of concentration downstream from an effluent discharge in a flowing stream. In almost all practical cases the flow will be much wider than it is deep; a typical channel dimension might be 30 m wide by 1 m deep, for example. If we take as a first estimate based

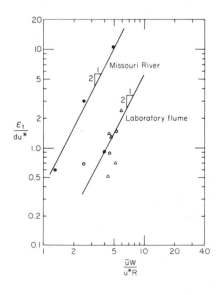

Figure 5.3 The effect of channel curvature on the transverse mixing coefficient. Data sources: ● Sayre and Yeh (1973) and Yotsukura *et al.* (1970), ○ Chang (1971), △ Fischer (1969). [After Yotsukura and Sayre (1976).]

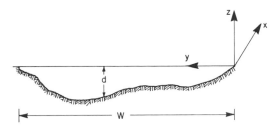

Figure 5.4 Definition of symbols and coordinate system.

on Eqs. (5.3) and (5.6) that $\varepsilon_t \approx 10\varepsilon_v$, and recall as mentioned in Chapter 1 that a mixing time is proportional to the square of the length divided by the mixing coefficient, we see that the transverse mixing time will be $30^2/10 = 90$ times the vertical mixing time. A result of this order is typical, so we can usually assume that vertical mixing is instantaneous compared to transverse mixing. Therefore in most practical problems we can start by assuming that the effluent is uniformly distributed over the vertical, or in other words we can analyze the two-dimensional spread from a uniform line source.

The coordinate system and nomenclature we will use for real streams is illustrated in Fig. 5.4. For the moment let us consider the case of a rectangular channel of depth d into which is discharged $\dot{M}$ units of mass of effluent per unit time in the form of a line source. A line source of $\dot{M}$ units into a flow of depth d is equivalent to a point source of strength $\dot{M}/d$ in a two-dimensional flow, for which Eq. (2.68) gives

$$C = [\dot{M}/\bar{u}d(4\pi\varepsilon_t x/\bar{u})^{1/2}]\exp(-y^2\bar{u}/4\varepsilon_t x) \tag{5.7}$$

if the channel is infinitely wide.[†] If the channel has width W the effect of the boundaries, $\partial c/\partial y = 0$ at $y = 0$ and $y = W$, can be accounted for by the method of superposition described in Section 2.3.2.4. It is convenient to define dimensionless quantities by setting

$$C_0 = \dot{M}/(\bar{u}dW), \qquad x' = x\varepsilon_t/\bar{u}W^2, \qquad y' = y/W. \tag{5.8}$$

If the source is located at $y = y_0$ ($y' = y_0'$), superposition gives the downstream concentration distribution as

$$\frac{C}{C_0} = \frac{1}{(4\pi x')^{1/2}} \sum_{n=-\infty}^{\infty} \{\exp[-(y' - 2n - y_0')^2/4x']$$
$$+ \exp[-(y' - 2n + y_0')^2/4x']\}. \tag{5.9}$$

[†] Equation (5.7) is exact only if the velocity u at all points in the channel is equal to the cross-sectional mean velocity $\bar{u} = \int_A u \, dA$, where A is the cross-sectional area; this restriction is acceptable in a wide rectangular channel because the tracer quickly averages the vertical velocity profile and there is no important transverse variation.

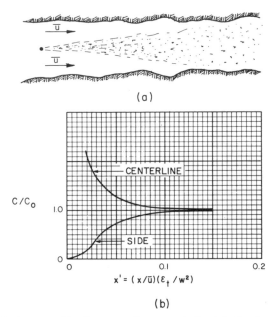

(a)

(b)

Figure 5.5 (a) A plume resulting from a continuous centerline injection into a stream of uniform depth and velocity. (b) A plot of concentrations at the centerline and along the sides as given by Eq. (5.9). The text describes how the same plot can be used to determine concentrations resulting from a side discharge.

Equation (5.9) is plotted in Fig. 5.5 for the case of a centerline discharge, $y_0' = \frac{1}{2}$. The figure shows the relative concentration on the centerline ($y' = \frac{1}{2}$) and on the side ($y' = 0, y' = 1$). For x' greater than about 0.1 the concentration is within 5 % of its mean value everywhere on the cross section. Thus a reasonable criterion for the distance required for what is often referred to as "complete mixing" from a centerline discharge is

$$L = 0.1\bar{u}W^2/\varepsilon_t. \tag{5.10}$$

Figure 5.6 illustrates the earliest stage of transverse mixing. In the figure the maximum value of x' is much less than 0.1.

If a pollutant is discharged at the side of a channel, the width over which mixing must take place is twice that for a centerline injection, but the boundary conditions are otherwise identical. The plume from a side discharge (neglecting effects of momentum and buoyancy of the discharge) is just one side of the plume from a centerline injection. For a given rate of injection of mass the concentration in the side plume is twice that of the centerline plume, but the rate of increase of the variance will be the same. This can be seen in the experimental results of a study in an irrigation canal near Albuquerque, New Mexico, by Fischer (1967b). (The test reach and injection apparatus for Fischer's experiment can be seen in

Figure 5.6   A plume resulting from a steady dye injection into the Columbia River near Hanford, Washington. (Photo courtesy of Battelle Northwest.)

Fig. 5.20.) The channel was nearly rectangular, straight for a distance of approximately 1000 ft above the injection point and through the measurement reach, and carried a flow approximately 2 ft deep. Two tests were performed under similar flow conditions and with identical rates of dye injection. In the first test, dye was injected onto the water surface at the centerline, and in the second the same injection was made approximately 1 ft out from the side. The concentration distributions observed during the two tests are shown in Fig. 5.7 and the variances of the distributions are plotted in Fig. 5.8. For the side injection the variance is computed with respect to the side of the channel, which is equivalent to the center of the channel for the centerline test. With this adjustment it can be seen that the variances for the two tests are similar. The test results show that the mathematical assumptions we have made throughout our discussion of how to treat boundaries are physically correct; solid boundaries really do lead to zero concentration gradients because of the requirement for zero mass flux, even in turbulent open channel flow.

The solution shown in Fig. 5.5 applies equally well to a side injection if W is replaced by $2W$. The curve marked "centerline" then gives the concentration

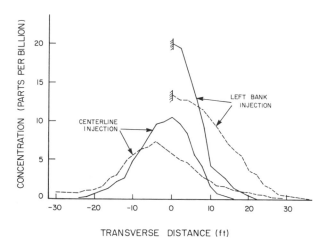

Figure 5.7   Observed concentrations downstream from centerline and side injections of dye in Fischer's (1967b) experiment. ——— 400 ft downstream from injection, ——— 1000 ft downstream from injection. Transverse distance is measured from the centerline for the centerline injection and from a point 1 ft from the left bank for the left bank injection.

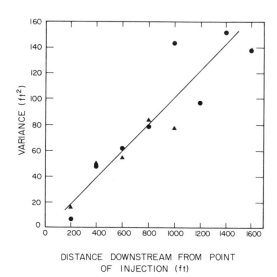

Figure 5.8   Variances of the concentration distributions shown in Fig. 5.7 (and others not shown) versus distance downstream from point of injection. ● Side injection. ▲ Centerline injection.

along the side where the injection takes place, and the curve marked "side" gives the concentration on the opposite side. The dimensionless distance for complete mixing is the same as for the centerline injection, but the real distance is four times as great because of replacing W by $2W$. Therefore the formula analogous to Eq. (5.10) for "complete mixing" from a side discharge is

$$L = 0.4 \bar{u} W^2 / \varepsilon_t. \tag{5.10a}$$

If the discharge has sufficient initial momentum and buoyancy that the initial mixing zone sketched in Fig. 5.1 ends in a plume over some fraction of the cross section, further turbulent mixing may have to be computed starting with a distributed source $C_i(y)$. In that case a superposition of line sources similar to the superposition illustrated in Fig. 2.5 yields:

$$C(y') = \int_0^1 \frac{C_i(y_0')}{(4\pi x')^{1/2}} \sum_{n=-\infty}^{\infty} \{\exp[-(y' - 2n - y_0')^2/4x']$$

$$+ \exp[-(y' - 2n + y_0')^2/4x']\} \, dy_0'. \tag{5.11}$$

Computations based on this equation are sometimes needed in the case of discharge of heated water into fast running rivers. A detailed description of such a calculation for thermal plumes in the Missouri River downstream from the Fort Calhoun and Cooper Nuclear Power Stations has been given by Sayre and Caro-Cordero (1977). Their computations also use the cumulative discharge method, which we discuss in Section 5.1.4.

Example 5.1: Spread of a plume from a point source.   An industry discharges 3 million gallons per day of effluent containing 200 ppm of a conservative substance near the center of a very wide, slowly meandering stream. The stream is 30 ft deep, the mean velocity of flow is 2 ft/sec, and the shear velocity is 0.2 ft/sec. Assuming that the effluent is completely mixed over the vertical, determine the width of the plume and the maximum concentration 1000 ft downstream from the discharge.

Solution.   The rate of input of mass is $\dot{M} = Qc = 4.65$ ft^3/sec [3 mgd (million gallons per day)] $\times$ 200 ppm $= 930$ ft^3/sec ppm (parts per million). From Eq. (5.6), estimate the transverse mixing coefficient to be approximately

$$\varepsilon_t = 0.6 d U^* = 0.6 \times 30 \text{ ft} \times 0.2 \text{ ft/sec} = 3.6 \text{ ft}^2/\text{sec}.$$

The width of the plume can be approximated by 4σ. Hence,

$$b = 4\sigma = 4\sqrt{2\varepsilon_t x/\bar{u}} = 240 \text{ ft}.$$

The maximum concentration, by Eq. (5.7), is

$$C_{max} = \frac{\dot{M}}{\bar{u}d} \frac{1}{\sqrt{4\pi\varepsilon_t x/\bar{u}}}$$

$$= \frac{930 \text{ ft}^3/\text{sec-ppm}}{2 \text{ ft/sec} \times 30 \text{ ft}(4\pi \times 3.6 \text{ ft}^2/\text{sec} \times 1000 \text{ ft/2 ft/sec})^{1/2}}$$

$$= 0.10 \text{ ppm.} \quad \blacksquare$$

Example 5.2: Mixing across a stream. An industrial plant discharges a conservative substance at the side of a straight rectangular channel. The channel is 2000 ft wide and carries a uniform flow 5 ft deep at a mean velocity of 2 ft/sec. The slope of the channel is 0.0002. Find the length of channel required for "complete mixing," as defined to mean that the concentration of the substance varies by no more than 5% over the cross section.

Solution. The shear velocity is given by

$$u^* = \sqrt{gdS} = \sqrt{32.2 \times 5 \times 0.0002} = 0.18 \text{ ft/sec.}$$

By Eq. (5.4) the transverse mixing coefficient is

$$\varepsilon_t = 0.15du^* = 0.15 \times d \times 0.18 = 0.135 \text{ ft}^2/\text{sec.}$$

Because the discharge enters the channel at the side, we must use twice the width in Eq. (5.10), hence the distance required for complete mixing is

$$L = 0.1\bar{u}(2W)^2/\varepsilon_t = 0.1 \times 2 \times (400)^2/0.135 = 240,000 \text{ ft.}$$

This is, of course, a very long distance for a real channel, and it is worth noting that according to Fig. 5.5 after even 25 miles of channel there will still be concentration variations across the channel of up to 30%. Of course in a real channel bends or changes in cross section would be likely within 25 miles, and would speed up the mixing. $\blacksquare$

Example 5.3: Blending of two streams. Sometimes we wish to compute the mixing of two streams which flow together at a smooth junction, and whose density is nearly the same so that the streams flow side by side until turbulence accomplishes the mixing. As an example, suppose that a city obtains its water supply partly by means of an aqueduct from another river basin and partly from local sources. The chemical contents of the two sources are quite different, so before processing in a water treatment plant it is necessary to blend the two supplies. The amount of water from each source is 50 ft^3/sec. It is proposed to blend the two supplies by running them together into a single rectangular channel 20 ft wide, having a slope of 0.001 and a Manning's "n" value of 0.030.

(a) Assuming the channel is straight, what length of channel is necessary to provide complete mixing between the two sources?

(b) To take advantage of the increased rate of mixing it is proposed to build the channel as a circular curve with a radius of 100 ft. In this case what length of channel is necessary?

Solution. The velocity and depth of flow, obtained by solving Manning's formula, are $d = 2.2$ ft and $\bar{u} = 2.28$ ft/sec. Hence, we have

$$u^* = \sqrt{gdS} = 0.268 \text{ ft/sec.}$$

To study the mixing of the two streams we may assume that there is a tracer whose concentration is C_0 in one stream and zero in the other. The initial condition at the upstream end of the channel is $C = C_0$ for $0 < y < W/2$ and $C = 0$ for $W/2 < y < W$, where $y = 0$ is at one bank. If the streams were mixed completely the concentration would be $\frac{1}{2}C_0$ everywhere on the cross section; in practice we would like to know when this condition is reached within, say, 5%.

Before actually solving the problem it is useful to notice an upper bound. The initial condition may be considered to consist of a uniform distribution of unit inputs in one-half of the channel. Since the unit inputs mix independently of each other, the longest mixing time will be that required for the unit input at the side of the channel to become uniformly mixed. According to Eq. (5.10) this requires a distance $L = 0.1u(2W)^2/\varepsilon_t$, which for the straight channel of part (a) is approximately 4000 ft. The actual distance required will be less than 4000 ft because the unit inputs closer to the channel centerline require less mixing time.

The exact solution can be obtained either from Eq. (5.11) or, more easily, by superposition of solutions for the step function in an unbounded system [Eq. (2.33)]. The initial and boundary conditions can be satisfied by superposing step functions from $-\frac{1}{2}C_0$ to $+\frac{1}{2}C_0$ at $y' = -\frac{1}{2} \pm 2n$ ($n = 1, 2, 3, \ldots$) and step functions from $+\frac{1}{2}C_0$ to $-\frac{1}{2}C_0$ at $y' = \frac{1}{2} \pm 2n$ ($n = 1, 2, 3, \ldots$), where $y' = y/W$ as before. The reader should verify that the sum of these step functions satisfies the initial condition; symmetry assures satisfaction of the boundary conditions. Hence, the exact solution is

$$C = \frac{C_0}{2} \sum_{n=-\infty}^{\infty} \left(\text{erf} \frac{y' + \frac{1}{2} + 2n}{\sqrt{4x'}} - \text{erf} \frac{y' - \frac{1}{2} + 2n}{\sqrt{4x'}} \right), \qquad (5.12)$$

where $x' = x\varepsilon_t/\bar{u}W^2$ as before. Values computed from Eq. (5.12) are shown in Fig. 5.9 for the centerline ($y' = \frac{1}{2}$) and the sides ($y = 1$ and $y = 0$).

The maximum deviation in concentration is 5% of the mean when x' is approximately 0.3. Therefore, to answer part (a) we take

$$\varepsilon_t = 0.15du^* = 0.15 \times 2.28 \times 0.268 = 0.0885 \text{ ft}^2/\text{sec}$$

and the required length of channel is

$$L = 0.3 \frac{\bar{u}W^2}{\varepsilon_t} = \frac{0.3 \times 2.28 \times (20)^2}{0.0885} = 3100 \text{ ft.}$$

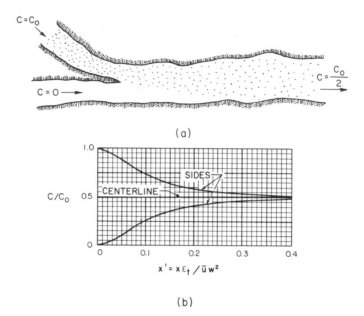

(b)

Figure 5.9 (a) Blending of two streams of equal discharge and (b) concentrations on the center-line and at each side downstream of the junction point, as given by Eq. (5.12).

For part (b), the case of a circular curve, we can use Fischer's (1969) laboratory result for a continuous curving channel, Eq. (5.5),

$$\varepsilon_t \approx 25 \frac{\bar{u}^2 d^3}{R^2 u^*} = 25 \times \frac{(2.28)^2 \times (2.2)^3}{(100)^2(0.268)} = 0.52 \text{ ft}^2/\text{sec}.$$

The required length of channel is

$$L = 0.3 \times \frac{2.28 \times (20)^2}{0.52} = 530 \text{ ft}. \quad \blacksquare$$

5.1.4   Complications in Real Streams; Use of the Cumulative Discharge Method

The plume formulas and analysis presented in Section 5.1.3 describe the initial mixing region in a river to a reasonable approximation, but in rivers there are some additional complications. The previous analysis assumed a uniform flow of constant velocity everywhere in the channel. In rivers the downstream velocity varies across the cross section, and there are irregularities along the channel. Some rivers even have groins built out perpendicular to the flow to retard sediment transport. Virtually all natural streams have a deep part of the cross section and shallow portions near the banks. The downstream velocity is fastest in the deepest section, and slows to near zero near the shores. Fig. 5.11 gives a typical cross section and transverse velocity distribution for a natural river (see below).

An approximate solution for near-field dispersal of an effluent in a river can be obtained by the methods given in the previous section. Simply assume that the river is of constant depth, and that the downstream velocity is equal to the mean value everywhere. The results so obtained may be adequate for many practical problems, especially since the actual shape of the cross section is often not well known nor is the value of the transverse mixing coefficient. If these quantities are accurately known, however, a better procedure is to use the "cumulative discharge" method described by Yotsukura and Cobb (1972) and Yotsukura and Sayre (1976). The cumulative discharge is defined by $q(y) = \int_0^y d\bar{u}^z \, dy$, where $\bar{u}^z(y) = 1/d \int_{-d}^0 u \, dz$ is the depth averaged velocity. It varies from zero at one bank to the total discharge Q at the other. The depth-integrated equation for transverse diffusion, assuming a steady-state concentration distribution, is

$$d\bar{u}^z \frac{\partial C}{\partial x} = \frac{\partial}{\partial y}\left(d\varepsilon_t \frac{\partial C}{\partial y}\right)$$

which can be transformed to obtain

$$\frac{\partial C}{\partial x} = \frac{\partial}{\partial q}\left(d^2 \bar{u}^z \varepsilon_t \frac{\partial C}{\partial q}\right). \tag{5.13}$$

Yotsukura and Sayre suggest that $d^2 \bar{u}^z \varepsilon_t$ can often be regarded as a constant "diffusivity," so that Eq. (5.13) is reduced to the form of Eq. (2.4) and reflected Gaussian solutions are obtained in the x-q coordinate system.

The value of the transformation is that a fixed value of q is attached to a fixed streamline, so that the coordinate system shifts back and forth within the cross section along with the flow. This greatly simplifies interpretation of tracer measurements in a meandering stream. Figure 5.10a shows the data obtained by Yotsukura et al. (1970) in the Missouri River near Blair, Nebraska (approximately 80 miles upstream from Omaha). The peak of the distribution moves from side to side as the river meanders. When the same data are plotted with respect to q (Fig. 5.10b), the peak remains at the injection location and the distribution spreads out as a decaying, reflected Gaussian distribution. The results in a meandering river, plotted against q, are indistinguishable from the results of a similar discharge into a straight canal plotted against y. In other words, the transformation from transverse distance to q as the independent variable essentially transforms a meandering river into an equivalent straight one. This statement is not exactly true if the meanders are sharp, because as Yotsukura and Sayre demonstrate it is theoretically necessary to introduce metric coefficients of an orthogonal curvilinear coordinate system, but for practical purposes the introduction of the metrics is usually not important.

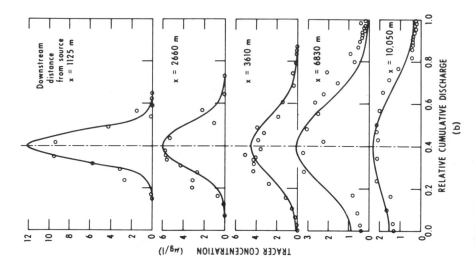

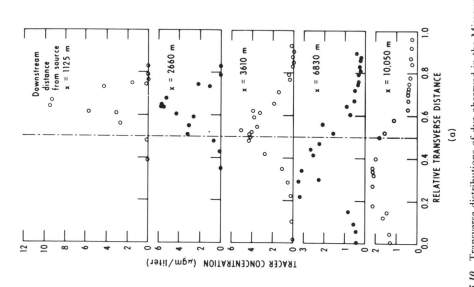

Figure 5.10 Transverse distributions of dye observed in the Missouri River near Blair, Nebraska, by Yotsukura *et al.* (1970), plotted (a) versus actual distance across the stream and (b) versus relative cumulative discharge. [After Yotsukura and Sayre (1976)].

5.1.5   Turbulent Mixing of Buoyant Effluents

If an effluent is lighter or heavier than the receiving water, the initial mixing may be strongly affected by the difference in densities. The most common example is the discharge of heated water from a power station. In the region close to the source we need an analysis of a buoyant plume interacting with a turbulent flowing stream, as discussed in Section 9.5. Sometimes, however, the residual buoyancy of the source continues to have an effect further downstream in the region where the mixing is caused primarily by the turbulence of the receiving water. To complete our analysis of turbulent mixing in rivers we would like to have some way of estimating how the turbulent mixing coefficients ε_v and ε_t are affected by an initial density difference.

Unfortunately not enough is known to permit confident engineering estimates. Oceanographers and boundary layer meteorologists have studied the effect of density stratification on vertical mixing, and some of this work is used or referenced in the following chapters on reservoirs and estuaries. There have been only a few laboratory studies directly relevant to mixing of effluents in rivers, however, and all of these have been in rectangular flumes. Extrapolation of the results to real rivers is difficult, partly because we are not certain that our scaling laws are accurate and partly because the shape and irregularity of the river cross section may play a role.

The parameters most likely to determine the density effects are the width, depth, and shear velocity of the receiving stream and the flux of buoyancy in the effluent (this concept is explained in detail in Chapter 9), defined by $B = (\Delta\rho/\rho)gQ_e$, where ρ is the density of the receiving water, $\Delta\rho$ the difference in density between receiving and effluent waters, g the acceleration of gravity, and Q_e the effluent discharge. These parameters make two dimensionless groups, B/du^{*3} and B/Wu^{*3}, which express, respectively, the ratio of the stabilizing influence of the effluent per unit depth and per unit width to the mixing power available in the receiving stream. We would expect that in the case of an effluent discharged near the side of a stream the ratio B/du^{*3} would determine how the density difference would influence transverse mixing, and that in the case of an effluent distributed over the surface of the stream (if lighter, or the bottom if heavier) the ratio B/Wu^{*3} would determine the result. In addition, however, the stream geometry, method of release of the effluent, and friction factor of the channel may have an effect, and there may be other conditions peculiar to a discharge site.

The laboratory results of Prych (1970) on transverse mixing and Schiller and Sayre (1973) on vertical mixing provide some approximate indications of when density effects come into play. Although neither study used the dimensionless numbers we have derived here, Prych's results can be interpreted to say that if B/du^{*3} is less than five, transverse spreading is reasonably well described by neglecting density effects, and Schiller and Sayre's results can be interpreted to say that if $B/Wu*^3$ is less than one, vertical mixing will be independent of density effects. We can also say qualitatively that if $B/du*^3$ is very large, an effluent will spread rapidly across a channel in the form of a density driven current and will be likely to form a layer at the surface (if lighter, or at the bottom if heavier). Also, if $B/Wu*^3$ is large, the vertical mixing of a surface or bottom layer will be slow. Mixing of layers is discussed in Chapter 6 with respect to the deepening of the epilimnion in reservoirs, but confident predictions for mixing of layers in rivers do not yet seem possible. Scale model studies may sometimes be useful. In most cases, however, both the ratios $B/du*^3$ and $B/Wu*^3$ should be the same in model and prototype, implying an undistorted model and a practical limitation that only a short reach of river can be modeled. The use of physical models is discussed in more detail in Chapter 8; whether a model should be built for a specific study is a matter for judgment, but given the lack of theoretical understanding or adequate experimental results there will be many cases in which a scale model can provide design information that is otherwise unobtainable.

5.2 LONGITUDINAL DISPERSION IN RIVERS

After a tracer has become adequately mixed across the cross section, the final stage in the mixing process is the reduction of longitudinal gradients by longitudinal dispersion. If an effluent is discharged at a constant rate into a river whose discharge is also constant, we will show in Section 5.5 that there is no need to be concerned about longitudinal dispersion; the reader interested only in that case may skip the present section. A common practice in sanitary engineering has been to assume immediate cross-sectional mixing and to neglect longitudinal dispersion; for example the derivation of the widely used Streeter–Phelps equation (see Fair *et al.*, 1971, or other sanitary engineering texts) depends on these assumptions.

There are, however, practical cases where longitudinal dispersion is important. The most immediately apparent is the accidental spill of a quantity of pollutant, such as a release of radioactive material from a riverside nuclear power station. A more common example is the daily cyclic variation of output from a sewage treatment plant; to compute downstream concentrations more accurately than on a daily average basis we must account for the effect of longitudinal dispersion of the streamwise gradients caused by the variation in effluent discharge. Therefore this section discusses how to estimate the longitudinal dispersion coefficient K in a natural stream for use in the one-dimensional dispersion equation derived in Chapter 4

$$\frac{\partial \overline{C}}{\partial t} + \bar{u}\frac{\partial \overline{C}}{\partial x} = K\frac{\partial^2 \overline{C}}{\partial x^2}. \tag{4.19}$$

As part of our discussion we will also note in Section 5.2.2 how Eq. (4.19) fails to account for all the observed characteristics of spreading in real streams, even after cross-sectional mixing would be expected to be complete, and we will give in Section 5.3 a method for analyzing dispersion before cross-sectional mixing is complete. Before reaching these practical matters, however, we begin with an explanation of how the shear flow dispersion theory developed in Chapter 4 gives a theoretical basis for the analysis of dispersion in streams.

5.2.1 Theoretical Derivation of the Longitudinal Dispersion Coefficient

The analysis of dispersion in natural streams is based on the shear flow dispersion analysis presented in Chapter 4. In Section 4.2 we presented Elder's analysis of dispersion due to the logarithmic velocity profile, which led to the result,

$$K = 5.93du^*.$$

Now consider the typical cross section of a natural stream sketched in Fig. 5.11. Henceforth, we will use a coordinate system as sketched in the figure, in which z is distance vertically upward from the water surface, y is measured across the stream from the right bank (looking downstream), and x is distance downstream. If an investigator were to go out in a boat, anchor at some fixed point, and measure the vertical profile of velocity $u(z)$ there, he would usually find that $u(z)$ follows approximately the logarithmic profile used by Elder, $u = \bar{u} + (u^*/\kappa)\{1 + \ln[(z + d)/d]\}$. Hence, as a first guess we might expect that Elder's analysis of shear flow dispersion might describe dispersion in natural streams.

That Elder's result does not describe dispersion in real streams, even to a first approximation, was first demonstrated by the measurements of Godfrey and Frederick (1970).† Godfrey and Frederick released slugs of solution of the radioactive tracer Gold-198 and measured the growth of the concentration of tracer versus time at various distances downstream from each release. Experiments were performed in four streams under a total of ten separate conditions, and observed values of K/du^* ranged from 140 to 500 (see Table 5.3). More recent experiments have confirmed that K/du^* varies widely but is almost always much greater than Elder's result. The largest value known to the writers is $K/du^* = 7500$ in the Missouri River observed by Yotsukura *et al.* (1970), and the smallest is $K/du^* = 8.6$ observed in the Yuma Mesa A canal in Arizona measured by Schuster (1965). The evidence makes clear beyond doubt that Elder's result does not apply to real streams.

---

† Although the publication is dated 1970, the experiments were performed in the period 1959–1961 and were made available shortly thereafter through an open file report.

TABLE 5.3

Experimental Measurements of Longitudinal Dispersion in Open Channels

Reference	Channel	Depth, d (m)	Width, W (m)	Mean velocity, $\bar{u}$ (m/sec)	Shear velocity, u^* (m/sec)	Observed dispersion coefficient, K (m²/sec)	$\dfrac{K}{du^*}$	K predicted by Eq. (5.16) (m²/sec)	K predicted by Eq. (5.19) (m²/sec)
Thomas (1958)	Chicago Ship Canal	8.07	48.8	0.27	0.0191	3.0	20		
State of California (1962)	Sacramento River	4.00		0.53	0.051	15	74		
Owens et al. (1964)	River Derwent	0.25		0.38	0.14	4.6	131		
Glover (1964)	South Platte River	0.46		0.66	0.069	16.2	510		
Schuster (1965)	Yuma Mesa A Canal	3.45		0.68	0.345	0.76	8.6		
Fischer (1967a)	Trapezoidal laboratory channel with roughened sides	0.035	0.40	0.25	0.0202	0.123	174	0.131	
		0.047	0.43	0.45	0.0359	0.253	150	0.251	
		0.035	0.40	0.45	0.0351	0.415	338	0.371	
		0.035	0.34	0.44	0.0348	0.250	205	0.250	
		0.021	0.33	0.45	0.0328	0.400	392	0.450	
		0.021	0.19	0.46	0.0388	0.220	270	0.166	
Fischer (1968b)	Green-Duwamish River, Washington	1.10	20		0.049	6.5–8.5	120–160	7.8	
Yotsukura et al. (1970)	Missouri River	2.70	200	1.55	0.074	1500	7500		3440

Godfrey and Frederick (1970) (predicted values of K from Fischer, 1968a)								
Copper Creek, Virginia (below gage)	0.49	16	0.27	0.080	20	500	6.0	
	0.85	18	0.60	0.100	21	250	28	
	0.49	16	0.26	0.080	9.5	245	11.4	
Clinch River, Tennessee	0.85	47	0.32	0.067	14	235	15	22
	2.10	60	0.94	0.104	54	245	86	73
	2.10	53	0.83	0.107	47	210	55	28
Copper Creek, Virginia (above gage)	0.40	19	0.16	0.116	9.9	220	2.8	
Powell River, Tennessee	0.85	34	0.15	0.055	9.5	200	9.1	
Cinch River, Virginia	0.58	36	0.21	0.049	8.1	280	30	
Coachella Canal, California	1.56	24	0.71	0.043	9.6	140	3.9	
Fukuoka and Sayre (1973)								
Sinuous rectangular laboratory channel, smooth sides, smooth and rough bottoms— 25 experiments	0.023 to 0.070	0.13 to 0.25		0.011 to 0.027				
McQuivey and Keefer (1974) (predicted values of K from Fischer 1975)								
Bayou Anacoco	0.94	26	0.34	0.067	33			13
	0.91	37	0.40	0.067	39			38
Nooksack River	0.76	64	0.67	0.27	35			98
Wind/Bighorn Rivers	1.10	59	0.88	0.12	42			232
	2.16	69	1.55	0.17	160			340
John Day River	0.58	25	1.01	0.14	14			88
	2.47	34	0.82	0.18	65			20
Comite River	0.43	16	0.37	0.05	14			16
Sabine River	2.04	104	0.58	0.05	315			330
	4.75	127	0.64	0.08	670			190
Yadkin River	2.35	70	0.43	0.10	110			44
	3.84	72	0.76	0.13	260			68

The reason that Elder's result does not apply was shown by Fischer (1966, 1967a) to be because of the transverse variation of velocity across the stream. Figure 5.11 shows the measured velocity distribution at one cross section of the Green-Duwamish River, Washington, the stream in which Fischer did his field experiments. Vertical profiles are approximately logarithmic, but it is also possible to derive a transverse profile by plotting the cross stream variation of the depth-averaged velocity,

$$\bar{u}^z(y) = \frac{1}{d(y)} \int_{-d(y)}^{0} u(y, z)\, dz, \tag{5.14}$$

as sketched in Fig. 5.12. $\bar{u}^z$ is a shear flow velocity profile extending over the width of the stream W, whereas $u(z)$, the profile used in Elder's analysis, extends only over the depth of flow d. In Chapter 4 we found in Eq. (4.47) that the longitudinal dispersion coefficient is proportional to the square of the distance over which the

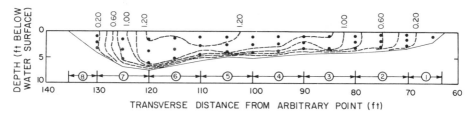

Figure 5.11   The cross-sectional velocity distribution looking downstream at Renton Junction in the Green-Duwamish River, Washington. Contours are lines of constant velocity, in increments of 0.2 ft/sec; dots indicate points where velocity was measured. The circled numbers show how the section is divided into stream tubes for the numerical model of the initial mixing period (Section 5.3) and to evaluate the longitudinal dispersion coefficient by the integration given by Eq. (5.16). Average velocities and discharges of the subareas are as follows:

Subarea (no.)	Area (ft)	Mean velocity in subarea (ft/sec)	Mean velocity in subarea relative to mean x-section velocity (ft/sec)	Relative discharge in subarea (col. 2 × col. 4) (cfs)
1	12.7	0.105	−0.799	−10.1
2	41.9	0.526	−0.378	−15.7
3	42.2	0.986	0.082	3.6
4	48.3	1.091	0.187	9.1
5	52.2	1.196	0.292	15.3
6	66.4	1.148	0.244	16.3
7	63.6	0.766	−0.138	−8.7
8	11.8	0.067	−0.837	−9.8
Totals	339.1			0.0

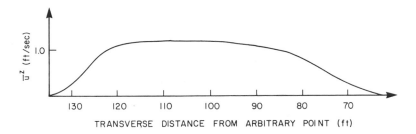

Figure 5.12 The transverse profile of the depth-averaged velocity corresponding to the velocity distribution shown in Fig. 5.11.

shear flow profile extends. Natural streams usually have width to depth ratios in the range of ten or greater. Thus we should expect to find that the transverse profile $u(y)$ should be 100 or more times as important in producing longitudinal dispersion as the vertical profile. This is in effect what Godfrey and Frederick discovered, although at the time they did not understand why.

A quantitative estimate of the dispersion coefficient in a real stream can be obtained by neglecting the vertical profile entirely and applying Taylor's analysis to the transverse velocity profile. Consider a slice of thickness dx moving at the mean flow velocity $\bar{u}$, as illustrated in Fig. 5.13, and let $u'(y) = \bar{u}^z(y) - \bar{u}$. The balance of diffusion and advection discussed in Section 4.1.2 is, in this case, a balance of transverse diffusive mass transport through the vertical face in the xz plane at the left edge of the slice versus the net advective mass transport through the vertical faces in the yz plane. The equivalent of the first integral of Eq. (4.35) is

$$\int_0^y u'(y)\, d(y)\, \frac{\partial \bar{C}}{\partial x}\, dy = d\varepsilon_t \frac{\partial C'}{\partial y}. \tag{5.15}$$

Implicit in this formulation is that $C'(y)$ and $u'(y)$ are assumed to vary only across the stream. Vertical variations are neglected because their effect, as mentioned

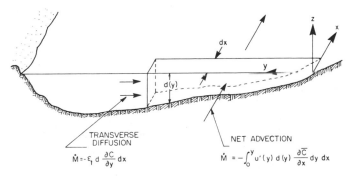

Figure 5.13 An illustration of the balance of advection and diffusion expressed by Eq. (5.15).

in the previous paragraph, is expected to be small compared to the effect of the transverse variations.

Equation (5.15) may be integrated to find $C'(y)$ and the result integrated over the cross section to give

$$K = -\frac{1}{A\,\partial\bar{C}/\partial x}\int_A u'C'\,dA = -\frac{1}{A}\int_0^W u'd\int_0^y \frac{1}{\varepsilon_t d}\int_0^y u'd\,dy\,dy\,dy. \quad (5.16)$$

This result is, of course, only an estimate because it is based on the concept of a uniform flow in a constant cross section. It differs from the result given by Eq. (4.36) only in the inclusion of the variable depth. If we define $d' = d/\bar{d}$, where $\bar{d}$ is the mean depth, introduce d' into the integral, and make the integral dimensionless as in Section 4.2, we find that the value of $\bar{d}$ cancels out and we have as the equivalent of Eq. (4.47),

$$K = IW^2\overline{u'^2}/\bar{\varepsilon}_t, \quad (5.17)$$

where I is the dimensionless integral given in Eq. (4.48), but with the dimensionless depth added in the places where the real depth appears in Eq. (5.16). In Section 4.2 we investigated several simple flows and found that I does not vary significantly because of variations in the velocity profile; the same result has been found for cross sections and velocity profiles typical of streams. In the following sections we will discuss the complications that arise in real streams, and then show how Eq. (5.17) can be used to obtain reasonably accurate predictions of the longitudinal dispersion coefficient.

Example 5.4   Use the cross-sectional distribution of velocity shown in Fig. 5.11 to estimate the longitudinal dispersion coefficient for the Green-Duwamish River at Renton Junction. Take $\varepsilon_t = 0.133$ ft^2/sec as given by Fischer (1968b).

~Solution.   A simple and reasonably adequate way of doing the triple integration required by Eq. (5.16) is shown in Table 5.4. The computation uses the eight subareas shown in Fig. 5.10. Numbers in the table are obtained as follows:

Column 1:   Transverse distance to the end of the subarea (the cross section starts at $y = 63$ and extends to $y = 136$; the numbers are with respect to an arbitrary datum).
Column 2:   The cumulative relative discharge to the end of the subarea, found by summing the last column underneath Fig. 5.10.
Column 3:   The mean depth of the subarea.
Columns 4 and 5:   The values of the integrals shown. Each value in column 4 is obtained by taking the value at the end of the previous subarea and adding the mean of the cumulative relative discharge multiplied by the width of the subarea and divided by the mean depth of the subarea and the transverse mixing

TABLE 5.4

Numerical Integration of Eq. (5.16)

(1)	(2)	(3)	(4)	(5)
y	$\int_0^y du'\,dy$	d	$\int_0^y \dfrac{1}{\varepsilon_t d}\int_0^y du'\,dy\,dy$	$\int_0^y u'd \int_0^y \dfrac{1}{\varepsilon_t d}\int_0^y du'\,dy\,dy\,dy$
63	0		0	0
		1.8		
70	-10.1		-148	750
		4.2		
80	-25.8		-469	5590
		4.2		
90	-22.2		-898	3130
		4.8		
100	-13.1		-1174	-6300
		5.2		
110	2.2		-1252	-24860
		6.6		
120	18.5		-1134	-44300
		6.4		
130	9.8		-967	-35170
		2.0		
136	0		-856	-26230

coefficient. Each value in column 5 is obtained by taking the value at the end of the previous subarea and adding the mean of column 4 in the subarea multiplied by the relative discharge in the subarea as given in the last column underneath Fig. 5.10.

The last figure in column 5 is the value of the integral in Eq. (5.16). Division by the total area gives $K = -(-26230)/339 = 77$ ft^2/sec. ∎

5.2.2 Dispersion in Real Streams

So far the analyses we have discussed have been limited to uniform channels because Taylor's analysis assumes that everywhere along the stream the cross section is the same. Real streams may have bends, sandbars, side pockets, pools, riffles, bridge piers, man-made revetments, and perhaps the occasional junked car or sunken barge. Every irregularity contributes to dispersion; some streams may be so irregular that no reasonable analysis can be applied. For instance, a mountain stream that consists of a series of pools and riffles is not a suitable place to apply Taylor's analysis. Nevertheless the majority of streams are uniform enough for an approximate analysis, if the limitations are borne in mind.

An immediate limitation, of course, is that Taylor's analysis cannot be applied until after the initial period described in Chapter 4. In a uniform channel the time scale for transverse mixing is W^2/ε_t and dimensionless distance downstream from a source can be defined as in Eq. (5.8) by $x' = (x/\bar{u})(\varepsilon_t/W^2)$. Numerical experiments have shown that in a uniform channel the variance of a dispersing cloud grows linearly with distance downstream for $x' > 0.2$. The skewed longitudinal distribution illustrated in Fig. 4.4 grows in the reach $0 < x' < 0.4$. For $x' > 0.4$ the skew decays towards a Gaussian distribution; the decay can usually be described by the routing procedure [(Eq. (5.20)] described in the next section. As mentioned in Section 4.1.2 the longitudinal distribution of concentration is expected to become approximately Gaussian for $x' > 1$. The various zones are illustrated in Fig. 5.14.

The irregularities in real streams increase the length of the initial period over that shown in the figure, and usually lead to the production of a long "tail" on observed concentration distributions. This tail is produced by the detention of small amounts of the effluent cloud. Figure 5.15 shows a cloud of Rhodamine dye dispersing in the Green River, Washington. Behind the main part of the cloud we can see pockets of dye retained in small, hardly noticeable irregularities along the sides of the channel. The dye is released slowly from these pockets, and causes measurable concentrations of dye to be observed long after the main portion of the cloud has passed.

One way to treat the tail is to ignore it; it may not contain much dye anyhow. The analysis of the Green River experiments (Fischer, 1968b) ignored the tail and showed that for the remainder of the dye cloud the application of Taylor's analysis worked well. On the other hand, Nordin and Sabol (1974) analyzed results of a number of dye dispersion studies and showed that in most cases the variance of the dye distribution increased faster than linearly with time through-

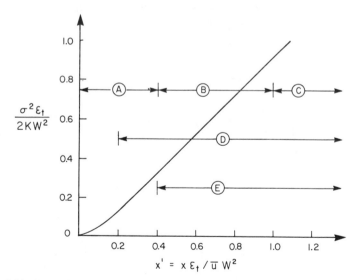

Figure 5.14 Stages in the evolution of a concentration distribution from a slug injection and expected growth of the longitudinal variance as given by numerical experiments in uniform channels. In real channels sidewall irregularities will increase the required values of x' as indicated in the text. Ⓐ Generation of a skewed distribution (see Fig. 4.4). Ⓑ Decay of the skewed distribution. Ⓒ Approach to a Gaussian distribution. Ⓓ Zone of linear growth of the variance. Ⓔ Zone where use of the routing procedure described in Section 5.2.3 is acceptable.

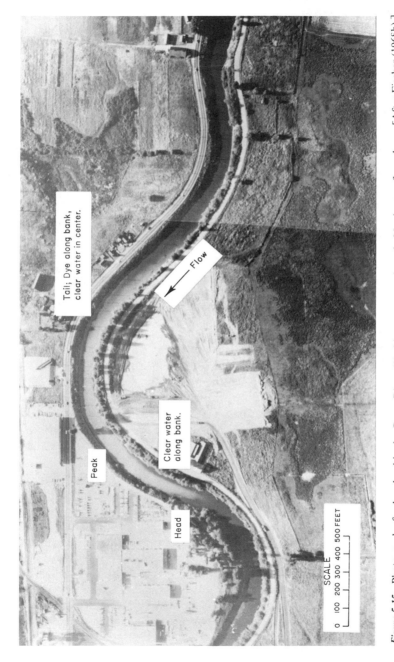

Figure 5.15 Photograph of a dye cloud in the Green River, Washington, approximately 35 minutes after release. [After Fischer (1968b).]

out the experiments (which is to say that Taylor's analysis never applied). The difficulty is that the computation of the variance weights the concentrations in the "tail" heavily, so that if the tail is not ignored the variance increases unreasonably.

Several writers have suggested that a complete analysis of dispersion in streams must include the effect of the side pockets, or as they are frequently called, "dead zones." Valentine and Wood (1977) and Valentine (1978) describe an analysis in which the stream is divided into a flowing zone, with a velocity profile, and a stagnant zone along the bottom or side. The exchange between the zones is postulated to depend on the difference in concentration between them and an exchange coefficient. The Aris moment method (Section 4.1.5) is used to compute the rate of increase of the variance and the coefficient of skew. The analysis shows that the dead zones increase the length of the initial period and the magnitude of the longitudinal dispersion coefficient. Valentine also describes laboratory experiments in which dead zones were produced by placing strips across the bottom of a flume. The depth of cavity produced by the strips was 0.5 cm. When the depth of flow over the top of the strips was 1 cm the "dead zones" seem to have increased the duration of the initial period by a factor of approximately two, and the magnitude of the dispersion coefficient by a factor of approximately five. Smaller ratios of dead zone volume to main flow volume had lesser effect. However, these results cannot be applied directly to natural streams, since the detailed geometry of the dead zone and its connection to the main flow undoubtedly determines the exchange properties. Valentine and Wood's and the other dead zone analyses make clear that side pocket and bottom irregularities do play an important role in determining the length of the initial period, the generation of the "tail," and the magnitude of the dispersion coefficient, but as yet it does not seem possible to quantify these results for application to real streams. Day (1975), for example, studied dispersion in five mountain streams in New Zealand and reported that even discounting the "tails" the dispersing clouds did not approach the Gaussian shape over the length of his experiments; this seems to have been because of "dead zones" in the bottom of the very rough stream he used, but the exact role of the roughness could not be quantified.

Mention should also be made of the effect of bends. Bends increase the rate of transverse mixing, and thereby to some extent reduce the dispersion coefficient. A much more important effect, however, is the tendency of bends to induce the sort of transverse velocity profile shown in Fig. 5.11. At any natural bend the high velocities will be concentrated toward the outside bank and the low velocities toward the inside one. Thus in a meandering stream the velocity differences across the stream are accentuated and the dispersion coefficient is likely to be much greater than in a stream which is artificially straightened.

A third effect, less obvious, comes about because in a natural channel the bends must be in alternating directions, i.e., first to one direction and then to the

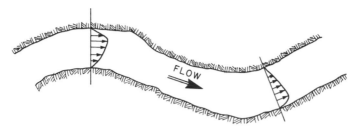

Figure 5.16 Typical velocity profiles in bends.

other. Recall that for any given velocity profile there is a steady-state concentration profile established through the process of transverse mixing. The dispersion coefficients given by Eqs. (5.16) and (5.17) depend on the steady-state profile being established; until it is, dispersion will be at a generally smaller rate. When a series of bends alternate in direction, the maximum of the velocity profile alternates from one side of the stream to the other, as shown in Fig. 5.16. The steady-state concentration profile also alternates, but if the bends come too close upon each other there may not be sufficient time for the steady-state profile to be reached. In that case the actual concentration profile will have smaller fluctuations, and the longitudinal dispersion coefficient will be less than computed from Eq. (5.16).[†]

Fischer (1969) quantified these remarks by writing a computer program to simulate an alternating series of bends. Whether the bends are sufficiently long for the steady-state concentration profile to be established depends on the ratio of the cross-sectional diffusion time to the time required for flow to round the bend, i.e., the ratio

$$\gamma = (W^2/\varepsilon_t)(\bar{u}/L), \tag{5.18}$$

where L is the length of the curve. In the Green-Duwamish River ($\gamma \approx 25$) the dispersion coefficient given by the computer program was almost identical to that given by Eq. (5.16), meaning that there was no effect due to the alternating direction and the dispersion was the same as would be given by one continuous curve. In the Missouri River ($\gamma \approx 130$), however, the dispersion coefficient was reduced by a factor of approximately eight from the results of Eq. (5.16). Both these results were verified by experiment. No other results were given, and if in another river γ were greater than 25 it would be necessary to use the computer program to accurately evaluate the effect of the curves. If, on the other hand, γ is less than 25 one may expect that all the effect of the curves is expressed by the observed velocity distribution and the result of Eq. (5.16) should be correct.

---

† See Section 4.3, in which a similar reduction is obtained in the case of oscillating velocity and concentration profiles in an unsteady flow.

5.2.3 Estimating and Using the Dispersion Coefficient in Real Streams

It makes little sense to try for too high accuracy in predicting dispersion coefficients in real streams. We cannot include the exact effects of irregularities, and
anyhow for most applications the results are insensitive to the exact value used
(for instance, the length of a dispersing cloud is proportional to the square root
of the dispersion coefficient). An approximate procedure given by Fischer
(1975) often works reasonably well. Fischer began with Eq. (4.47), $K = I\overline{u'^2}h^2/E$, and selected $I = 0.07$ as corresponding to a reasonable approximation
of the velocity profile in a real stream. The characteristic length h should be
somewhere between the half-width for a symetric velocity profile and the full-
width for a completely assymetric one; a reasonable choice is $h = 0.7W$, where
W is the full-width. Laboratory experiments (Fischer, 1966) have given the
ratio of $\overline{u'^2}/\bar{u}^2$ to vary between 0.17 and 0.25, with a mean of 0.2; similar ratios
are found in streams. Substituting these values, and $E = \varepsilon_t = 0.6du^*$ from
Eq. (5.6), into Eq. (4.47) gives

$$K = 0.011\bar{u}^2 W^2/du^*. \qquad (5.19)$$

This result is obviously quite approximate since it depends on a rough estimate
of ε_t and does not explicitly reflect the presence of "dead zones." Nevertheless,
it has the advantage of predicting the dispersion coefficient from the usually
available quantities of mean depth, width, mean velocity, and surface slope.

In most studies predictions by Eq. (5.19) have been found to agree with
observations within a factor of four or so, as can be seen in Table 5.3. Calculation
of "observed" values from field data is itself not simple and is usually accurate
only within a factor of two or so, so agreement of observation and prediction
within a factor of four is reasonably good. Observed values are usually computed
by one of two methods, the "change of moment" method in which the rate of
growth of the variance is computed and Eq. (2.36) applied, or a "routing
procedure" developed by Fischer (1968a). The difficulty with the change-of-
moment method is that the long "tails" on observed distributions make it
difficult to compute a meaningful value of the variance. The routing procedure
avoids this problem by matching a downstream observation of passage of a
tracer cloud to the prediction based on an upstream observation. Figure 5.17
illustrates the method. Curve 1 is a plot of tracer concentration versus time from
Godfrey and Frederick's experiment in Copper Creek, Virginia, at a discharge
of 300 ft^3/sec, and is used as the upstream observation. Curve 2, the curve of
downstream concentration versus time should, according to the solution of the
one-dimensional dispersion equation, be given by

$$C(X_2, t) = \int_{-\infty}^{\infty} C(X_1, \tau) \frac{\exp\left|\dfrac{-\{\bar{u}(\bar{t}_2 - \bar{t}_1 - t + \tau)\}^2}{4K(\bar{t}_2 - \bar{t}_1)}\right|}{\sqrt{4\pi K(\bar{t}_2 - \bar{t}_1)}} \bar{u}\, d\tau, \qquad (5.20)$$

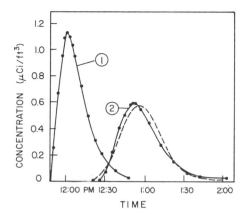

Figure 5.17 An illustration of Fischer's (1968a) routing procedure applied to run 1-60 of Godfrey and Frederick (1970). A release of a slug of radioactive tracer was made at 11:08 a.m. ——— Observed data: curve 1, 7870 ft downstream of the release; curve 2, 13,550 ft downstream of the release. ——— Result obtained by Eq. (5.20) using curve 1 and $K = 230$ ft^2/sec to predict curve 2.

where $\bar{t}_1$ and $\bar{t}_2$ are the mean times of passage at the upstream and downstream stations located, respectively, at X_1 and X_2.[†] The value of K can be varied to obtain the best fit between observed and predicted downstream curves, and the "best fit" value regarded as the observed longitudinal dispersion coefficient.

The routing procedure can also be used to predict downstream distributions once the value of K is known. This procedure has been found to be reasonably accurate if the dimensionless time of the upstream observation is greater than 0.4; if the dimensionless times are less than 1 the actual downstream observation usually is more skewed than the prediction, but the error is small enough to be of little significance.

Example 5.5: Dispersion of a slug. Ten pounds of Rhodamine WT dye are dropped onto the surface of the stream whose cross section is shown in Fig. 5.11. Estimate the value of the longitudinal dispersion coefficient using Eq. (5.19), and the length of the initial zone in which Taylor's analysis does not apply. Estimate the peak concentration that will be observed 20,000 ft downstream of the injection, and the length of the dye cloud at the time that the peak passes that point.

[†] Equation (5.20) is obtained from Eq. (2.30), the superposition integral, by making what has been termed the "frozen cloud" assumption, i.e., that the dispersing cloud changes its shape only slightly during the time taken to pass the observation point. Thus $C(X_i, t) \cong C(x, \bar{t}_i)$ with $x - X_i = \bar{u}(\bar{t}_i - t)$. This assumption is not exactly correct but is reasonably accurate for large dimensionless times. The properties of $C(X_i, t)$ have been studied by Tsai and Holley (1978), who showed that the variance of the concentration versus time curve increases linearly for times corresponding approximately to $x' > 0.4$.

Assume that the cross section shown in the figure is representative of the entire stream, that the shear velocity is $\frac{1}{10}$ the mean velocity, and that the stream is gently meandering with $\varepsilon_t = 0.4du^*$.

Solution. From Fig. 5.10 the mean velocity, width, and depth may be found to be

$$\bar{u} = 0.90 \text{ ft/sec}, \qquad W = 73 \text{ ft}, \qquad \bar{d} = 4.65 \text{ ft}$$

and we estimate the shear velocity to be $u^* = 0.09$ ft/sec. By Eq. (5.19)

$$K = 0.011\bar{u}^2 W^2/du^* = 0.011(0.9)^2(73)^2/(4.65)(0.09) = 113 \text{ ft}^2/\text{sec}.$$

The length of the initial period is determined by $x' = 0.4$. The transverse mixing coefficient is

$$\varepsilon_t = 0.4du^* = 0.167 \text{ ft}^2/\text{sec}$$

so the initial period ends when

$$x = 0.4\bar{u}W^2/\varepsilon_t = (0.4)(0.9)(73)^2/(0.167) = 11{,}500 \text{ ft}.$$

At the observation station where $x = 20{,}000$ ft, $x' = x\varepsilon_t/\bar{u}W^2 = (20{,}000)(0.167)/(0.9)(73)^2 = 0.7$, which means that the shape of the cloud will be approaching Gaussian. For practical purposes we can assume that it is Gaussian. In Fig. 5.14 the straight line portion of the curve intersects the x' axis at approximately $x' = 0.07$, which serves as a "virtual origin" for a cloud which would have followed the diffusion equation throughout its growth. The length of the real cloud can be found by assuming that the cloud started with zero variance at $x' = 0.07$ and grew steadily at the rate $d\sigma^2/dt = 2K$. Thus

$$\sigma^2 = 2K(W^2/\varepsilon_t)(x' - 0.07) = 4.54 \times 10^6 \text{ ft}^2.$$

Adopting the practical view that the length of the cloud is approximately 4σ we have

$$\text{"length"} = 4\sigma = (4)(2130) = 8500 \text{ ft}.$$

The peak concentration is given by

$$C_{max} = \frac{M}{A\sqrt{4\pi Kx/\bar{u}}} = \frac{10 \text{ lb}}{(339.1)\sqrt{4\pi(113)(20{,}000)/(0.9)}}$$

$$= 5.52 \times 10^{-6} \text{ lb/ft}^3$$

$$= 88.4 \text{ ppb} \quad \text{(parts per billion)}.$$

The observed value of the peak concentration should be somewhat less than this estimate because some material is trapped in the dead zones to make up the "tail" on the concentration distribution, and also because some of the dye will probably decay or be adsorbed on sediments. ∎

5.3 A NUMERICAL ANALYSIS FOR THE INITIAL PERIOD

The purpose of this section is to describe a simple but often adequate numerical procedure for computing concentrations in the initial reach downstream from a point of discharge in which a slug discharge is transformed into a longitudinally skewed concentration distribution by the combined action of transverse mixing and longitudinal advection. As discussed previously, this reach is defined approximately by $0 < x' < 0.4$, where $x' = xW^2/\bar{u}\varepsilon_t$. The one-dimensional dispersion equation is not applicable in this reach, so a numerical approach is needed.

The approach followed here is that of Fischer (1968b). It is not the most elegant, in the sense of using sophisticated techniques of numerical analysis, but it is easy to program, gives rapid computations, and has given adequate agreement with observed data. It uses observed velocities at only one cross section, in accord with the requirement that Taylor's analysis can be used only in a uniform flow, but an extension accounting for varying cross-sectional shape and using a more accurate numerical technique, but at the expense of substantial added programming complication, is described by Harden and Shen (1979).

The technique divides the total flow by vertical lines into n stream tubes of area $A_1, \ldots, A_n$, as shown in Fig. 5.11. Each stream tube is assigned a relative velocity, $u_1', \ldots, u_n'$, based on actual velocity measurements, care being taken that

$$\sum_{j=1}^{n} u_j' A_j = 0. \tag{5.21}$$

A computer mesh for concentration values $C(I, J)$ is established, where I refers to longitudinal distance in a coordinate system moving at the mean flow velocity and J to the jth stream tube. A time step Δt is selected, subject to conditions given below; the computer longitudinal distance step is taken as

$$\Delta x = u_{j\,\text{max}}' \Delta t, \tag{5.22}$$

in which u_j' is the mean velocity of the jth stream relative to a coordinate system moving at the overall cross-sectional mean velocity. Thus, the average flow in the stream tube of maximum relative velocity $u_{j\,\text{max}}'$ is moving at plus or minus one computer mesh point per time step.

Each time step is assumed to consist of two parts: first, the concentration distribution within each stream tube is advected up or downstream according to the velocity of that tube; second, at each cross section, transfer is accomplished between adjoining stream tubes according to the predetermined mixing coefficients.

In the advective part, an entire new set of mesh point values $D_t(I, J)$, is generated from the values $C_t(I, J)$, where the subscript t indicates the values after t time steps. The advective velocities are converted to units of mesh points per time step by the relation

$$U_j = u_j'(\Delta t/\Delta x). \tag{5.23}$$

A concentration which is advected part way between two computer mesh points is proportioned between them, inversely as the distance from each. Thus, the $D_t(I, J)$ are obtained from the relation

$$D_t(I, J) = C_t(I, J) + H(U_j)U_j[C_t(I - 1, J) - C_t(I, J)]$$
$$+ H(-U_j)U_j[C_t(I, J) - C_t(I + 1, J)], \tag{5.24}$$

in which H is the heaviside step function (which equals $+1$ if the argument is positive and zero otherwise). In other words, the second term on the right is used if U_j is positive and the third term if U_j is negative. In Chapter 8 we will discuss how the proportioning between grid points introduces a "numerical diffusion" whose coefficient lies between 0 and $0.125(\Delta x)^2/\Delta t$ depending on the ratio of the distance the concentration is advected to Δx. It is sometimes possible to choose values of Δx and Δt such that the numerical diffusion approximately equals the dispersion caused by the vertical velocity gradient, which we have otherwise neglected. The only important requirement, however, is that the induced numerical diffusion be much less than the longitudinal dispersion caused by the transverse velocity gradient, i.e., $0.125(\Delta x)^2/\Delta t \ll K$ where K is given by Eq. (5.16); if so the numerical diffusion will have little effect on the result given by the model.

For the transverse mixing part the following quantities are defined

h_j is the area of surface dividing stream tubes j and $j + 1$, per unit downstream length;

s_j is the distance between centroids of stream tubes j and $j + 1$;

ε_j is the mixing coefficient between stream tubes j and $j + 1$;

ΔC_j is the difference in concentration between stream tubes j and $j + 1$, $[C(I, J + 1) - C(I, J)]$.

The mass transport between stream tubes per time step is computed by assuming that for the duration of the step the concentration gradient at the dividing surface equals the difference in convected concentrations at the mesh points divided by the distance between them; that is

$$\Delta M_{j, j+1} = h_j \varepsilon_j (\Delta C_j/s_j)\Delta x \Delta t. \tag{5.25}$$

Since the mesh-point concentration is meant to represent the concentration within the entire stream tube, the change in concentration $\delta c(I, J)$ at mesh point (I, J) is given by

$$\delta c(I, J) = \frac{1}{A_j \Delta x} (\Delta M_{j, j+1} - \Delta M_{j-1, j}). \tag{5.26}$$

A new set of C net values for the $t + 1$ time step is calculated from D net values of the t step using the relation

$$C_{t+1}(I, J) = D_t(I, J) + \left\{ \frac{h_j \varepsilon_j}{s_j} [D_t(I, J + 1) - D_t(I, J)] \right.$$

$$\left. - \frac{h_{j-1} \varepsilon_{j-1}}{s_{j-1}} [D_t(I, J) - D_t(I, J - 1)] \right\} \frac{\Delta t}{A_j}. \qquad (5.27)$$

The scheme is stable if, for all value of j,

$$\varepsilon_j \Delta t / (s_j)^2 < 0.5; \qquad (5.28)$$

in practice it is well to keep this ratio less than approximately 0.2.

One advective step followed by one diffusive step completes the computation for one time step. The method can accept any desired initial distribution; for example, a slug source can be modeled by setting all initial concentrations equal to zero except for one, a plane source can be modeled by setting all initial concentrations zero except for the concentrations on one line, which are all set to some constant value, or any two-dimensional initial distribution can be introduced. A constant or variable rate of discharge can also be modeled, although a source at a fixed point must be moved through the grid since the program uses a moving coordinate system.

Fischer (1968b) demonstrated the accuracy of this model by simulating a two-dimensional flow with a logarithmic velocity profile and obtaining approximately the analytical result of Elder (1959; see Section 4.2). The computer program was also run to simulate the Green–Duwamish River using the eight stream tubes shown in Fig. 5.10, $\Delta x = 30$ ft and $\Delta t = 35.7$ sec. Two runs were made, one using a line source and the other a point source inserted only in stream tubes 4 and 5. The results are shown in Fig. 5.18, which also shows the

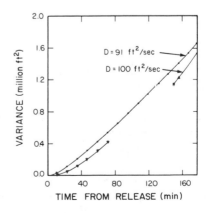

Figure 5.18   Results of Fischer's (1968b) numerical simulation of dispersion in the Green–Duwamish River. Legend: ─●─●─●─● dispersion from a line source; ─▼─▼─▼─ , dispersion from a point source; ─■─────▲───── , measured values and trend of observed results during prototype experiments on August 3 and September 9, 1965.

variances observed at Renton Junction during two field experiments. The computer program gave a value of $K = 91$ ft^2/sec, as compared with results of the field experiments of 100 ft^2/sec and the predictions by Eq. (5.16) of 77 ft^2/sec and by Eq. (5.19) of 83 ft^2/sec. The computer results show that the details of how the source is injected have only a small effect on the longitudinal variance, once the cloud is well mixed across the stream.

5.4 MEASUREMENT OF STREAM DISCHARGE BY TRACER TECHNIQUES

A relatively inexpensive way of measuring stream discharge in places where a gaging site is difficult to construct is to use a tracer technique. There are two methods: release of a single "slug" of a concentrated tracer, and continuous release over an extended time period at a constant rate of a diluted mixture of tracer. The slug release method is easier because it is not necessary to have a device capable of a continuous release at constant rate, but the continuous release method is probably more accurate and requires many fewer samples.

In the slug release method a known quantity of fluorescent dye or radioactive isotope is dropped into the stream. Radioactive tritium, for instance, can be obtained in a sealed glass vial which can be held beneath the water surface and crushed by tongs. Somewhere far enough downstream to allow complete cross-sectional mixing samples are taken at equal time intervals throughout passage of the tracer. The total mass passing the sampling station during a sampling period of length T_s is $M = \int_0^{T_s} CQ \, dt$, and if the stream discharge Q is constant this can be written $M = QT_s(1/T_s \int_0^{T_s} C \, dt)$. The quantity in parentheses is the average concentration, which can be found by one measurement if samples of equal volume taken at equal time increments are simply placed in one container and mixed. The stream discharge can then be computed by the formula,

$$Q = M/C_{av} T_s, \qquad (5.29)$$

where M is the amount of tracer injected and C_{av} the average concentration of all the samples taken together.

In the continuous release method a solution containing dye at concentration C_i is injected into a stream at discharge rate Q_i. The easiest way to do this is to construct a mariotte bottle, as shown in Figs. 5.19 and 5.20. The tank containing the solution is sealed, so that the only entry of air is through the bottom of the tube shown in the sketch. Hence the head on the orifice is always the distance from the bottom of the vent tube to the orifice, no matter what the level of fluid in the tank (so long as it is above the bottom of the vent). The discharge of the stream is found by going a sufficient distance downstream to allow adequate cross-sectional mixing (essentially the same distance as is required for the slug release method), allowing sufficient time for the concentration in the stream to

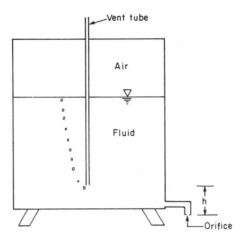

Figure 5.19 Schematic drawing of a mariotte bottle. The tank is airtight except for the vent tube. The fluid exits through a sharp-edged orifice at a rate controlled by orifice size and the constant head h. h is the vertical distance from the bottom of the vent tube to the orifice.

reach equilibrium, and taking a few samples (in theory one is sufficient). The stream discharge is given by

$$Q = Q_i C_i / C_s, \qquad (5.30)$$

where C_s is the concentration of the sample.

If the release is made at the side of the stream the distance required for adequate mixing is approximately $0.4\bar{u}W^2/\varepsilon_t$. If the release is made at the center of discharge (i.e., roughly where the discharge is equal on both sides of the release) the required distance is $0.1\bar{u}W^2/\varepsilon_t$, and if the release is divided into equal parts and made at the two quarter points the required distance is only $0.025\bar{u}W^2/\varepsilon_t$. Moreover, if the slug release method is used division of the slug in half and injection at the two quarter points reduces the amount of tracer required and the sampling time by factors of four as compared to a side release. Clearly it is worth the trouble to divide the release over the cross section as much as possible. In the absence of any access to or over the water even a simple slingshot might be used to inject a distributed slug. It is also worth noting that Florkowski et al. (1969), who tested the slug release method using tritium in several streams in Kenya, obtained reasonable accuracy using substantially smaller mixing lengths than those given here.

Example 5.6: Stream discharge measurement by the slug release method. It is desired to use the method to measure the discharge of the Missouri River near Omaha in a reach which is 600 ft wide and averages 10 ft deep. Approximate estimates are that the mean velocity is 6.0 ft/sec, the transverse mixing coefficient is 1.5 ft/sec, and the longitudinal dispersion coefficient is 16,000 ft^2/sec.

(a)

(b)

Figure 5.20 A mariotte bottle in use in a mixing study. The photographs were taken during the transverse mixing study described by Fischer (1967b). (a) Close-up of the vessel. (b) Dye being injected at the channel centerline.

(a)   How far downstream of the injection should the measurements be taken?

(b)   How long should sampling be continued?

(c)   If the tracer being used is rhodamine WT 20% dye solution and the desired concentration of the average sample is approximately 1.0 ppb, how much tracer should be injected?

Solution.   (a) If the tracer is inserted as a single slug on the stream centerline Eq. (5.18) gives the required distance for complete cross-sectional mixing as

$$L = 0.1\bar{u}W^2/\varepsilon_t = \frac{0.1 \times 6 \times (600)^2}{1.5} = 145,000 \text{ ft.}$$

(b)   The length computed in (a) corresponds to a travel time of 24,000 sec during which the variance of the cloud, assuming Fickian dispersion throughout the period, grows to a magnitude,

$$\sigma^2 = 2Dt = 2 \times 16,000 \times 24,000 = 7.7 \times 10^8 \text{ ft}$$

giving for the standard deviation

$$\sigma = 28,000 \text{ ft.}$$

Sampling a length of 6σ will encompass more than 99 % of the cloud, with some allowance for error. Hence, a reasonable sampling duration would be

$$t_s = \frac{6\sigma}{\bar{u}} = \frac{6 \times 28,000}{6} = 28,000 \text{ sec.}$$

(c)   The amount of tracer required is based on the estimate of the discharge, and is computed by

$$M = QC_{av}T_s = 36,000 \text{ ft}^3/\text{sec} \times 62.4 \text{ lb/ft}^3 \times 10^{-9} \times 28,000 \text{ sec}$$
$$= 63 \text{ lb of dye.} \quad \blacksquare$$

5.5   DISPERSION OF DECAYING SUBSTANCES

Many applications of mixing theory are to nonconservative substances, such as biochemical oxygen demand in a sewage effluent or heat in a power station discharge. Since many of these substances can be considered to be undergoing a first-order decay it is useful to see the effect of the decay term on the analysis. Consider a substance whose rate of decay in a stagnant water body is given by

$$dC/dt = -kC \tag{5.31}$$

and suppose that $\dot{M}$ units of mass per unit time of this substance are discharged into a river whose discharge is Q. Downstream of the initial mixing zone the diffusion equation becomes

$$\bar{u}\frac{\partial C}{\partial x} = K\frac{\partial^2 C}{\partial x^2} - kC \tag{5.32}$$

with a boundary condition that $C \to 0$ as $x \to \infty$. A solution is

$$C = C_0 \exp\{-(kx/\bar{u})[(2/\alpha)(\sqrt{\alpha + 1} - 1)]\}, \tag{5.33}$$

$$= C_0 \exp[-(\bar{u}x/2K)(\sqrt{\alpha + 1} - 1)], \tag{5.33a}$$

where

$$\alpha = \frac{4Kk}{\bar{u}^2} \tag{5.34}$$

and C_0 is a constant of integration. $\alpha = 0$ corresponds to neglecting the dispersion term in the equation, because for α tending to zero the quantity in the square brackets in Eq. (5.33) tends to one and the solution is that for simple first-order decay.

The constant of integration must be determined from the condition that the rate at which mass in entering the stream must equal the rate at which it is being removed by decay. If the source is located at $x = 0$ and the initial mixing distance has length X_m, we can write for the reach in which the one-dimensional equation applies that

$$\int_{X_m}^{\infty} kCA\, dx = \dot{M}(X_m), \tag{5.35}$$

where C is given by Eq. (5.33) and $M(X_m)$ is the mass transported through a cross section at $x = X_m$. Substitution of Eq. (5.33) into (5.35) permits computation of C_0 given $\dot{M}(X_m)$; however $\dot{M}(X_m)$ is not the same as the rate of release of mass at the source because some decay occurs during the initial mixing process. How much that is can only be computed from a detailed investigation of the concentration distribution in the initial zone, so an analytical expression for C_0 based on the one-dimensional theory is not possible. Most practical studies have avoided the problem by neglecting the existence of the initial zone, in effect setting $X_m = 0$ in Eq. (5.35). The integration then gives

$$C_0 = \frac{\dot{M}}{Q}\left[\frac{2}{\alpha}(\sqrt{\alpha + 1} - 1)\right], \tag{5.36}$$

in which $\dot{M}$ is the rate of addition of mass at the source.

Although the reader may find uses of the combination of Eqs. (5.33) and (5.36) in the engineering literature, an interesting contradiction should be noted.

For steady flow in a river we have previously noted that the distance required for cross-sectional mixing is approximately

$$X_m = 0.4\bar{u}W^2/\varepsilon_t. \tag{5.37}$$

The distance required for decay of the substance to a factor of e^{-1} is

$$X_d = \bar{u}/k \tag{5.38}$$

and the ratio is $X_d/X_m = 2.5\varepsilon_t/kW^2$. The quantity α may be written approximately, using Eqs. (5.19) and (5.6), as

$$\alpha = 4Kk/\bar{u}^2 = 0.024W^2k/\varepsilon_t, \tag{5.39}$$

which means that in order for the decay distance to exceed the cross-sectional mixing distance $(X_d > X_m)$ α must be less than approximately 0.06. The corresponding value of the bracketed term in Eq. (5.33) is 0.985, which means that the solution is almost exactly that for first order decay neglecting longitudinal dispersion. Thus there are two possibilities: either the material decays before it mixes across the cross section, in which case Eq. (5.32) is not a suitable model, or else the longitudinal dispersion term in Eq. (5.32) has a negligible effect and can be dropped. If the former, $X_d < X_m$, the concentration distribution must be computed numerically. The scheme described in Section 5.3, with the modification that a decay step is added after each diffusive step, will often be satisfactory. If $X_m < X_d$, and if the river and effluent discharges are steady as assumed in dropping the time derivative from Eq. (5.32), the downstream concentration is reasonably well given by the first-order decay solution $C = (\dot{M}/Q)\exp(-kx/\bar{u})$. It should be noted, however, that effluent discharges are hardly ever steady. The typical daily fluctuation in output from a sewage treatment plant, for example, leads to gradients of concentration of the discharged material along the river, and these gradients are subsequently leveled out by the process of longitudinal dispersion. Although unsteady solutions of the one-dimensional diffusion equation are available (see Bennett, 1971, for example), practical problems involving daily variations are usually handled most easily by numerical models of the sort described in Chapter 8.

Chapter 6

Mixing in Reservoirs

Reservoirs are man-made lakes usually constructed to store water for later release, flood control, or power generation. Reservoirs can also be used to control the temperature of the outflowing water and sometimes also the salinity, turbidity, and other water quality parameters. It is now common to provide outlets at different levels so that the reservoir operator has some control over the quality of the water released; for example a variable depth outlet is provided in the dam for the Oroville Reservoir, California, so that the operator can select colder bottom water when needed for fish or warmer surface water when needed for irrigation. It has also become apparent, however, that while controlling the quality of the outflow the operator must be concerned about the quality of the water that remains, and this requires an understanding of the dynamics of mixing and the internal flows within the reservoir.

There is no generally accepted method of classifying reservoirs, but in dealing with any particular reservoir it is helpful to have a general idea of some gross parameters. Some indication of the effect on water quality is given by the mean residence time. In a large, deep reservoir like Lake Mead the residence time, defined as the volume of the reservoir divided by the mean inflow rate, is often several years, whereas in some run-of-the-river reservoirs, formed by small dams mostly for power generation, the residence time may be only a week. If the residence time is short, water quality in the reservoir may be determined primarily by that of the inflow, but if the residence time is long we may expect significant effects from surface or bottom inputs or from biological activity.

148

Another important parameter is the climate. Reservoirs in cold climates usually have an annual cycle of stratification. During the summer a warm surface layer (the epilimnion) floats on the colder main body (the hypolimnion). This heating phase is followed by cooling of the surface during the fall and a general overturning before the surface once again freezes. In warm climates the complete overturning may not occur. The strength of local winds is also important, especially in wider reservoirs in which the entire surface layer is sometimes blown to one side.

The first attempts to model the internal structure of reservoirs were generally concerned with predicting the temperature of the outflow. Raphael (1962) described a method of dividing a reservoir into horizontal slabs; his approach formed the basis for numerical programs developed by Water Resources Engineers, Inc., in the 1960s and the numerical method described by Markovsky and Harleman (1971). Numerical models of this type have been used extensively and have often given adequate predictions of outflow temperatures. Many of the temperature models are limited, however, in that they use empirical eddy diffusivities. Since the major input of heat is at the water surface, an adequate verification for temperature is possible without necessarily modeling the internal dynamics correctly. Recent modeling studies have concentrated on improving the modeling of the full range of mixing processes, so that the model can better describe a wider range of water quality problems. Turbidity is one example. Turbidity in the water increases the extinction of light below the water surface and concentrates the heat input distribution at the surface, thereby affecting the stratification and creating a feedback loop in which water quality parameter has an effect on the temperature distribution which, in turn, affects the parameter. Another important example is the distribution of nutrients and chemicals. Thornton and Lessen (1976) found in a study using a model that while the value of the mixing coefficient used within the reservoir had only a small effect on the predicted outflow temperature, the availability of nutrients was often controlled almost entirely by mixing. A small change in the mixing coefficient led to a large change in the population of algae. The distribution and control of salinity is also an important problem in some reservoirs. In all these examples an adequate model requires an understanding of a wide range of processes occurring throughout the reservoir.

The plan of this chapter is to present a detailed, separate discussion of each of the important mixing processes, and then in the final section to show how all of them are represented in what we believe to be the most up-to-date (at this writing) model for a particular class of reservoir. We begin with a brief exposition of the annual cycle of the Wellington Reservoir in Western Australia as a way of introducing the processes to be discussed in later sections, and also because the Wellington is the reservoir we will use for our final example of how a model can be applied. Then we discuss the specific processes of mixing and deepening on the epilimnion, vertical and horizontal mixing in the hypolimnion, mixing driven by the inflow, and mixing driven by the outflow. Our final example shows how all these processes must be accounted for in a strategy to control salinity in the Wellington. It should be understood that the focus of this chapter is on studies of the type of reservoir exemplified by the Wellington, which has an annual cycle of stratification and is large enough to have more than a one-year retention time, but small enough to allow the use of a one-dimensional analysis, wherein all properties are assumed to be a function only of depth. Our focus is broad enough to include many of the world's important reservoirs, and much of what we say is relevant to mixing processes in natural lakes as well. It should be understood, however, that many reservoir problems may involve special considerations not completely treated herein, and that in particular we exclude very large lakes in which the currents are strongly influenced by the earth's rotation.

6.1   RESERVOIR BEHAVIOR

A reservoir is exposed to three disturbing influences. First, the meteorological conditions in the area will determine the strengths of any energy transfers across the air–water interface. These may consist of thermal exchanges due to radiation, sensible heat transfer and evaporation, or mechanical energy transfers due to wind stirring. Second, the water from the inflowing streams may impart kinetic and potential energy, and third, some of the energy of the outflowing water may be transformed to kinetic energy of the reservoir water.

6.1.1   The Annual Cycle

In order to fix our ideas a little more firmly let us concentrate on the behavior of a particular lake. We have chosen the Wellington Reservoir in Western Australia (Fig. 6.1) because of our intimate knowledge of its behavior and also because it displays a particularly simple annual cycle; each major mixing mechanism is clearly distinguishable.

In Fig. 6.2 we have summarized the relevant data for the water year 1975. Throughout this chapter we will use a five digit number rather than the date. The first two digits refer to the calendar year and the last three give the day within that year.

We start the year in winter (June 1975) when the lake is homogeneous. The surface water continually cools, becomes heavier and sinks. The mixing is increased by the winds of this season keeping the reservoir in a homogeneous state throughout the month of June. In such a state there is little resistance to the stirring action of the wind and the dissolved oxygen data suggest active turbulent mixing throughout the depth of the reservoir. At the end of June the first inflows bring down the salt accumulated in the catchment during the

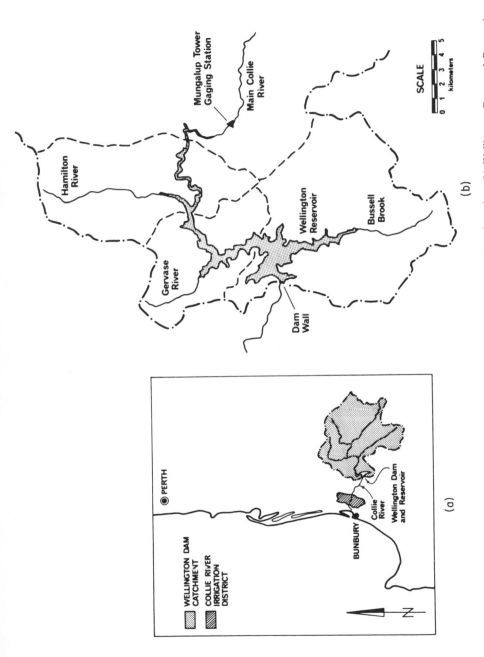

Figure 6.1 Geographic location of the Wellington Reservoir, Western Australia. (a) Location plan. (b) Wellington Dam and Reservoir area enlarged. —·— Wellington Dam catchment boundary. —·— Subcatchment boundaries of the four inflowing streams.

151

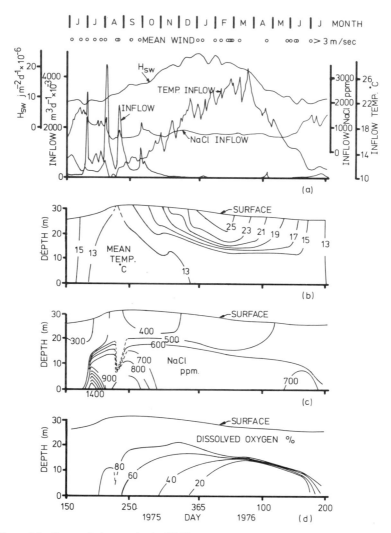

Figure 6.2 Seasonal changes in the Wellington Resevoir. (a) Inflow characteristics and solar radiation. (b) Mean temperature field. (c) Mean salinity field, expressed as NaCL parts per million. (d) Mean percentage saturation of dissolved oxygen.

summer months. The cold, salty water flowing into the reservoir is much heavier than that resident in the reservoir and so it flows to the bottom. During its downward path it is, however, considerably diluted.

The activity of the wind and the inflows combat the stability introduced by this density layering and we see appreciable mixing over the whole depth, leading to a rapid decrease in the peak salinities of the water lodged in the bottom of the reservoir.

By September the inflows have dropped and the windy season is over, allowing the solar heating to take effect and build an increasing degree of stratification over the months of October, November, and December. During the months of January and February there is little change in any of the variables, suggesting that the reservoir has reached equilibrium with its surroundings.

Strong, persistent winds are a feature of late summer and fall. These stir the surface layers and a well defined three tier structure results. A surface mixed layer called the epilimnion retains the active mixing nature of the homogeneous reservoir. Underneath this surface layer there usually is a sharp temperature transition separating the deeper waters from this active surface layer. This transition is termed the thermocline. The deep waters, protected from the wind by the stable nature of the thermocline, form the hypolimnion. Here, the water motion is generally sluggish and disturbed only by sporadic mixing events. The cumulative effect of the winds in January and February is to deepen the thermocline by about 6 m.

In March and April the solar radiation begins to diminish sharply, leading to strong cooling of the upper layers. From the oxygen data the very sharp nature of the transition between the mixed upper layers and the anaerobic hypolimnetic lower water is particularly noticeable. There is essentially no transfer of oxygen across this transition zone. The cycle is completed in July when the lake mixes to the bottom.

Lakes are found in a great variety of geographical locations and meteorological climates. It is not surprising, therefore, that large variations are observed in their behavior. However, as we shall show, the response to outside forces will always be similar and the differences in lake behavior are really only in the frequency with which the above cycle is repeated. This may range from very many times per year to only partial deepening for a particular year.

6.1.2   The Water Density Structure and Its Effects on the Motion within a Reservoir

The isotherms shown in Fig. 6.2b demonstrate the strong degree of temperature stratification induced by the solar radiation. Indeed, the reservoir was at least partially stratified for nearly ten months of the year. Quite apart from the stratification due to salinity, in these months the water at the bottom was always considerably colder than that at the surface and one would expect horizontal planes of constant density in periods of weak winds, inflows or outflows. Figure 6.3 shows a longitudinal cross section of the Wellington Reservoir with mean isotherms taken at a time of stable stratification. The horizontality of the isotherms is remarkable! Individual profiles, however, often show well defined periodic oscillations or seiches and time averages must always be taken if the horizontal structure is to be revealed. In periods of high winds or inflow, the constant density lines may oscillate strongly about their mean level. If the

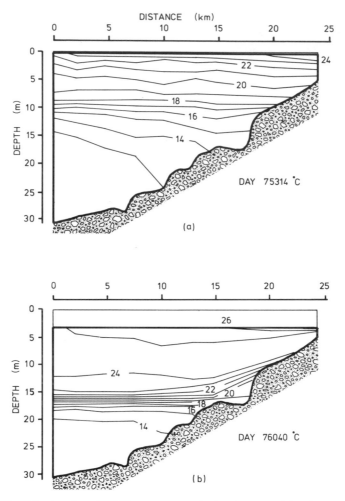

Figure 6.3   Side elevation of Wellington Reservoir showing typical longitudinal mean temperature structures. (a) November 10, 1975: Thermocline structure extends all the way to the surface during the beginning of summer. (b) February 9, 1976: A well defined mixed layer had formed to a depth of 15 m. It was not until June of that year that the mixed layer deepened to the bottom of the reservoir.

disturbances persist the departure from horizontality may also remain. An example of such longer departures occurs when the reservoir experiences long periods of high inflows. In this chapter we take the view that although there may be periodic departures from the vertical density stratification, nearly all reservoirs return to a horizontal state during calm conditions and it is useful to think of the processes causing these deviations as inducing perturbations to this horizontal structure.

As a preliminary to the study of motions which cause mixing, let us first examine the gentle seiching initiated by a weak external disturbance. In order to do this let us partition the ambient water density ρ_a such that:

$$\rho_a = \rho_0 + \rho_e(z) + \rho(x, z, t), \tag{6.1}$$

where ρ_0 is the density of water at the mean temperature, $\rho_e(z)$ the density above ρ_0 when motion is absent and the structure is horizontal, $\rho(x, z, t)$ the density changes induced by any motion. The horizontal and vertical coordinates (x, y) have their origin at the bottom of the dam wall, and the horizontal distance is measured upstream and z vertically up is taken as positive, and t is the time. A measure of the scale of the density gradients is given by the parameter

$$\varepsilon' = \frac{-1}{\rho_0} \frac{d\rho_e}{dz}. \tag{6.2}$$

In a homogeneous fluid, $\rho_a = \rho_0$ everywhere and gravity merely induces a passive hydrostatic pressure. The gravitational body force acting on a small spherical particle of fluid is directed through the center of mass which coincides

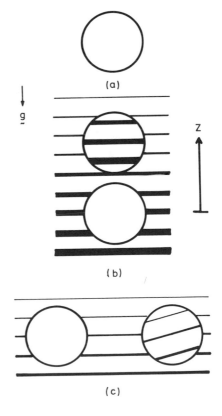

Figure 6.4   Generation of buoyancy forces on a sphere of fluid. (a) Homogeneous fluid in which buoyancy and pressure forces both act through the center of volume. (b) A vertical displacement in a stratified fluid leads to a net vertical buoyancy force. (c) A rotational displacement in a stratified fluid leads to a body-force moment on the spherical particle tending to return it to its original position.

with the center of volume through which the pressure forces act (see Fig. 6.4a). Thus there is no torque applied to the particle and far from boundaries the motion can be considered to be irrotational. Indeed for motions which do not involve surface waves the only effect of the body force is to introduce a hydrostatic pressure equal to $\rho_0 g(H - z)$, where H is the depth of the lake and g the acceleration due to gravity. Thus by defining an active pressure $p^* = p - \rho_0 g(H - z)$ the effect of the body force may be completely removed.

On the other hand when a free surface is present as in lakes, then we must account for motions at the free surface and retain the pressure as an unaltered variable since the boundary condition at the free surface contains the absolute atmospheric pressure.

The influence of the free surface may be gauged by examining the pressure balance that must exist if the free surface is to remain at a constant atmospheric pressure. Suppose fluid traveling with a velocity u impinges on the sidewalls of the reservoir and is brought to rest. The stagnation pressure will be of order $\rho_0 u^2$. This pressure increase must be offset by a rise in the free surface. If H is the depth of the reservoir and g the acceleration due to gravity, then the maximum hydrostatic pressure increase possible will be of the order $\rho_0 gH$. The square root of the ratio of these pressures, called the Froude number,

$$F_r = u/\sqrt{gH} \tag{6.3}$$

will yield a measure of the relative influence of the free surface. If E_r is large then the surface displacements may be expected to be large.

We can also attach a rather simple kinematic interpretation to this statement. In hydraulics texts it is shown that long surface waves travel with a speed proportional to $\sqrt{gH}$, hence the ratio given by Eq. (6.3) compares the fluid velocity with the wave speed. If the fluid moves slower than the wave speed ($F_r < 1$) then the free surface is able to adjust by wave action to a displacement and remain nearly horizontal. Conversely, if $F_r > 1$ then the advective changes occur faster than the speed with which waves can adjust the free surface and large surface displacement will be observed. Such cases may occur near glory hole spillways.

In a stratified reservoir (as shown in Fig. 6.3) where the density varies, two major differences arise. Both may be usefully described by considering small spherical fluid particles undergoing motions similar to that encountered at the free surface described above.

First consider two such spheres of volume δV separated vertically by a distance Z. The gradient in density means that these two spheres will have masses different by order $\varepsilon \rho_0 Z \delta V$. If a certain pressure field begins to accelerate these two particles, the upper, lighter particle will accelerate faster for the same applied pressure gradient. Indeed, if the applied pressure gradient is β, then the acceleration of the upper particle is of order $\beta/(\rho_0 - \varepsilon \rho_0 Z)$ and that of the lower will be order β/ρ_0. Thus the ratio of lower to upper accelerations is $1 - \varepsilon Z$.

In a reservoir the maximum vertical separation is the depth H and the parameter

$$B = \varepsilon H, \qquad (6.4)$$

usually attributed to Boussinesq, is a measure of the influence of density variations on the inertia of the fluid. In a 200 m deep lake with a stratification gradient ε' of 10^{-4}/m, the value of B is 2×10^{-2}, equivalent to a 2% correction. It is therefore reasonable to neglect the variation of ρ_a in calculating the inertial response of the water.

The second and more important difference a variable density introduces is a nontrivial gravitational body force. Once again this is most easily seen by considering the motion of a small spherical particle of fluid. Suppose the particle has volume δV and is displaced, as shown in Fig. 6.4b, a vertical distance Z, then its density anomaly with the surroundings will again be of the order $\varepsilon \rho_0 Z$. This induces a body force, in the direction of the particle's original position, of order $\varepsilon \rho_0 Z g \delta V$. This force, which now actively enters any vertical momentum balance, tends to inhibit vertical motion. If the stratification is strong, this force will dominate the other prevailing forces, and the fluid will avoid vertical motions and flow mostly horizontally.

Disregarding other forces, the equation of motion of the spherical particle displaced vertically becomes

$$(d^2Z/dt^2) + N^2Z = 0, \qquad (6.5)$$

where the buoyancy frequency

$$N = \sqrt{\varepsilon'g} \qquad (6.6)$$

determines the frequency of oscillation.

Furthermore, as the sphere has more of its mass concentrated in the lower half, the center of mass will not coincide with the center of volume and the sphere will tend to rotate if given an angular displacement (Fig. 6.4c). Motions with such an associated rotation are called baroclinic and the usual notions about inviscid fluids executing irrotational motion must now be discarded. The frequency of rotation of the sphere shown in Fig. 6.4c may be computed by considering angular momentum. If θ is the angular displacement then angular momentum leads to the same simple harmonic equation.

$$(d^2\theta/dt^2) + N^2\theta = 0. \qquad (6.7)$$

Hence it is possible for a particle in a stratified fluid to execute translational and rotational simple harmonic motion or a simple linear combination thereof. Figure 6.5a shows how such motions may be combined to yield what are known as standing internal waves. At the crest of these waves the motion is one of pure vertical translation and at the node the motion is oscillatory shear.

If the effects of viscosity of the water are neglected the water to the left of the

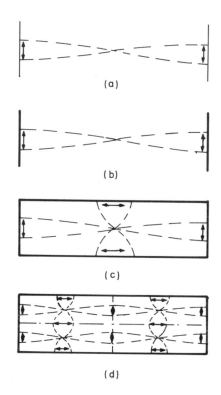

(a)

(b)

(c)

(d)

Figure 6.5 Standing wave patterns in a stratified fluid. (a) Simple vertical oscillations in a stratified fluid of infinite extent. (b) Vertical boundary may be introduced at the end points without changing the motion. (c) Containment of the fluid in a cavity requires horizontal motions at the top and bottom of the cavity. A horizontal pressure field must be set up to accelerate the fluid across the cavity. (d) Second harmonic oscillations in a cavity may be visualized as first harmonic oscillations in four subcavities.

point A and to the right of B may be replaced by a solid wall as shown in Fig. 6.5b. The motion represents vertical sloshing with the isotherms moving vertically up and down. If, as depicted in Fig. 6.5c, the tank has a top and bottom boundary, then the vertical sloshing leads to oscillatory horizontal motions in the central part of the tank. For a tank of length L conservation of mass requires the magnitude of this velocity to be $(L/H)\,dz/dt$. The horizontal acceleration of the fluid will thus be $(L/H)(d^2Z/dt^2)$. As shown in Fig. 6.5c when the fluid moves to the right at the top of the box it must return to the left in the lower half and vice versa. Hence, the pressure field will cause a vertical pressure force on our oscillating water particle equal to $\rho_0(L^2/H^2)(d^2Z/dt^2)$ and this must now be included in the force balance so that Eq. (6.5) becomes

$$\left(1 + \frac{L^2}{H^2}\right)\frac{d^2Z}{dt^2} + N^2Z = 0. \tag{6.8}$$

This leads to a simple harmonic motion with a frequency

$$\omega = N\Big/\left(1 + \frac{L^2}{H^2}\right)^{1/2}, \tag{6.9a}$$

indicating that the standing wave frequency now depends on the geometry of the container. In the limit $H \rightarrow \infty$ we recover our previous result.

Since this rectangular tank may itself be divided into imaginary subrectangular compartments as shown in Fig. 6.5d, we see that harmonic waves can exist with a frequency

$$\omega = N \Big/ \left(1 + \frac{m^2}{k^2} \frac{L^2}{H^2}\right)^{1/2}, \tag{6.9b}$$

for the case of $k \times m$ compartments.

Summarizing, in a stratified fluid a disturbance will cause a pressure wave to set up an initial response similar to that in a homogeneous reservoir. The pressure field will induce a surface wave which sets up the free surface in the time it takes a surface wave to travel the length L of the reservoir, i.e., order $L/\sqrt{gH}$. This flow displaces the fluid and brings buoyancy forces into play. The transition to steady state is realized by internal waves moving through the water continually adjusting the constant density lines so that they remain horizontal on the average.

If a typical velocity in the water is u then a particle will travel a distance H in a time of order H/u. On the other hand, we see from Eq. (6.9a) that constant density lines adjust to a vertical displacement in a characteristic time N^{-1}. The ratio of the adjustment time to the displacement time

$$F = u/NH, \tag{6.10}$$

called the internal Froude number, is thus a measure of whether internal waves can adjust the constant density lines as quickly as they are distorted by the advection. If $F > 1$ then the fluid will have sufficient momentum to overcome the buoyancy force and the effect of any stratification will be negligible; the fluid will behave as if it were homogeneous. On the other hand, if $F < 1$ then internal waves can continually adjust the density structure to keep it oscillating about a horizontal plane.

Often the stratification is more like a two-layer system, for example, as shown in Fig. 6.3b, where a sharp thermocline separates the surface epilimnion from the hypolimnion. In such cases the effective frequency of the internal seiches is

$$Ne = \frac{m}{2L} \left(g \frac{\Delta\rho}{\rho_0} \frac{h(H-h)}{H}\right)^{1/2}, \tag{6.11}$$

where the top layer has a density ρ_0, the bottom has a density $\rho_0 + \Delta\rho$, h is the depth of the epilimnion, and m the number of nodal lines (Turner, 1973, p. 20). The corresponding internal Froude number is

$$F_i = u \Big/ \left(\frac{\Delta\rho g H}{\rho_0}\right)^{1/2}. \tag{6.12}$$

Once waves are set up they will persist, even if the disturbances have stopped, until dissipation has robbed the waves of their energy or equivalently until the momentum variations have been diffused. It was seen in Chapter 2 that the time a substance takes to diffuse a distance H is equal to H^2/κ. Here we are dealing with momentum so that the diffusivity should be replaced by the kinematic viscosity v. Since in the vertical, the largest wavelength possible is $2H$ and the fastest period is N^{-1}, the ratio H^2/vN^{-1} is the ratio of the dissipation or diffusion time to the internal wave period. The square of this ratio

$$Gr = N^2 H^4/v^2 \qquad (6.13)$$

is, by convention, called the Grashof number. If the Grashof number is greater than one, the wave field will decay only very slowly. If Gr is less than one, viscous dissipation dampens the waves as fast as they are formed.

There is a second more subtle damping mechanism. Waves can only exist if there is a density gradient in the fluid. Suppose temperature is responsible for the stratification and the thermal diffusivity κ is large, then the density anomaly will diffuse away quickly. The ratio of momentum diffusion to thermal diffusion is called the Prandtl number

$$Pr = v/\kappa. \qquad (6.14)$$

If Pr is greater than one then momentum diffuses faster than heat and if Pr is less than one then momentum diffuses slower than heat. The two most common stratifying agents in a reservoir are heat and salt with respective Prandtl numbers 7 and 900.

All reservoirs are located on a rotating earth so that as the earth rotates about its axis with a frequency $\Omega_0 \approx 1$ revolution per day, the lake rotates as on a turntable with an angular velocity $\Omega = \Omega_0 \sin \theta$, where θ is the latitude of the lake. At the equator $\Omega = 0$ and at the poles $\Omega = 7.3 \times 10^{-5}$ rad/sec. Thus there is a further time scale to consider in our problem. This naturally leads to further dimensionless numbers. The most common is the Rossby Number

$$R_0 = u/L\Omega, \qquad (6.15)$$

being the ratio of the period of rotation to the time of advection. If R_0 is large then rotational effects are overcome by the fluid's momentum; if $R_0 < 1$ then we may expect strong influences of rotation. The magnitude of u will depend on whether the current is driven by wind, outflow, inflow, or other disturbing forces.

For periodic motions the time scales for comparison are the period of the motion and the period of rotation. Thus a long surface wave will be influenced by the earth's rotation if

$$\sqrt{gH}/L\Omega = R_0/F_r < 1. \qquad (6.16a)$$

On the other hand, internal wave motions are affected if the ratio

$$\omega/\Omega < 1, \tag{6.16b}$$

where ω is given by Eq. (6.9b). If this ratio is larger than one then the water motion can adjust by internal wave motion faster than it takes the lake to rotate appreciably and the motion will be independent of the rotation. Waves where the two time scales are nearly equal are important in large lakes (see Mortimer, 1974). Some care must, however, be taken that the motion is not comprised of waves with a very small vertical and long horizontal wavelength, since as seen from Eq. 6.9b such motions possess very small frequencies ω.

6.2 EXTERNAL ENERGY SOURCES FOR MIXING

Before proceeding with a discussion of mixing models we must first determine the magnitudes of the disturbances caused by the external energy inputs. These may conveniently be partitioned into transfers across the lake surface, those which are advected in by the streams and that energy which is introduced by the outflows.

6.2.1 Surface Momentum and Mechanical Energy Transfer

Wind exerts a drag force on the surface of a lake. The actual stress felt by the water surface is influenced by the wind strength, the stability of the meteorological boundary layer over the water surface, the variability of the wind speed over the lake, the length of fetch, the degree of wave development, and the amount of wave energy dissipation at the shores of the lake.

Of these influences the wind speed is the dominant factor determining the stress τ and it is customary to write

$$\tau = C_D \rho_A U^2 \tag{6.17}$$

where U is the wind speed at 10 m above the water surface, ρ_A the density of the air, and C_D the drag coefficient incorporating all the variability induced by the other influences.

A great deal of research effort has been expended in determining the value of C_D for different atmospheric and lake conditions. For a neutral atmospheric boundary layer, Hicks (1972) found that lake size was an insignificant factor and small lakes had similar coefficient to those found in ocean experiments. On the other hand, C_D is a weak function of the wind speed. Hicks (1972) found that $C_D = 1.0 \times 10^{-3}$ for wind speeds up to 5 m/sec and then it rose linearly to a value of 1.5×10^{-3} for winds speeds of 15 m/sec. These values have been confirmed by a great many researchers. However, it must be remembered that in writing Eq. (6.17) we are assuming that the wind stress is transmitted directly to the upper layers of the water and that none of the stress is lost by surface wave radiation and subsequent dissipation at the lake boundaries.

As the water becomes very shallow, less than 2.5 m, the longer waves will not be able to develop fully and the water surface will remain smoother. Hicks *et al.* (1974) showed that for such conditions C_D remains close to 1.0×10^{-3} for all wind speeds.

The stability of the air column also has a strong influence on the value of C_D. Warm winds blowing over a cold water body are stabilized by the temperature difference which in turn results in less friction. The value of C_D can be reduced by as much as 40% for stable conditions and increased equally by up to 40% for very unstable air flows. Hicks (1975) suggested an iterative procedure designed to correct for the effects of stability, without requiring sensors at more than one elevation.

McBean and Paterson (1975) and Bean *et al.* (1975) give data for the variation of surface stresses over Lake Ontario. Great variability was observed both spatially and temporally, many times greater than the variation introduced by the changes in C_D due to stability, wind speed, or lake depth.

Thus from a practical point of view, since the normal lake site usually has only a remotely located land based meteorological station, the only consistent action is to assume C_D equal to a constant. For a station with a wind sensor located 10 m above the ground or above the lake's surface, an average value of 1.3×10^{-3} is appropriate for most engineering calculations.

The rate of working W by the wind on the water surface is given by

$$W = \tau u_w, \qquad (6.18)$$

where u_w is the drift velocity of the water near the surface and the stress τ is given by Eq. (6.17). The drift velocity has rarely been measured directly, but as in boundary layer flow u_w may be expected to scale with shear velocity $u^* = (\tau/\rho_0)^{1/2}$. How much of the work given by Eq. (6.18), appears as kinetic and potential energy of surface waves, as kinetic energy of a mean drift current, or as kinetic energy of eddy motions below the wave zone, is still poorly understood. We shall return briefly to this matter in Section 6.3.2, but the reader may also wish to consult the book by Phillips (1977) for a more penetrating discussion of this subject.

6.2.2   Surface Thermal Energy Transfer

Once again specification of the thermal transfer of heat by evaporation, conduction, and radiation is complicated by the great variability of meterological conditions encountered and by the usually poor availability of reliable meteorological data. An excellent summary of the formulas pertaining to transfers from a lake surface is given in the TVA publication (1972).

In general the heat flux due to conduction, called the sensible heat transfer H_s, and the loss of heat due to evaporation H_L, may be written in a form similar to that for the surface stress.

In particular, the dominant influences are incorporated if we write

$$H_s = C_s \rho_A C_p U(T_0 - T) \tag{6.19}$$

and

$$H_L = C_L \rho_A L_W U(Q_0 - Q), \tag{6.20}$$

where C_s and C_L are coefficients incorporating the same variability influences as described for the wind stress, C_p is the specific heat of air, T_0 the water surface temperature, T the temperature at 10 m, L_W the latent heat of evaporation, Q_0 the saturation specific humidity (kilograms of water moister/kilograms of air–water moisture) at T_0, and Q the specific humidity at 10 m.

For neutrally stable boundary layers Hicks (1972) recommends $C_s \approx C_L = 1.45 \times 10^{-3}$. Friehe and Schmitt (1976) collated all available ocean and lake data and determined an average value of 1.4×10^{-3} for C_s and 1.3×10^{-3} for C_L, but it should be noted that the stability (see Hicks, 1975) has a large effect on these values. Normal meteorological changes over a lake could easily cause variations from 0.8×10^{-3} for very stable conditions to 1.6×10^{-3} at very unstable conditions. The TVA report (1972) recommends values of $C_s \approx C_L = 1.5 \times 10^{-3}$ for estimates from remote land based meteorological stations. As a rule of thumb, it is always better to use a value verified for local meteorological and atmospheric conditions rather than ones which are perhaps more general, but which have not been verified locally.

The long wave radiation from the water vapor in the sky is given by TVA (1972)

$$H_1 = -5.18 \times 10^{-13}(1 + 0.17C^2)(273 + T)^6 \tag{6.21}$$

and the back radiation from the water surface

$$H_2 = 5.23 \times 10^{-8}(273 + T_0)^4, \tag{6.22}$$

where C is the fraction of the sky covered by cloud.

These formulas are based on average values of reflectance and the reader is referred to the TVA report for further details.

The direct short wave radiation H_{sw} incident at the lake's surface may again be estimated from a knowledge of the radiation outside the atmosphere and the conditions of the atmosphere (see TVA, 1972), but since radiometers are relatively cheap and reliable it is better to measure this flux directly.

Example 6.1. Table 6.1 gives values of meteorological variables recorded on 9 February (day 76040) 1976 near the Wellington Reservoir, Western Australia. Use these values to compute the instantaneous heat fluxes at the water surface. Assume that the sky was cloudless for the whole period, the values

TABLE 6.1

Meteorological Variables Recorded on February 9, 1976, near the Wellington Reservoir, Western Australia

	Given Quantities						Derived Quantities		
Time (h)	Short wave, H_{sw} (W/m^2)	Air temp. at 10 m T (°C)	Water temp. T_0 (°C)	Vapor conc. at 10 m, $Q \times 10^2$	Vapor conc. (saturation), $Q_0 \times 10^2$	Wind speed at 10 m, U (m/sec)	H_L (W/m^2)	H_S (W/m^2)	$H_1 + H_2$ (W/m^2)
1.0	0	17.0	24.4	1.0	1.9	0.2	7.6	2.6	101
2.0	0	17.0	24.5	1.0	1.9	0.5	19.1	6.6	101
3.0	0	16.9	24.4	1.1	1.9	0.1	3.4	1.3	101
4.0	0	16.8	24.4	1.0	1.9	0.4	15.3	5.3	101
5.0	0	16.4	24.4	1.0	1.9	0.8	30.5	11.2	105
6.0	0	17.0	24.3	1.1	1.9	0.3	10.2	3.8	101
7.0	−18	18.6	24.3	1.0	1.9	0.4	15.3	4.0	90
8.0	−230	23.0	24.3	0.9	1.9	0.1	4.2	0.3	60
9.0	−444	28.8	24.3	0.7	1.9	0.9	45.8	−7.1	17
10.0	−628	29.3	24.4	0.7	1.9	0.2	10.2	−1.7	14
11.0	−769	29.8	24.4	0.7	1.9	0.1	5.1	−0.9	10
12.0	−859	31.8	24.4	0.7	1.9	0.8	40.7	−10.4	−6
13.0	−889	34.6	24.5	0.7	1.9	1.0	50.8	−17.7	−30
14.0	−863	35.0	24.7	0.8	1.9	0.5	23.3	−9.0	−31
15.0	−720	34.2	24.9	0.9	1.9	3.2	135.6	−52.2	−23
16.0	−650	33.4	25.1	0.9	1.9	2.3	97.5	−33.5	−15
17.0	−403	32.2	25.2	0.8	1.9	4.1	191.2	−50.3	−6
18.0	−180	30.3	25.2	1.0	1.9	3.4	129.2	−30.4	10
19.0	−23	28.6	25.1	0.9	1.9	3.2	135.6	−19.6	23
20.0	0	25.0	24.6	1.0	1.9	2.8	106.8	−2.0	47
21.0	0	23.9	24.5	0.9	1.9	1.4	59.3	1.5	55
22.0	0	20.2	24.5	1.0	1.9	1.8	68.7	13.6	81
23.0	0	18.1	24.5	1.0	1.9	1.1	37.3	12.4	95
24.0	0	18.0	24.5	1.0	1.9	0.8	27.1	9.1	95

of the coefficients $C_s = C_L = 1.45 \times 10^{-3}$, $\rho_A = 1.20$ kg/m^3, $L = 2.4 \times 10^6$ J/kg, and $C_p = 1012$ J/kg °C.

Solution. See Table 6.1 and Fig. 6.6. ■

The heat fluxes calculated in Example 6.1 are all immediately absorbed by the water at the surface, except the short wave radiation which is able to penetrate to appreciable depths. Depending on the color and clarity of the water some of the incident radiation will be felt at depths as far as 10–15 m below the surface. There are again numerous empirical formulas which correlate the light attenuation with the turbidity of the water. Generally a relationship of the form

$$H_z/H_{sw} = e^{-\eta(H-z)} \tag{6.23}$$

is found to fit the data, where H_z is the radiation remaining at a height z and η is a coefficient of extinction. The TVA (1972) report summarizes available data and it is seen that η ranges from 0.2 m^{-1} for very clear lakes to 4.0 m^{-1} for turbid eutrophied lakes. The actual value obviously depends greatly on the biological growth in the lake. The exact extinction coefficient is not important for mixing models in the cooling phase since wind stirring and convective overturn at night usually mix the heat gained from the sun to a depth greater than the penetration depth. However, on the heating part of the cycle the value of η controls the gain in stability of the upper layers and thus strongly influences the depth of the thermocline.

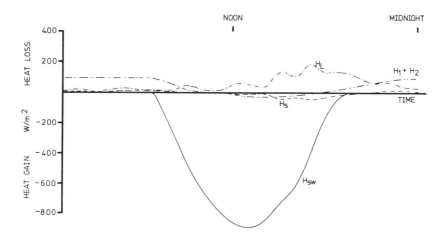

Figure 6.6   Graph showing shortwave radiation H_{sw}; net longwave radiation H_1 and H_2; sensible heat exchange H_s and the latent heat transfer H_L for February 9, 1976, at the Wellington Reservoir.

Example 6.2.   Given the incident energies of the previous examples, calculate the change in temperature in the top 10 m of the water column. Use one meter intervals in your calculation and assume that $\eta = 0.35 \text{ m}^{-1}$.

Solution.   The long wave radiation, sensible heat transfer, and the evaporative heat losses all effect only the upper 1 m. The short wave radiation will be distributed according to Eq. (6.23). Measuring from the surface down the net heat input over the whole day may be tabulated as shown in Table 6.2.

This calculation indicates that cooling in the evening leads to cold surface waters overlying warmer water. Such a situation is unstable and we might expect to see rather vigorous "convective" stirring or mixing as evening approaches. Such mixing occurs very often and is responsible for stirring the heat gained during the day into the epilimnion. ■

TABLE 6.2
Absorption of Thermal Energy with Depth

Depth $(H - z)$ (m)	Short wave energy at z (J/m²) × 10^{-7}	Net short wave energy absorbed in slab $z - 1$ to z (J/m²) × 10^{-7}	Surface transfers (J/m²) × 10^{-7}	Temp. change[a] (°C)
0	−2.4			
1	−1.7	−0.7	0.8	−0.3
2	−1.2	−0.5		+1.2
3	−0.8	−0.4		+0.9
4	−0.6	−0.2		+0.5
5	−0.4	−0.2		+0.5
6	−0.3	−0.1		+0.3
7	−0.2	−0.1		+0.3
8	−0.15	−0.05		+0.1
9	−0.10	−0.05		+0.1
10	−0.007	−0.03		+0.07

[a] This column shows the corresponding temperature changes in the water column induced by the energy absorption.

6.2.3   Inflow Energy Available for Mixing

A stream enters a reservoir with a finite velocity and is often cooler than the resident water in the reservoir which has been warmed by the sun. Thus apart from bringing in a mass of water into the reservoir, the stream also introduces kinetic and potential energy.

The flux of kinetic energy K_1 introduced is given by the integral

$$K_1 = \int_A \frac{1}{2} \rho_1 u^3 \, dA, \qquad (6.24)$$

where A is the cross-sectional area of the entering river at a sufficient distance upstream to be away from the influence of the reservoir itself, ρ_1 the density of the stream water, and u the velocity of the water.

The introduction of potential energy from a stream is most conveniently illustrated by considering a very much simplified model of a reservoir inflow. Figure 6.7 shows a reservoir with a length of river equal to say one day's inflow. Let us assume that the reservoir water has a constant density $\rho_0 < \rho_1$. The energy due to buoyancy differences released and possibly available for mixing may be estimated by calculating the decrease in mean potential energy as the river water flows to the bottom exchanging without mixing its place with reservoir water (Fig. 6.7).

The potential energy P_i of a system before the inflow has taken place is given by

$$P_i = \rho_0 V_0 g \bar{z}_0 + \rho_1 V_1 g \bar{z}_1, \qquad (6.25)$$

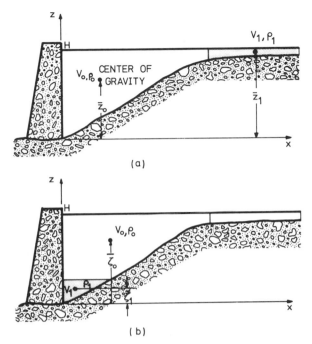

Figure 6.7 Potential energy released as an inflow of heavier water enters a reservoir. (a) An inflow volume V_1 about to enter reservoir. (b) Heavier inflow water has surged to the bottom and exchanged position with reservoir water.

where V_0 and V_1 are the reservoir and river water volumes and $\bar{z}_0$, $\bar{z}_1$ are the heights of the centers of mass of the reservoir and river volumes. The potential energy P_f of the system after the inflow has taken place is given by

$$P_f = \rho_0 V_0 g\bar{z}_0 + \rho_0 V_1 g\bar{z}_1 + \Delta\rho V_1 g\bar{\zeta}_1, \qquad (6.26)$$

where $\bar{\zeta}_1$ is the height of the center of mass of the river water lodged in the bottom of the reservoir and $\Delta\rho = (\rho_1 - \rho_0)$. Hence the energy released by the inflow is

$$P_i - P_f = \Delta\rho V_1 g(\bar{z}_1 - \bar{\zeta}_1). \qquad (6.27)$$

The processes by which this energy is dissipated are discussed in Section 6.4.

6.2.4 Outflow Energy Available for Mixing

One feature that may be conveniently used to differentiate a reservoir from a lake is that reservoirs normally have large through flows and thus retention times usually less than one or two years.

From first courses in hydraulics we learn that if a hole is punched into the bottom of a container and the fluid is allowed to drain out then Torricelli's law gives the velocity at the hole as $\sqrt{2gH}$. This formula is obtained by using Bernoulli's law and presupposes that all of the potential energy lost due to the fluid surface falling is completely converted to kinetic energy of the fluid leaving through the hole.

This same idea may be generalized to a stably stratified reservoir. Suppose the stratification is sufficiently strong so that the vertical velocity remains constant over a horizontal plane as the water drains out the outlet. Under such conditions the ambient density ρ_a is only a function of z and t and the time variation is

$$(\partial \rho_a / \partial t) = -w(z)(\partial \rho_e / \partial z) \tag{6.28}$$

since the water is merely falling vertically with a velocity $-w(z)$.

The potential energy P of the reservoir water is given by

$$P = \int_0^H gz A(z) \rho_a(z, t) \, dz, \tag{6.29}$$

where $A(z)$ is the plan area of the reservoir at the elevation z.

The rate of change of potential energy

$$\frac{dP}{dt} = gH A_s \rho_s \frac{dH}{dt} + \int_0^H gz A(z) \frac{\partial \rho_a}{\partial t} \, dz, \tag{6.30}$$

where A_s is the surface area of the reservoir and ρ_s the density at the surface. By conservation of volume the discharge

$$Q = -A_s(dH/dt) = -A(z)w(z) \tag{6.31}$$

so that Eqs. (6.28) and (6.30) may be combined to yield

$$\frac{dP}{dt} = -gH\rho_s Q + gQ \int_0^H z \frac{\partial \rho_e}{\partial z} \, dz$$

$$= -gHQ(\rho_s - \rho_e(H)) - gQ \int_0^H \rho_e(z) \, dz. \tag{6.32}$$

Under the assumption of the Boussinesq approximation the rate at which kinetic energy K is being lost at the outlets is given by

$$dK/dt = -\tfrac{1}{2}\rho_0 Q u^2, \tag{6.33}$$

where u is the mean velocity at the outlet.

Equating the expressions (6.32) and (6.33) yields the velocity

$$u = [2g(H + He)]^{1/2}, \tag{6.34}$$

where $He = 1/\rho_0 \int_0^H \rho_e(z) \, dz$. We recover the Torricelli result when $\rho_e \equiv 0$.

In Section 6.1.2 we showed that whenever a stratified body of water is disturbed, as would be the case when the outflow is initiated or changed, internal waves are generated. Hence each time the outflow Q is changed a small fraction of the potential energy normally available as exit kinetic energy will be taken up to generate internal wave kinetic and potential energy. This becomes particularly important in pump storage schemes if the dominant frequency of discharge variation corresponds to the natural internal frequency of the thermocline seiche. In such cases resonance could lead to very large heaving motions of the interface with associated mixing (see Section 6.4).

6.3 VERTICAL MIXING IN THE EPILIMNION

Most reservoirs undergo a period of stratification and it is not uncommon for the water of a lake to be temperature stratified for the major part of the year. Under such conditions the mean isotherms are horizontal surfaces, their position being disturbed only by surface winds and stream inflows. Such disturbances will cause internal wave motions which oscillate the structure. However, for mild winds and inflows these oscillations do not induce mixing and we may time average the temperature to obtain a stable vertical temperature structure independent of the oscillations. Under such conditions mixing is confined to surface penetrative convection in the epilimnion. This case is treated first in Section 6.3.1. In Section 6.3.2 we assess the mixing induced by mild winds "stirring" the epilimnion layers through the formation of waves. Storms with severe winds greatly increase mixing in the whole reservoir. The increased surface stress accelerates the epilimnion and forms large shears at the thermocline, leading to active production of turbulent kinetic energy. Strong winds thus directly lead to severe mixing in the epilimnion and the associated internal wave activity also enhances vertical mixing in the hypolimnion. In Section 6.3.3 we define various time scales of motion and advance a mixing model based on a simple energy budget.

6.3.1 Penetrative Convection

A reservoir sheltered from surface winds will normally be characterized by a well defined thermal structure. The lake will possess a stably stratified hypolimnion, a sharp thermocline, and an epilimnion which undergoes diurnal temperature fluctuations due to daytime heating and nighttime cooling. The depth of penetration of the short wave radiation depends on the water clarity but, in the absence of wind, there is always an identifiable temperature rise and stratification in the surface layers (see Example 6.2) during the sunlight hours. As night falls and radiative heat losses begin to dominate the thermal exchange at the surface, the surface layer cools and convective motions mix the upper layers. Often these convective motions proceed until they reach the mature thermocline where they begin to erode the stable structure. This diurnal cycle is illustrated by the data from the Wellington reservoir reproduced in Fig. 6.8.

Let us now examine in some detail this mixing process. For simplicity consider first the agitation induced by surface cooling in a lake which has already overturned and which is homogeneous with a uniform density ρ_a.

Suppose that the surface is cooling at a rate $\tilde{H}$(W/m). A thin cool surface layer will form, become buoyantly unstable and release negatively buoyant thermal plumes which penetrate into the body of the lake. This process is generally called penetrative convection. The surface temperature drop induced by the cooling adjusts so that the sustained thermal currents are vigorous enough to carry away just the right amount of coldness $\tilde{H}$ from the surface to the deeper waters. Usually this temperature difference is less than 0.2°C.

Conservation of heat, applied to the uniform water of the lake as a whole, leads to a rate of mean temperature change,

$$dT/dt = -\tilde{H}A_s/C_p\rho_a V, \qquad (6.35)$$

where C_p is the specific heat of water, V the volume of the lake, A_s the lake surface undergoing cooling, and T the average lake temperature.

The rate of shrinking of the lake volume due to this overall cooling may be calculated from the associated density change,

$$dH/dt = -\alpha\tilde{H}/C_p\rho_a, \qquad (6.36)$$

where it must be remembered that an upward heat flux is taken as positive and α is the thermal coefficient of expansion of the water.

Now the potential energy P of the lake was shown in Section 6.2.4 to be given by Eq. (6.29),

$$P = \int_0^H \rho_a gzA(z)\,dz, \qquad (6.37)$$

so that the rate of mechanical energy released as the cooling proceeds is

$$\frac{dP}{dt} = \rho_a gHA_s\frac{dH}{dt} + \frac{1}{\rho_a}\frac{d\rho_a}{dt}P. \qquad (6.38)$$

Substituting for dH/dt and defining the elevation of the centroid by $\bar{z} = P/V\rho_a g$ yields a rate of change

$$\frac{dP}{dt} = -\frac{\alpha g\tilde{H}}{C_p}A_s(H - \bar{z}). \qquad (6.39)$$

In other words, the potential energy of the lake decreases at a rate directly proportional to the loss of heat at the surface, and when the surface cools there is

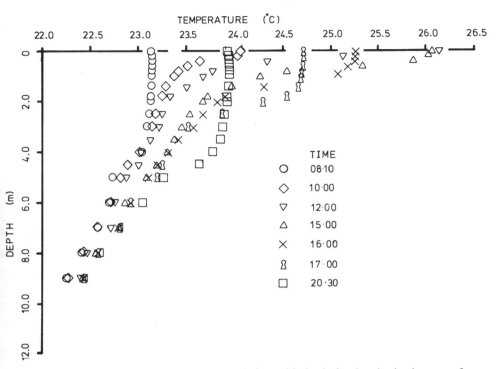

Figure 6.8 Graph of mean temperature variations with depth showing the development of a diurnal mixed layer. The data was taken by Rayner (unpublished manuscript) in the Wellington Reservoir on January 15, 1976. The deepening between 12.10 and 16.00 was strongly controlled by the surface buoyancy flux, but at about 16.30 the wind increased to about 6.00 m/sec deepening the surface mixed layer to 5 m by 20.30.

a transfer of energy at the surface from the mean potential energy to the turbulent kinetic energy budget where it then becomes available for the generation of turbulent penetrative convection.

The diurnal cycle in such a reservoir may thus be conveniently viewed as a heat engine. One might define its "useful work" or "useful effect" as the transport of O_2 or other things which would otherwise languish in the upper layers, down to depths where they are needed by animals and bacteria. The sun puts in heat during the day; the night sky serves as the cooling reservoir. Curiously, the mechanical effect is produced during the cooling, rather than the heating, phase. The useful work is not, to a great degree, exported for consumption outside the engine, but is performed internally. (Greedy little engine!)

The conversion process (thermal expansion or contraction, followed by the mechanical collapse of a gravitationally unstable structure), is an exotic one and not very efficient.

The velocity induced by such cooling may again be estimated from a very simple model. Consider a small parcel of cold surface water falling under the action of gravity. Neglecting viscous and diffusive effects the parcel will have a velocity scale u_f of order $(\Delta\rho/\rho_a)gh$ by the time it reaches a depth h. The density anomaly $\Delta\rho$ is, however, still unknown. It seems plausible that the activity will adjust itself such that the heat being carried away just balances $\tilde{H}$. If u_f is too small then $\Delta\rho$ will increase, increasing u_f which, in turn, increases the heat being carried away. At equilibrium, $u_f\Delta T$ must thus be order $\tilde{H}/C_p\rho_0$. Combining these two estimates yields the order of magnitude of the free fall velocity of the thermals under a cooling water surface

$$u_f = \left(\frac{\alpha g H\tilde{H}}{C_p\rho_a}\right)^{1/3}. \tag{6.40}$$

As the penetrative convection proceeds, a tracer introduced would mix rapidly over the depth. An estimate of the effective diffusion coefficient ε_z may be obtained from Eq. (3.40) by assuming that the velocity scales with the plume velocity u_f and the length with the depth H so that we may write

$$\varepsilon_z = ku_f H. \tag{6.41}$$

Unfortunately, there appears to be little data to verify Eq. (6.41) or to estimate the magnitude of the coefficient k.

So far we have only considered cooling of an overturned lake with a uniform density profile. Now let us apply this generalization to a lake with a characteristic hypolimnetic structure, a sharp thermocline with a temperature differential ΔT, and a well mixed epilimnion of depth h, as shown in Fig. 6.9. As the cooling commences, the cold turbulent plumes form, and begin to descend into the epilimnion.

The front of the penetration may be conveniently defined as the position of the leading billows of the individual thermal plumes originating at the surface. Once the plume fronts reach the thermocline they meet the resistance of the stable stratification. Any further propagation can only proceed by entrainment of the colder water into the turbulent epilimnion. This is equivalent to lifting the entrained fluid through a distance $h/2$. If the entrainment proceeds at a rate of dh/dt, then the rate of work required is equal to $-\alpha g\Delta T\rho_0(h/2)(dh/dt)$. Hence, it is now possible for some of the turbulent kinetic energy locked in the convective motions to be used to deepen the mixed layer. Once the thermocline has deepened to the bottom, the energy is once again completely dissipated internally.

For a uniformly distributed $\tilde{H}$, the deepening process will be one dimensional and it is sufficient to write the turbulent kinetic energy budget for a column of fluid 1 m^2 in area and extending from $z = 0$ to $z = H$. If K is the total turbulent kinetic energy of the motion in the epilimnion then we may write,

$$dK/dt = -(dP/dt) - \rho_0\Phi, \tag{6.42}$$

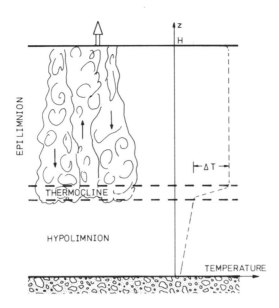

Figure 6.9 Pictorial representation of rising and falling plumes eroding the thermocline during penetrative convection.

where P now represents the total change in potential energy and Φ represents all the internal losses.

Adding the entrainment work contribution described above to Eq. (6.39) leads to

$$dP/dt = -(\alpha g h \tilde{H}/2C_p) + \alpha g \Delta T \rho_0 (h/2)(dh/dt). \tag{6.43}$$

The rate of change of kinetic energy will be proportional to the rate at which quiescent hypolimnion water is being entrained and accelerated to the turbulent state of the epilimnion. Thus we can write

$$dK/dt = \tfrac{1}{2}C_T{}^f \rho_0 u_f{}^2 (dh/dt), \tag{6.44}$$

where $C_T{}^f$ is a coefficient to be determined by calibration and u_f is defined by Eq. (6.40) except that h must be substituted for H since now the thermals descend only a distance h.

Substituting Eqs. (6.43) and (6.44) into Eq. (6.42) leads to the required balance:

$$(C_T{}^f u_f{}^2 + \alpha \Delta T g h)\frac{dh}{dt} = u_f{}^3 \left(1 - \frac{2\Phi}{u_f{}^3}\right). \tag{6.45}$$

A great deal of research has been carried out in an effort to determine the losses Φ. These arise from internal dissipation, leakage of turbulent kinetic energy through the thermocline into the hypolimnion, and the transfer of heat, and thus buoyancy, through the thermocline by mixing and, at low Peclet numbers $u_f h/\kappa$, by molecular diffusion.

Sherman et al. (1978) reviewed numerous convective penetration experiments and found that internal dissipation accounts for about a 60% loss of energy and mechanical energy leakage can consume another 20% if the lower fluid is stratified. Denton (1978) carried out a set of very careful experiments designed to estimate the losses of energy due to thermal (buoyancy) leakage. He found that this transfer was dominated by molecular diffusion at low Peclet numbers and that in general the thermal leakage is a strong function of both the P_e and R_i^f numbers, where $R_i^f = \alpha \Delta T g h / u_f^2$.

However, this leakage is strictly not a loss term since leakage of heat through the base of the epilimnion will cause a temperature rise in the hypolimnion immediately below the thermocline, thus reducing the stability of the thermocline which in turn enhances the penetration speed of the entraining front. Denton (1978) advanced a multiple layer model which is successful at describing convective penetration at all P_e numbers.

For our purposes of computing thermocline behavior in reservoirs under widely varying conditions these specialized models are not warranted. It is best to retain the assumption originally due to Kraus and Turner (1967), that the losses Φ are directly proportional to the buoyancy flux input u_f^3. We shall choose the efficiency $C_\kappa^f = (1 - 2\Phi/u_f^3)$ so that the thermocline behavior is modeled under normal conditions. The justification for this simplified approach lies in the fact that cooling seldom occurs without wind stress and until the detailed coupling of both mechanisms (see Section 6.3.2) is better understood there is little rationale for using anything but the simplest model.

Studies on the Wellington Reservoir have shown that good comparisons of the temperature and salinity structure are achieved with $C_T^f = 0.5$ and $C_\kappa^f = 0.13$. The value for C_κ^f lies well within the range of values observed in laboratory and oceanic experiments, provided it is remembered that here C_κ^f is the net efficiency including all losses. The value of 0.38 given in Sherman et al. (1978) is an efficiency which includes only loss due to internal dissipation; these authors account for mechanical energy losses separately.

The entrainment of cold water will cool the epilimnion at a rate equal to $(\Delta T/h)dh/dt$ so that Eq. (6.35) must be modified to read

$$\frac{dT}{dt} = \frac{-\tilde{H}}{C_p \rho_0 h} - \frac{\Delta T}{h} \frac{dh}{dt}, \tag{6.46}$$

where T is the temperature of the epilimnion water.

Example 6.3.   Consider a reservoir with a well mixed epilimnion at 25°C and with a thickness of 10 m. The temperature drop across the thermocline is 3°C and the temperature variation in the hypolimnion is 0.1°C/m. If the net cooling at the surface over one day is at a rate of 290 W/m², calculate the new depth of the thermocline.

Solution. The solution to this problem is obtained by simultaneously solving Eqs. (6.45) and (6.46). This is most easily done with an explicit scheme with a time step of four hours. The physical properties of water used are assumed constant and equal to the values at 25°C: $\alpha = 2.57 \times 10^{-4}/°C$, $C_p = 4179$ J/kg/°C, $\rho_0 = 997$ kg/m³, $C_\kappa{}^f = 0.13$, and $C_T{}^f = 0.5$.

The solution is set out in Table 6.3, but we shall derive the first line of Table 6.3 step by step, and leave the remainder to be verified by the reader.

Step 1: $h = 10$ m, $\Delta T = 3°C$.
Potential energy per unit rate of deepening,

$$\alpha \Delta T g h = 2.57 \times 10^{-4} \times 3 \times 9.8 \times 10$$
$$= 7.56 \times 10^{-2} \text{ m}^2/\text{sec}^2 \text{ (column 4)}.$$

Step 2: $h = 10$ m, $\tilde{H} = 290$ W/m²
Plume velocity,

$$u_f = \left(\frac{\alpha g h \tilde{H}}{C_p \rho_0}\right)^{1/3} = \left(\frac{2.57 \times 10^{-4} \times 9.8 \times 10 \times 290}{4179 \times 997}\right)^{1/3}$$
$$= 1.21 \times 10^{-2} \text{ m/sec (column 5)}.$$

Step 3: $u_f = 1.21 \times 10^{-2}$ m/sec, $\Delta t = 1.44 \times 10^4$ sec.
Work done by penetrative convection,

$$C_\kappa{}^f u_f{}^3 \Delta t = 0.13 \times (1.21 \times 10^{-2})^3 \times 1.44 \times 10^4$$
$$= 3.28 \times 10^{-3} \text{ m}^3/\text{sec}^2 \text{ (column 6)}.$$

Step 4: $u_f = 1.21 \times 10^{-2}$ m/sec, $\Delta T = 3°C$, $h = 10$ m.
Total energy required per unit rate of deepening,

$$C_T{}^f u_f{}^2 + \alpha \Delta T g h = 0.5 \times (1.21 \times 10^{-2})^2 + 7.56 \times 10^{-2}$$
$$= 7.56 \times 10^{-2} \text{ m}^2/\text{sec}^2.$$

Step 5: From Eq. (6.45) we get that

$$\Delta h = \frac{C_\kappa{}^f u_f{}^3 \Delta t}{(C_T{}^f u_f{}^2 + \alpha \Delta T g h)}$$
$$= \frac{3.28 \times 10^{-3}}{7.56 \times 10^{-2}}$$
$$= 0.044 \text{ m} \quad \text{(column 8)}.$$

TABLE 6.3

Sample Calculation for Computing Mixed Layer Deepening by a Simple Penetrative Convection Model

Time (h)	$\bar{H}\Delta t/C_p\rho_0 h$ (°C) (×10²) 1	$\Delta T\Delta h/h$ (°C) (×10²) 2	$\Delta T = \Delta T + 0.1\Delta h$ ① − ② (°C) 3	$\alpha\Delta T g h$ (m²/sec²) (×10²) 4	$u_f = (\alpha g h\bar{H}/C_p\rho_0)^{1/3}$ (m/sec²) (×10²) 5	$C_\kappa^f u_f^3 \Delta t$ (m³/sec²) (×10³) 6	$C_T^f u_f^2 + \alpha\Delta T g h$ (m²/sec²) (×10³) 7	Δh (m) 8	h (m) 9
0	10.0	1.30	3.00	7.56	1.21	3.28	7.56	0.044	10.00
4	9.98	1.30	2.89	7.31	1.21	3.30	7.31	0.045	10.04
8	9.93	1.29	2.78	7.07	1.21	3.31	7.07	0.047	10.09
12	9.88	1.29	2.67	6.82	1.21	3.33	6.82	0.049	10.14
16	9.83	1.28	2.56	6.57	1.21	3.34	6.57	0.051	10.19
20	9.77	1.28	2.45	6.32	1.21	3.36	6.32	0.053	10.24
24			2.34						10.29

Step 6: The new depth of the epilimnion will be

$$10 + 0.044 = 10.044 \text{ m} \quad \text{(column 9)}.$$

Step 7: The temperature drop due to heat lost in the four hours

$$= \frac{\tilde{H}\Delta t}{C_p \rho_0 h}$$

$$= \frac{290 \times 1.44 \times 10^{+4}}{4179 \times 997 \times 10}$$

$$= 1.00 \times 10^{-1} \, {}^\circ\text{C} \quad \text{(column 1)}.$$

Step 8: The temperature drop due to entrainment of colder water

$$= \frac{\Delta T \Delta h}{h}$$

$$= \frac{3 \times 0.044}{10} = 1.30 \times 10^{-2} \, {}^\circ\text{C} \quad \text{(column 2)}.$$

Step 9: The new temperature difference between the epilimnion and the hypolimnion

$$= \Delta T(\text{old}) + 0.1\Delta h - \frac{\tilde{H}\Delta t}{C_p \rho_0 h} - \frac{\Delta T \Delta h}{h}$$

$$= 3.00 + 0.004 - 0.10 - 0.013 = 2.89 \quad \text{(column 3)}.$$

The results listed in Table 6.3 show that the thermocline would have deepened by about 29 cm over that particular day. The flux 290 W/m^2 is typical for a cool winter day at latitudes of 30–40°. Table 6.3 also clearly illustrates that the temporal term $C_T{}^f u_f{}^2$ is very small compared to $\alpha \Delta T g h$ and may usually be neglected in reservoir work. ■

6.3.2 Mixing Due to Weak Winds

A wind blowing over a lake exerts a stress on the water surface that causes waves to form, break and transfer momentum to the water. The wave motion, especially when waves are breaking, produces turbulence in the upper layers. This turbulence then interacts with the mean shear in the upper few meters to produce further turbulent kinetic energy. Often this interaction produces a secondary motion as well as a mean windward drift. Such secondary motions are called Langmuir cells and they are distinguishable to an observer by the characteristic slick pattern associated with the regions of convergence. The net turbulent kinetic energy produced in these upper few meters is then exported to the lower parts of the epilimnion by turbulent diffusion or by the advective motion associated with the Langmuir circulation.

The total power input from the wind is given by Eq. (6.18). The fraction available as turbulent kinetic energy is that which is produced directly by the wind pressure fluctuations working on the water surface, that which is introduced by the breaking waves and that which is generated in the top one or two meters of the water column. These surface processes have so far defied analysis, but from Eq. (6.18), with $u_w \sim u^*$, the introduced power is seen to be proportional to $\rho_0 u^{*3}$, where u^* is the water shear velocity. This follows by noting that the Reynolds stresses near the surface are of order $\rho_0 u^{*2}$ and the mean drift, as well as the velocities associated with Langmuir circulation, are all of order u^*.

In addition to this stirring of the surface layers the wind will also cause the water to accelerate, so that after short times the whole epilimnion will have a mean motion with a velocity $u(z)$. The shear associated with this mean motion may then contribute further to the production of turbulent kinetic energy.

There are very few detailed measurements of the velocity distribution throughout the epilimnion and the thermocline. However, the few available observations by Csanady (1972), Gregg (1976), Thorpe (1977), and Kraus (1977) suggest that the vertical lake structure may be pictured as follows:

(a)   A comparatively thin surface layer near the stirring agent. The dominant activity here is the production of turbulent kinetic energy which is then exported to the fluid below.

(b)   A uniform central layer in which the energy exported from the surface layer is used to homogenize the fluid.

(c)   A thin front separating the turbulent interior from the quiescent fluid below. Here the remainder of the eddy energy from the surface, plus any which may be locally generated by shear, less that which is locally dissipated or radiated downward by internal waves, is expended to entrain quiescent fluid into the central layer above.

(d)   A hypolimnion which is the storage volume for the deeper water. Normally it is stabilized by a weak temperature gradient with mixing being sporadic and confined to isolated patches.

For the moment we shall neglect the mean velocity structure and the associated production of turbulent kinetic energy within the epilimnion, and account only for the stirring action of the wind. As we shall see in Section 6.3.3, this requires that the wind stress be not too great or that the lake be not too large.

If we define C_κ^* as the efficiency of the wind energy utilization and further if we assume that the input power from the penetrative convection and from the wind stress are cumulative, then Eq. (6.45) may be written as

$$(C_T q^2 + \alpha \Delta T g h)(dh/dt) = C_\kappa^f u_f^{\,3} + C_\kappa^* u^{*3}, \qquad (6.47)$$

where q is a measure of the turbulent root mean square velocity resulting from the combination of cooling and wind stirring and C_T the kinetic energy coefficient. There is some doubt how we should define q, but consistency suggests

$$q = (u_f{}^3 + \eta^3 u^{*3})^{1/3}, \tag{6.48}$$

where $\eta = (C_\kappa{}^*/C_\kappa{}^f)^{1/3}$ so that Eq. (6.47) reduces to the simple form:

$$(C_T q^2 + \alpha \Delta T g h)(dh/dt) = C_\kappa{}^f q^3, \tag{6.49}$$

analogous to Eq. (6.45). This equation was first derived by Kraus and Turner (1967) for the description of ocean mixed layers.

The experiments of Halpern (1974) indicate that $\eta^3 C_\kappa{}^f$ should be 6.5 and the original estimates of Turner (1969) suggests a value of 16.5. Laboratory data reviewed by Sherman et al. (1978) requires a value $\eta^3 C_\kappa{}^f$ of 2.2. The wind stress experiments of Wu (1973) on the other hand yielded a value of only 0.23 and correlation of predictions from Eq. (6.49) with profile data from the Wellington Reservoir led Imberger et al. (1978) to fix $\eta^3 C_\kappa{}^f$ at 0.15 for thermoclines at a depth of approximately 20 m. However, these authors also included a wind reduction correction for the sheltering effect of the surrounding hills. Without this, $\eta^3 C_\kappa{}^f$ would have had a value closer to 0.03.

The reasons for this large variation in the value of $\eta^3 C_\kappa{}^f$ may be explained by noting that in Eq. (6.49) $\eta^3 C_\kappa{}^f$ represents the net efficiency of the surface introduction of kinetic energy. In any particular experiment this energy is the sum of turbulent kinetic energy introduced at the surface, the shear production in the epilimnion and the mechanical and thermal energy losses through the thermocline. The relative importance of these contributions varies greatly from one experiment to the next and until these are parameterized separately, large variations of the value $\eta^3 C_\kappa{}^f$ may be expected.

Furthermore, the energy introduction depends on u^{*3}, so that using different time steps for the wind records will lead to very large errors. For instance a 5-m/sec daily average wind would introduce 16 times less energy than a 20-m/sec wind blowing for only six hours, even though the daily averages are identical.

Leakage losses are relatively unimportant in reservoirs, but as Spigel (1978) has shown shear production may have a dominating role in periods of high wind. For this reason it is recommended that $\eta^3 C_\kappa{}^f$ be given the value of 0.23 and that the turbulent kinetic energy budget be augmented by a separate parameterization of the shear production (Section 6.3.3). This recommendation is based on Wu's (1973) experiment where shear effects were negligible.

We may not have first expected the efficiency $\eta^3 C_\kappa{}^f$ to be constant, independent of the depth h. However, the flow in the epilimnion may be approximated by a fully turbulent shear flow. In Tennekes and Lumley (1972) it is shown that for such flows the dissipation per unit volume is proportional to $\rho_0 u^3/l$, where u is the velocity scale of the turbulent fluctuations and l the length scale of the larger eddies. In the epilimnion we have seen that $u \sim u^*$ and $l \sim h$, so that the dissipation in the entire epilimnion water column per unit area will be proportional to $\rho_0 u^{*3}$ and thus the efficiency $\eta^3 C_\kappa{}^f$ may be expected to be constant.

Lastly, great care must be exercised when dealing with reservoirs situated in hilly terrain. Uneven wind sheltering may lead to differential deepening and rather complicated secondary motions may arise (see Imberger *et al.*, 1978).

Example 6.4. Consider the lake structure defined in Example 6.3. Suppose that as well as the surface cooling there is a 7.1 m/sec wind blowing over the lake surface. Repeat the calculations of the previous example including the wind power and determine the depth to which the thermocline will deepen in one day.

Solution. From Eq. (6.17) and the definition of the shear velocity it follows that

$$u^* = \left| 1.3 \times 10^{-3} \frac{\rho_A}{\rho_w} U^2 \right|^{1/2}$$

$$= 8.88 \times 10^{-3} \text{ m/sec.}$$

Using a value $\eta^3 C_\kappa{}^f = 0.23$ and $C_\kappa{}^f = 0.13$, the value of $\eta = 1.23$. The deepening is found by integration of Eqs. (6.46) and (6.49). The details of calculations are identical to those in Example 6.3 and we give only the results in Table 6.4. ∎

6.3.3 Reservoir Behavior under Severe Wind Conditions

So far we have been able to account for mixing in the epilimnion without regard to the motion of the water in the reservoir. The importance of this simplification must be stressed since simulations can be made of temperature and other water quality parameters without computing the momentum balance.

However, the wind stress will initiate motion and move the water in the epilimnion in the direction of the wind. If the water surface is to remain nearly horizontal, as it does, then the water in the hypolimnion must counter this flow and move in the reverse direction. As shown in Fig. 6.10, a shear will develop across the thermocline which will increase with time until the thermocline has tilted sufficiently to set up a hydrostatic pressure gradient which just balances the surface stress. At this stage the motion changes from a whole basin circulation to two closed gyres, one each in the epilimnion and the hypolimnion and the shear at the interface will decrease to a very small value. All the work done by the wind is then either dissipated internally or used to deepen the epilimnion. However, as indicated in Section 6.2, the set up time is proportional to the seiching period of the thermocline and this may be as much as two or three days, giving the wind stress ample opportunity to develop an appreciable shear across the thermocline.

TABLE 6.4

Sample Calculation for Computing Mixed Layer Deepening by a Krause and Turner (1967) Type Model

Time (h)	$\bar{H}\Delta t/C_p\rho_0 h$ (°C) ($\times 10^2$)	$\Delta T \Delta h/h$ (°C) ($\times 10^2$)	ΔT (°C)	$\alpha\Delta Tgh$ (m²/sec²) ($\times 10^2$)	u_f (m/sec) ($\times 10^2$)	u^* (m/sec) ($\times 10^2$)	$[C_\kappa^f u_f^3 + \eta^3 C_\kappa^f u*^3]\Delta t$ (m³/sec²) ($\times 10^3$)	$C_T q^2 + \alpha\Delta Tgh$ (m²/sec²) ($\times 10^2$)	Δh (m)	h (m)
0	1	2	3	4	5	6	7	8	9	10
0	10.0	2.22	3.00	7.56	1.21	8.88	5.60	7.58	0.074	10.00
4	9.96	2.21	2.89	7.33	1.21	8.88	5.63	7.35	0.077	10.07
8	9.88	2.17	2.79	7.11	1.21	8.88	5.66	7.13	0.079	10.15
12	9.81	2.15	2.67	6.87	1.22	8.88	5.68	6.89	0.082	10.23
16	9.72	2.13	2.56	6.64	1.22	8.88	5.70	6.66	0.086	10.31
20	9.64	2.10	2.45	6.42	1.22	8.88	5.72	6.44	0.089	10.40
24			2.34							10.49

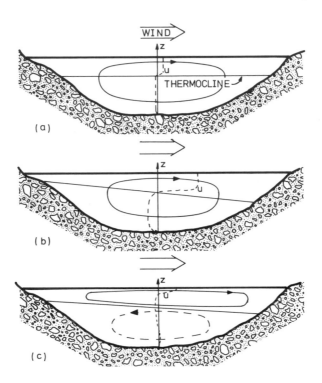

Figure 6.10 Formation of baroclinic motions in a lake exposed to wind stresses at the surface. (a) Initiation of motion. (b) Position of maximum shear across the thermocline. (c) Steady state baroclinic circulation.

As deepening proceeds the entrained hypolimnion water, initially moving with a small velocity in the reverse direction, mixes with the water in the epilimnion and assumes the same velocity. In order to conserve momentum such entrainment must then decrease the velocity in the epilimnion by the fraction $h/(h + \Delta h)$. The associated rate of decrease of mean kinetic energy will be proportional to $\rho_0(\Delta u)^2 \, dh/dt$, where Δu is the difference in velocity between the water of the epilimnion and that of the hypolimnion. Some of this energy will become available to the mixing processes and we should add this production contribution to the right-hand side of Eq. (6.49). Again the efficiency C_s of the production must be determined experimentally. So far no direct measurements have been made of C_s. However, Sherman *et al.* (1978) estimated indirectly from available laboratory data that C_s was approximately equal to 0.5. The complete turbulent kinetic energy balance thus becomes

$$(C_T q^2 + \alpha \Delta T g h)\frac{dh}{dt} = C_\kappa{}^f q^3 + C_s(\Delta u)^2 \frac{dh}{dt}. \qquad (6.50)$$

To determine the deepening of the thermocline we must now simultaneously solve Eqs. (6.46), (6.50), and the momentum equations. Niiler (1975) has carried out such a computation for the ocean mixed layer and the reader is referred to the book edited by Kraus (1977) for an excellent summary of this and other ocean mixed layer work.

Pollard et al. (1973) simplified Eq. (6.50) and balanced only the second and last terms. This yielded the deepening law

$$h = C_s \frac{(\Delta u)^2}{\alpha \Delta T g}, \tag{6.51}$$

which may be combined with Eq. (6.46) and the horizontal momentum equations to calculate Δu and ΔT. They applied their model to ocean mixed layers where the model predictions proved excellent.

The situation in a lake is complicated by the boundaries of the lake basin. These must be accounted for in any momentum balance. However, because of the simplicity inherent in the formulation of Section 6.3.2 it is useful to first investigate the domain of validity of this simpler approach. We shall follow the work of Spigel (1978) who dissected the interactions between the mixing and the motion by introducing the concept of competing time scales.

We shall now define the time scales of mixing and motion in a reservoir with a well defined thermocline at a depth h. If q is the measure of the turbulent velocity, then the time taken to mix a substance over the depth of the epilimnion will be of order T_τ, where

$$T_\tau = h/q. \tag{6.52}$$

If the mixing encounters a thermocline, then it must erode the water in the hypolimnion and entrain it into the upper turbulent waters. We have seen that when shear production is absent, Eq. (6.49) may be applied. Neglecting the correction due to the unsteady term, deepening the epilimnion to twice the original depth will take a time of order T_e^q;

$$T_e^q = T_\tau R_i^q, \tag{6.53}$$

where $R_i^q = \alpha \Delta T g h / q^2$.

On the other hand if the wind stress induces large shears, then shear production will dominate and Eq. (6.51) is a useful approximation. For such vigorous mixing $\tilde{H}$ becomes unimportant in Eq. (6.46) and the solution to the thermal equation will become $\Delta T h = $ constant. Lastly, since the shear is a direct consequence of the applied stress, the rate of change of momentum $\rho_0 \, d(hu)/dt$ will be equal to $\rho_0 u^{*2}$, so that

$$\Delta u \sim u = u^{*2} t / h \tag{6.54}$$

For deep thermoclines the reverse motion in the hypolimnion must also be accounted for. However, since the shear mechanism only dominates for h small we will neglect this correction.

Combining these three results yields a time of deepening to twice the original depth of

$$T_e = (2\eta/C_s^{1/2})T_\tau R_i^{*\,1/2},\qquad(6.55)$$

where $R_i^* = \alpha g \Delta T h / u^{*2}$.

Shear production contributes to the deepening if F_i is order $C_s^{-1/2}$, where F_i is evaluated with the maximum velocity Δu achieved for a particular shear velocity u^*. Now, we have already seen that the speed of an internal wave at a density discontinuity is order $\sqrt{\alpha \Delta T g h}$, when the thermocline is close to the surface, so that the half-period of oscillation T_i, which is the time for the wave to travel the length of the reservoir, will be

$$T_i = L/\sqrt{\alpha \Delta T g h},\qquad\text{for}\quad h \ll H.\qquad(6.56)$$

The maximum velocity occurs when the thermocline has achieved the maximum tilt so that an estimate of the maximum velocity will be obtained by substituting the time for one quarter of the period into Eq. (6.54),

$$\Delta u_{max} = u^{*2}L/2h\sqrt{\alpha \Delta T g h}.\qquad(6.57)$$

Beyond this time Eq. (6.54) is no longer valid. The velocity shear will quickly return to order u^*. Hence the maximum internal Froude number F_i will increase to $C_s^{-1/2}$ before a time $T_i/2$ has elapsed provided that

$$R_i^* \le LC_s^{1/2}/2h.\qquad(6.58a)$$

Shear production will dominate if it is sustained long enough to deepen the layer to $2h$ by this mechanism. Setting T_e equal to $T_i/2$ in Eq. (6.54) leads to the secondary boundary

$$R_i^* = LC_s^{1/2}/4h.\qquad(6.58b)$$

For a value of R_i^* below this, the layer will deepen predominantly with energy from the mean shear.

The velocity in the epilimnion will cause a reverse velocity in the hypolimnion. Assuming that the interface is planar at its maximum excursion allows an estimate for the tilt Δh at the end of the reservoir. By applying conservation of mass we must have that

$$\Delta h = \tfrac{1}{2}L/R_i^*,\qquad(6.59)$$

where the equal sign should be interpreted only as an approximation since the estimate assumes a very elongated reservoir.

By comparing Eq. (6.59) with (6.58a) we see that the thermocline may surface whenever shear productions enters the entrainment process at the thermocline. Imberger *et al.* (1978) suggested the bounds given by Eq. (6.58a) as offering a means for differentiating between large and small lakes on the grounds that small lakes are ones in which shear production is negligible. We will now show that Eq. (6.58a) has an even more central role. So far we have concentrated on the energy budget and the nature of the deepening has remained somewhat hidden. We saw in Section 6.3.1 how penetrative convection may depend on the details of the interface at the thermocline. The same is true for wind mixing.

Introduction of turbulent kinetic energy at the surface leads to deepening by erosion and as a rule results in a very sharp well-defined thermocline. By contrast mixing which is energized by shear production is characterized by vigorous "billows" at the thermocline (Thorpe, 1977) resulting in a smeared interface. The billowing is a result of Kelvin–Helmholz instabilities described in detail in Sherman et al. (1978). Here we are only concerned with the interface thickness after the billows have subsided, the local shear reduced, and the flow returned to a laminar state.

Corcos and Sherman (1976) have analyzed the nature of the instability and they concluded that the thickness of the interface δ_B, after the mixing has subsided, is given by

$$\delta_B = 0.3(\Delta u)^2/(\alpha \Delta T g) \tag{6.60}$$

and the time T_B required for this to occur would be given by

$$T_B = 20(\Delta u)/(\alpha \Delta T g). \tag{6.61}$$

Both estimates are independent of molecular processes, but of course the ultimate mixing associated with the billows will depend on molecular diffusivities.

Substituting the expression for the maximum velocity from Eq. (6.57) into Eq. (6.60) shows that when $R_i^* \leq L/2h$

$$\delta_B/h \geq 0.3, \tag{6.62}$$

or in other words, the thermocline is smeared over essentially the whole epilimnion.

If $R_i^* > L/2h$ then $u_{max} \leq u^* L/2hR_i^{*\,1/2}$, so that

$$T_B/T_\tau \leq (10\eta R_i^{*\,-3/2}L)/h. \tag{6.63}$$

On the other hand if $R_i^* < L/2h$, then $u_{max} \leq u^{*2}T_c/h$ and

$$T_B/T_\tau \leq 20\eta C_s^{-1/2}R_i^{*\,-1/2}. \tag{6.64}$$

With severe tilting of the thermocline ($R_i^* < L/2h$) deepening results in a longitudinal variation in density. For one-dimensional models to be useful it is important that these are quickly dispersed. There are essentially two mechanisms by which this may be accomplished. The first is shear dispersion where the shear profile is established by the wind stress itself. In Chapter 4 we saw that the time T_m required to mix a substance over a distance L in a fully turbulent layer of depth h is given by

$$T_m \sim T_\tau L^2/h^2. \tag{6.65}$$

Second, if the wind is weak, the differential deepening may itself build up a longitudinal density differential in the epilimnion, sufficient to drive a gravitational circulation. This circulation may then lead to longitudinal shear dispersion. Imberger (1976) has derived a time scale T_g for such mixing with

$$T_g = T_r(L^4/h^4)(1/R_i^{g^2}).$$ (6.66)

Restoration to a one-dimensional vertical structure after the wind ceases is achieved by damping of the internal waves. Once again we can arrive at a simple model for the damping time. Kalkanis (1964) has shown that at the boundary of an oscillating fluid a boundary layer forms in which there is energetic dissipation. The thickness of this turbulent layer δ was determined experimentally to be given by

$$\delta = u_{max} T^{1/2} e/471 v^{1/2},$$ (6.67)

where u_{max} is the maximum velocity due to the oscillations, e the boundary roughness, v the coefficient of kinematic viscosity, and T the wave period.

The work done in the oscillatory boundary layer is equal to its kinetic energy content per period. Hence the work done in time t is equal to order $\frac{1}{2}\delta A\rho_0 u^2 t/2T_i$; where A is the boundary area. The waves will be dissipated when the total kinetic energy order $\frac{1}{2}\rho_0 u^2 V$ locked in the internal wave motion has been dissipated by the layers at the boundary. The dissipation time scale T_d may thus be derived by equating these two energies:

$$T_d = (V/\delta A)T.$$ (6.68)

Example 6.5. Heaps and Ramsbottom (1966) have analyzed internal seiching in Lake Windermere. The approximate dimensions of this lake are such that $L = 6.6 \times 10^3$ m, $H = 65$ m, $A = 2.5 \times 10^7$ m^2, and $V = 6.6 \times 10^8$ m^3. In September 1951 the lake was partitioned by a sharp thermocline with the temperature of the epilimnion being 14.5°C and that of the hypolimnion being 7°C. If the epilimnion had a thickness $h = 12$ m, calculate the thickness of the turbulent boundary layer for a wind which induced a shear velocity of 1.6×10^{-2} m/sec. What is the characteristic wave damping time?

Solution. The density of the hypolimnion water will be 999.95 kg/m^3 and the water in the epilimnion has a density of 999.15 kg/m^3 so that the effective acceleration due to gravity g' will be 7.8×10^{-3} m/sec^2. Equation (6.56) gives the period of oscillation

$$T = 2T_i = \frac{2 \times 6.6 \times 10^3}{(7.8 \times 10^{-3} \times 12)^{1/2}}$$

$$= 4.31 \times 10^4 \text{ sec}$$

$$= 12.0 \text{ hr.}$$

This may be compared to an observed period of approximately 12.7 h.

The order of magnitude of the maximum velocity induced by the internal seiching is given by Eq. (6.57),

$$u_{\max} = \frac{(1.6 \times 10^{-2})^2 \times 6.6 \times 10^3}{2 \times 12 \times (7.8 \times 10^{-3} \times 12)^{1/2}} = 0.23 \text{ m/sec.}$$

The thickness of the turbulent boundary layer is given by Eq. (6.67). However, this equation requires an estimate of the roughness height e. There is some uncertainty, but a reasonable number would be the same as for a smooth river channel. Chow (1959, p. 196) gives an average value $e = 6 \times 10^{-2}$ m. With this value and taking $v = 1.3 \times 10^{-6}$ m^2/sec

$$\delta = \frac{2.3 \times 10^{-1} \times (4.31 \times 10^4)^{1/2} \times 6 \times 10^{-2}}{471 \times (1.3 \times 10^{-6})^{1/2}}$$

$$= 5.3 \text{ m.}$$

Substituting this result into Eq. (6.68) yields the required estimate for damping time scale

$$T_d = \frac{6.6 \times 10^8 \times 4.31 \times 10^4}{5.33 \times 2.5 \times 10^7}$$

$$= 2.13 \times 10^5 \text{ sec} \approx 59 \text{ hr.}$$

Heaps and Ramsbottom (1966) developed a precise theory of forced oscillations with a linear damping term. Calibration of the decay of the oscillations of their model with the observed seiching in Lake Windermere gave a characteristic damping time of 55 hr. We thus see that the above choice of e which represents the roughness length of a smooth mud bottom was reasonable. ■

The estimates from Eqs. (6.67) and (6.68) may be combined to yield an expression of T_d in terms of R_i^*:

$$\frac{T_d}{T_\tau} = \frac{1333 \eta V}{eA \text{ Gr}^{1/4}} \left(\frac{h}{L}\right)^{1/2} R_i^{*1/2}. \tag{6.69}$$

The value of $\eta V/eA \text{ Gr}^{1/4}$ does not vary greatly from lake to lake, and may be assumed constant when comparing the time scale T_d with the other times of motion and mixing.

Figure 6.11 is reproduced from Spigel (1978) and shows the variation of the various time scales as a function of R_i^*. For the purposes of this discussion we have assumed that $q \approx u^*$ and $C_s^{1/2} \approx 1$. Also included in Fig. 6.11 is the estimate for the tilt $2\Delta h/L$ and the broadening δ/h of the interface by the Kelvin–Helmholtz mechanism. Using this diagram Spigel (1978) has defined four distinct deepening regimes. These are schematically illustrated in Fig. 6.12 and possess the following properties.

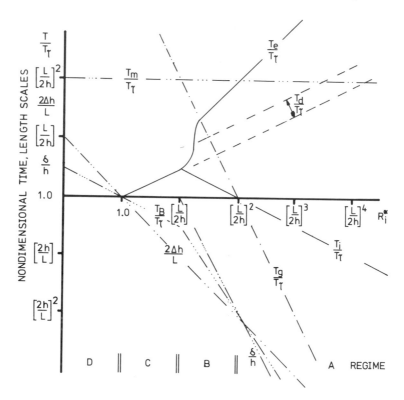

Figure 6.11 Time scales for mixed layer deepening as a function of Richardson number $R_i^* = \alpha \Delta T g h / u^{*2}$. The time scales T_τ, T_e, T_i, T_B, T_m, T_g, and T_d are defined by Eqs. (6.52), (6.55), (6.56), (6.61), (6.65), (6.66), and (6.69). The nondimensional tilt height $\Delta h / L$ is given by Eq. (6.59) and the billowing length scale δ_B / h is defined by Eq. (6.60). [After Spigel (1978).]

(a) $R_i^* > (L/2h)^2$. The deepening process proceeds very slowly by turbulent erosion. The interface set up is small and internal waves persist for long times. Any longitudinal gradients are quickly obliterated by longitudinal gravitationally driven mixing. Mixing and internal motions are completely uncoupled and the interface remains sharp.

(b) $(L/2h) < R_i^* < (L/2h)^2$. Internal waves are the prominant feature of this regime. Entrainment and billowing have minor effects on the internal waves, yet the wave amplitude can be quite severe. Erosion keeps the interface sharp and it is energized almost exclusively by surface stirring. Billowing should be observable with the interface becoming less defined the closer R_i^* is to $L/2h$. At this boundary the thermocline surfaces at one end, becomes smeared and essentially deepens to the bottom at the downwind end all in much the same time. Furthermore, at the boundary $R_i^* = L/2h$ the mean kinetic energy being

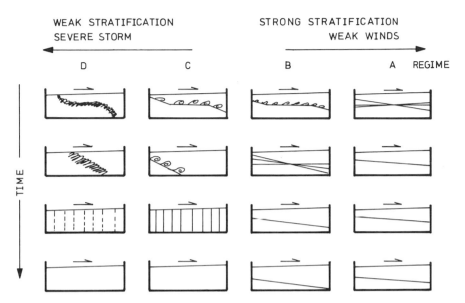

Figure 6.12 Pictorial representation of the mixing sequences for the different regimes of mixed layer deepening. [After Spigel (1978).] Regime A: internal waves form and decay; very slow deepening with sharp thermocline. Regime B: internal waves with slight billowing; slow deepening of tilted interface. Regime C: large billows with severe thermocline tilt; vertically mixed with a longitudinal gradient. Regime D: severe vertical mixing leading to a weak longitudinal gradient.

utilized by the billowing process is approximately equal to that used for deepening. The billowing will be rapid, but the presence of the turbulence in the upper layer may hide the organized structure associated with the billowing. Internal waves are heavily damped.

(c) $1 < R_i^* < L/2h$. Throughout this regime the thermocline will be diffuse and steeply inclined. The process deepens the epilimnion rapidly to the bottom, the mixing being predominantly energized by shear production. The final state will be a longitudinal temperature gradient which will be mixed horizontally on a time scale T_m.

(d) $R_i^* < 1$. Deepening is now so rapid and chaotic that the interface will not be well defined. The flow will appear a little like the reverse of a lock exchange flow.

Unfortunately, there does not appear to be lake data available which would allow the whole classification to be tested. Spigel (1978) has, however, summarized all available data (Table 6.5) and concluded that all available observations conform with the above classification scheme. Hence by estimating the regime of a particular lake it should be a relatively simple matter to predict the gross mixing features to be expected in the lake.

TABLE 6.5

Classification of Phenomena Observed by Previous Investigators[a]

Reference(s)	Predicted regime	Sharp interface	Sustained seiching	Billows	Diffuse thermocline	Thermocline surfaces	Response in hypolimnion	Longitudinal gradient
Mortimer (1974): Lake Windermere, 26 October 1949	C	no	no	—	yes	yes	yes	yes
Thompson (1969): Laboratory wind wave tank	C	no	no	—	yes	yes	yes	yes
Mortimer (1952): Lake Windermere, 9 June 1947	C	no	no	—	yes	yes	no	no
Thorpe and Hall (1977): Loch Ness	C	yes	—	yes	no	no	no	no
Mortimer (1953): Lake Windermere, 18–19 July 1951	C	yes	yes	—	no	no	no	no
Heaps and Ramsbottom (1966): Lake Windermere	B	yes	yes	—	no	no	no	no
Wedderburn (1912): Loch Earn	B	yes	yes	—	no	no	no	
Thorpe (1971, 1974): Loch Ness	B	yes	yes	—	no	no	no	no
Thompson (1969): Laboratory wind wave tank	B	yes	yes	—	no	no	no	no
Thorpe (1977): Loch Ness	B	yes	yes	—	no	no	no	no
Wu (1973), (1977): Laboratory wind wave tank	A	yes	no	—	no	no	no	no

[a] After Spigel (1978).

Example 6.6. Table 6.6 gives the temperature structure of the upper meters on the morning of February 29 in the Wellington Reservoir.

TABLE 6.6

Temperature Structure in the Upper Meters of the Wellington Reservoir on the Morning of February 29[a]

Depth (m)	Temp. (°C)	Depth (m)	Temp. (°C)	Depth (m)	Temp. (°C)
5.40	25.71	6.79	25.48	8.33	25.15
5.84	25.57	7.29	25.44	8.89	24.99
6.31	25.52	7.80	25.39	29.29	12.40

[a] The temperature at depth is included for comparison.

The recorded wind speed at 10 m was 6.3 m/sec and the rate of heat loss due to all sources was 199.4 W/m². The depth $\bar{h}$ of the median temperature was approximately 10 m, the overall reduced acceleration due gravity $g' = \Delta\rho/\rho_0 g$ was equal to 1.18×10^{-2} m/sec², the effective length of the reservoir at that depth is approximately equal to 2.1×10^4 m, the reservoir volume is 1.86×10^8 m³, and its boundary area is 4.05×10^7 m². Investigate the various time scales and determine the likely regime of deepening. Calculate the deepening over the first three hours.

Solution. The shear velocity

$$u^* = \left(1.3 \times 10^{-3} \frac{\rho_A}{\rho_w} U^2\right)^{1/2}$$

$$= 7.87 \times 10^{-3} \text{ m/sec.}$$

To determine the various time scales we must compute an equivalent two layer Richardson number

$$R_i^* = \frac{g'\bar{h}}{u^{*2}} = \frac{1.18 \times 10^{-2} \times 10}{(7.87 \times 10^{-3})^2} = 1905.$$

The aspect ratio of the thermocline

$$\frac{L}{2\bar{h}} = \frac{21 \times 10^3}{2 \times 10} = 1050,$$

so that the deepening process falls into the regime (B) with vigorous mixing deepening and tilting the thermocline. We may expect roughly equal contributions for all sources of energy.

Now

$$u_f = \left(\frac{\alpha g \bar{h} \tilde{H}}{C_P \rho_0}\right)^{1/3}$$

$$= \left(\frac{2.57 \times 10^{-4} \times 9.8 \times 10 \times 199.4}{4175 \times 998}\right)^{1/3}$$

$$= 1.06 \times 10^{-2} \text{ m/sec}$$

from which we may obtain,

$$q = (u_f^3 + \eta^3 u^{*3})^{1/3}$$
$$= [(1.06 \times 10^{-2})^3 + (1.23)^3(7.87 \times 10^{-3})^3]^{1/3}$$
$$= 1.28 \times 10^{-2} \text{ m/sec.}$$

The Richardson number

$$R_i^q = \frac{g'\bar{h}}{q^2} = \frac{1.18 \times 10^{-2} \times 10}{(1.28 \times 10^{-2})^2} = 718.$$

Hence we may compute the time scale as follows:

$$T_\tau = \frac{\bar{h}}{q} = \frac{10}{1.28 \times 10^{-2}} = 781 \text{ sec} = 13.1 \text{ min,}$$

$$T_e^q = T_\tau R_i^q = 781 \times 718 = 5.6 \times 10^5 \text{ sec} = 156 \text{ hr,}$$

$$T_e = \frac{2\eta}{C_s^{1/2}} T_\tau R_i^{*1/2} = \frac{2 \times 1.23 \times 781 \times (1905)^{1/2}}{(0.5)^{1/2}} = 1.2 \times 10^5 \text{ sec} = 32.9 \text{ hr,}$$

$$T_i = \frac{L}{\sqrt{g'\bar{h}}} = \frac{21 \times 10^3}{(1.8 \times 10^{-2} \times 10)^{1/2}} = 6.1 \times 10^4 \text{ sec} = 17.0 \text{ hr,}$$

$$T_m = T_\tau \frac{L^2}{\bar{h}^2} = \frac{781 \times (21 \times 10^3)^2}{10^2} = 3.4 \times 10^9 \text{ sec} = 3.8 \times 10^4 \text{ day,}$$

$$T_g = T_\tau \frac{L^4}{\bar{h}^4} \frac{1}{(R_i^q)^2} = \frac{781 \times (21 \times 10^3)^4}{10^4 \times 718^2} = 2.9 \times 10^{10} \text{ sec}$$

$$= 3.4 \times 10^5 \text{ day.}$$

The maximum velocity to be expected will be of the order

$$u_{max} = \frac{u^{*2} T_i}{2\bar{h}} = \frac{(7.87 \times 10^{-3})^2 \times 6.1 \times 10^4}{2 \times 10}$$

$$= 0.189 \text{ m/sec,}$$

but since $h < \bar{h}$ the actual layer may move somewhat faster (see Table 6.7 where we calculate the actual velocity).

TABLE 6.7

Sample Calculation for Computing Mixed Layer Deepening by Including Shear Production of Turbulent Kinetic Energy

Time 1	$\dfrac{\bar{H}\Delta t}{C_p\rho_0 h}$ ($\times 10^2$) 2	$\dfrac{\Delta T \Delta h}{h}$ ($\times 10^2$) 3	T Epilimnion 4	T Hypolimnion 5	$\alpha\Delta Tgh$ ($\times 10^3$) 6	$u_f = \left(\dfrac{\alpha gh\bar{H}}{C_p\rho_0}\right)^{1/3}$ ($\times 10^3$) 7	u^* ($\times 10^3$) 8	$q = (u_f^3 + \eta^3 u^{*3})^{1/3}$ ($\times 10^2$) 9	$C_\kappa^f q^3 \Delta t$ ($\times 10^4$) 10
0.0			25.71	25.72	1.90	8.67	7.87	1.16	
1.0	3.19	1.14	25.67	25.52	2.21	8.90	7.87	1.17	7.30
2.0	2.95	1.20	25.63	25.48	2.38	9.13	7.87	1.19	7.50
2.0	0	1.14	25.62	25.44	3.08	9.35	7.87	1.20	0
3.0	2.54	1.33	25.58	25.39	3.49	9.58	7.87	1.21	8.09
3.0	0	1.33	25.57	25.15	8.25	9.80	7.87	1.23	0
4.0	2.21	2.85	25.54	24.99	11.12	10.02	7.87	1.25	8.71

Time	$\Delta u = \Delta u + \dfrac{u^{*2}\Delta t}{h}$ ($\times 10^2$) 11	$C_s(\Delta u)^2 \Delta h$ ($\times 10^3$) 12	$10+12+16$ Available mixing energy ($\times 10^3$) 13	Required energy $(C_T q^2 + \alpha\Delta Tgh)\Delta h$ ($\times 10^3$) 14	Action $13 \gtreqless 14$, $16 \gtreqless 14$ 15	Excess energy ($\times 10^3$) 16	Velocity after deepening $= \dfrac{h}{h+\Delta h}\Delta u$ ($\times 10^2$) 17	Δh 18	h 19
0.0									5.40
1.0	4.13	0.38	1.11	0.87	Mix	0.25	3.82	0.44	5.84
2.0	7.64	1.37	2.36	1.07	Mix next layer	1.29	7.07	0.47	6.31
2.0	7.07	1.20	2.49	1.17	Mix	1.32	6.57	0.48	6.79
3.0	9.85	2.43	4.56	1.58	Mix next layer	2.98	9.17	0.50	7.29
3.0	9.17	2.15	5.13	1.82	Mix	3.31	8.57	0.51	7.80
4.0	11.43	3.46	7.64	4.41	Mix	3.23	10.70	0.53	8.33

The expected tilt of the thermocline

$$\Delta h = \frac{1}{2}\frac{L}{R_i{}^*} = \frac{21 \times 10^3}{2 \times 1905} = 5.5 \text{ m},$$

and the thickness associated with Kelvin Helmholz billowing

$$\delta_B = \frac{0.3(u_{max}^2)}{g'} = \frac{0.3 \times 0.189^2}{1.18 \times 10^{-2}} = 0.91 \text{ m}.$$

This will occur in times of order

$$T_B = \frac{20(u_{max})}{g'} = 320 \text{ sec},$$

indicating that the billowing occurs much faster than the acceleration of the layer.

The turbulent wall boundary layer depends on the wall roughness. Let us choose $e = 2 \times 10^{-2}$ m, so that

$$\delta = \frac{u_{max} T^{1/2}}{471 \nu^{1/2}} = \frac{0.189 \times (1.2 \times 10^5)^{1/2} \times 2 \times 10^{-2}}{471 \times (1.3 \times 10^{-6})^{1/2}} = 2.4 \text{ m}.$$

The damping time

$$T_d = \frac{V}{\delta A}T = \frac{186 \times 10^6 \times 2 \times 6.1 \times 10^4}{1.8 \times 40.5 \times 10^6}$$

$$= 2.55 \times 10^5 \text{ sec} = 70.8 \text{ hr}.$$

In summary we see that the wind stress signal is communicated to the mixed layer in about 15 min. Hence any integration scheme shorter than this would have to account for changes in the mean velocity profile. The epilimnion may be expected to deepen considerably in one day, the mixing energy being mainly derived from the mean kinetic energy of the layer. The thermocline will remain diffuse as $T_B \ll T_i$ and although the tilt is appreciable it will not invalidate the one dimensional assumption $\Delta h \leq h$.

In Table 6.7 we have set out a very simple scheme for integrating Eqs. (6.46) and (6.50), analogous to that presented in Example 6.3. The momentum at any instant is estimated from Eq. (6.57) up to a time $T_i/2$, after which time it is assumed that the interface has set up and the interfacial shear has decreased to zero. A more detailed description of the momentum of the upper layers is not warranted, unless the lake behavior should fall into regimes C and D, which is not the case in most medium sized lakes. Normally, the calculations shown in Table 6.7 are carried out with the help of a computer and the approximation of allowing heat transfer only at the surface is easily overcome. The last column in Table 6.7 gives the depth to which the epilimnion has deepened for the times shown in the first column. It will be noted that after two hours there was an excess energy, which allowed deepening without further surface kinetic energy input. This is a consequence of the explicit integration scheme. In practice the front would merely deepen faster and accommodate the larger mixing energies.

■

6.4 VERTICAL MIXING IN THE HYPOLIMNION

In the previous sections we have shown how the vigorous mixing encountered in the epilimnion may be parameterized and how the deepening of the thermocline may be calculated provided the effect of the earth's rotation is small. By implication we have seen how a stable thermocline acts as a protective cover for the hypolimnetic waters against the disturbances created by wind on the surface. In addition there is usually a weak, but definite density gradient stabilizing the hypolimnion so that it may be expected that vertical mixing in lakes is quite small, perhaps even only on a molecular level.

There have been many studies in which the evolution within the hypolimnion of concentration distributions of natural or artificially introduced tracers have been measured. In lakes with a well developed vertical thermal structure the following conclusions regarding vertical mixing may be made:

(a) Vertical diffusivities of heat range from molecular diffusivities up to values of 10^{-4} m^2/sec.

(b) Vertical diffusivity of heat and other tracers are about the same.

(c) The diffusivity is generally higher at periods of strong winds and inflow (outflow).

(d) The diffusivity ε_z generally decreases with increasing N. There have been many correlations of ε_z versus N in the form

$$\varepsilon_z = a(N^{-2})^n, \tag{6.70}$$

where n ranges from 0.2 to about 2.0, depending on the lake considered.

(e) The vertical profiles of temperatures, tracer or dye are usually full of structure with scales ranging from a few centimeters to 10 m. Indeed Kullenberg (1974) distinguished between four types of vertical steppiness in their dye profiles.

Thus even though the hypolimnion is overall very stable with rather large average Richardson numbers, there is nevertheless sometimes relatively vigorous vertical mixing. The only explanation for this apparent contradiction is that although overall there is not sufficient kinetic energy to cause mixing there are areas in the lake at any particular time where the energy density has been increased by some type of concentrating mechanisms allowing a local breakdown and mixing of the structure.

The mixing is thus patchy and intermittent and quickly collapses under the action of buoyancy once the input energy has been expended. Upon collapse the mixed patches elongate and interleave with themselves and their surroundings, leading to steplike vertical density structures. Thorpe (1977) has given the first documentation of such localized mixing. The existence of the concentrating mechanisms thus allows extraction of energy for mixing even when overall there is very little mean kinetic energy compared to the potential energy of the system.

We can follow Ozmidov (1965b) and estimate the vertical extent of the mixed regions before their collapse in terms of the internal dissipation ε (see Chapter 3). If N characterizes the density gradient the buoyancy and inertia forces roughly balance for motions with vertical scales l such that the internal Froude number is roughly one. In other words,

$$l = u/N. \tag{6.71}$$

Now, since below this size the turbulence is not effected by buoyancy, we can use the Kolmogorov estimate of the velocity u in terms of the dissipation ϵ per unit mass. From dimensional analysis

$$u = (\epsilon l)^{1/3}. \tag{6.72}$$

Substituting this into Eq. (6.71) yields the ratio

$$l/H = \epsilon^{1/2}/HN^{3/2} \tag{6.73}$$

as the dimensionless scale of mixing events within the reservoir.

To date there have been no conclusive studies made which parameterize the average or effective vertical mixing gained by summing over all the random localized mixing events resulting from random external inputs. Later in this section we will once again use energy arguments to arrive at a global parameterization, but before doing so let us briefly review possible mechanisms which could lead to a local concentration of energy for mixing. In Fig. 6.13 we have given a schematic representation of the mechanisms to be described and the reader may find it helpful to refer to this sketch when reading the next paragraphs.

(a)   Probably the most important mechanism in lakes of small and medium size ($R_i^* > L/2h$) is that of boundary mixing. We have already seen in the estimation of the damping time for thermocline seiching that an oscillatory boundary layer forms at the reservoir basin boundaries, having typically a thickness of 2–6 m and peak a velocity of 0.02–0.2 m/sec. The period of oscillation is equal to $2T_i$ which depends on the degree of stratification and the basin shape. If the roughness is assumed to be made up of rocks and tree stumps then a reasonable value for the roughness height e is $0.01 - 0.2$ m. Hence a typical roughness Reynolds number is of order 10^2–10^5 which is large enough to yield a fully turbulent boundary layer. Buoyancy will have little influence on the layer structure since the layer is only slightly larger than the mixture length l.

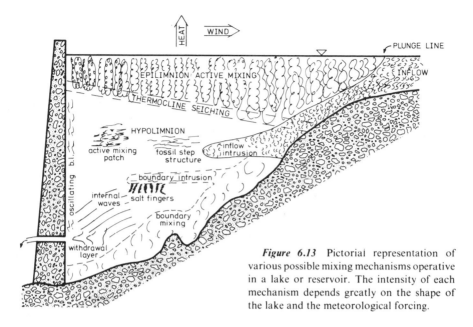

Figure 6.13 Pictorial representation of various possible mixing mechanisms operative in a lake or reservoir. The intensity of each mechanism depends greatly on the shape of the lake and the meteorological forcing.

Now the layer dissipates kinetic energy by mixing so that it will also mix potential energy and at least at the boundary the density gradient will be continually annihilated. The fluid in the boundary layer then assumes a mean density which differs from that at the same level in the hypolimnion. The mean horizontal density gradient then causes slow intrusion flows into the hypolimnion. In Section 6.7.3 we shall return to this problem when dealing with river inflows.

An effective vertical transport is thus realized by vigorous mixing at the boundary with the mixed boundary fluid intruding gravitationally into the hypolimnion. In this way energy from the seiching motion will be extracted and used to transport mass vertically even though the Richardson number associated with the seiching motion itself may be very large. Such boundary mixing has recently been documented by Caldwell *et al.* (1978) in Lake Tahoe in California. Cacchione and Wunsch (1974) showed that under extreme conditions internal waves may even shoal and break at the boundaries providing an even greater input. Lastly, Robinson and McEwan (1975) illustrated how an oscillatory boundary layer in a stratified fluid may itself become unstable and mix even on smooth walls. This could be the case on the face of the dam wall.

(b)   Decreases in local Richardson number may be achieved by a variety of internal wave instability and interactions. Since internal waves are generated at the underside of the thermocline and at the lake boundaries, energy may be transferred from the wind through the thermocline to internal waves which then interact with each other in a variety of possible instability mechanisms to produce a local mixing event. Alternatively seiching of the thermocline will lead to internal wave generation at the boundaries, adding further to the internal wave energy spectrum. The possibilities for local mixing via these mechanisms are numerous and should always be kept in mind when dealing with vertical transfers in the hypolimnion. However, as yet no detailed measurements exist and we can only speculate as to their effect.

(c)   Lastly, if the lake is stratified by both heat and total dissolved solids and if one of these has an unstable distribution, then some of this energy may be available for mixing via double diffusion mechanisms (see Sherman *et al.* 1978).

Let us now turn to a very simple parameterization of the net mixing from all the mechanisms shown in Fig. 6.13.

Ozmidov (1965) postulated, as mentioned above, that in a stratified flow local energy concentrations will generate mixed patches with a scale l given by Eq. (6.73) and further, that inside these patches the turbulence is isotropic and the mixing is described approximately by Kolmogorov's $\frac{4}{3}$ law given in Chapter 3. He reasoned that, since l is limited by Eq. (6.73), the effective transport of mass or heat would be given by substituting Eqs. (6.71) and (6.72) into Eq. (3.40) to give an effective vertical transport coefficient proportional to (ϵ/N^2). However, some care must be exercised here since we have really implicitly assumed a uniform density of concentration of mixing bursts throughout the water column.

Thorpe (1977) measured l directly in Loch Ness and determined that the above is a very good description of the local vertical transfer of mass. However, to apply this result in a model we must still somehow calculate l, the frequency of occurrence and the spatial distribution of the mixing patches. In principle this would require a detailed analysis of all the above listed mechanisms which would then allow l to be determined given the energy inputs to the reservoir. This has so far not been attempted.

An alternative approach may be implemented by assuming that on average the whole reservoir is in "equilibrium" with the external energy inputs and that ϵ may be estimated by the equation

$$\epsilon = (WA_s + dP/dt)/V, \tag{6.74}$$

where W is obtained from Eq. (6.18) and dP/dt is calculated using Eqs. (6.27) and (6.30). This then leads to a vertical diffusion coefficient

$$\varepsilon_z \sim \left(\frac{WA_s + dP/dt}{V}\right)\frac{1}{N^2}. \tag{6.75}$$

Introducing a characteristic mixing time $T_m = P/[WA_s + (dP/dt)]$ and a stability $S = (d\rho_a)/(dz)(H/\Delta\rho)$, where $\Delta\rho$ is the density difference between the surface and bottom layers, Eq. (6.75) may be written in the form

$$\varepsilon_z = CH^2/T_m S. \tag{6.76}$$

The coefficient C, equal to $cPS/(\rho_0 H^2 V N^2)$, is basically constant as may be seen by writing P in the form $\Delta \rho g V \bar{z}$, where $\bar{z}$ is the height of the effective center of volume. This leads to $c = \bar{z}/H$ and even for reservoirs with widely different geometries this coefficient may be expected not to vary widely. A numerical integration of the vertical component of the diffusion Eq. (3.37), with the above diffusion coefficient and $C = 0.048$ was in excellent agreement with the salinity distribution measured over a period of three years in the Wellington Reservoir.

The remaining problem is to determine the vertical mixing of momentum. Even a superficial examination of the energy transfer mechanism outlined above shows that momentum and mass will be transferred at different rates. This is so because internal waves are able to transfer momentum but do not transfer mass or heat. A guide may be obtained from the work of Ivey and Imberger (1978) who found from direct measurements of the thickness of withdrawal layers a value of $\varepsilon_z^v/\varepsilon_z^T = 20$. However, much more detailed measurements are required before a firm recommendation can be given.

6.5 HORIZONTAL MIXING IN RESERVOIRS

We have seen that generally a reservoir is partitioned by the thermocline. Above we find a vertically well mixed column of water agitated by the wind stress and by natural convection and moving with horizontal motions ranging from little eddies to large gyres. Below, stratification usually dominates and the mean motions consists of a set of interleaving density currents of great horizontal and small vertical extent. Let us now examine some of the consequences of this picture with respect to horizontal diffusion.

6.5.1 Horizontal Mixing in the Epilimnion

In large open lakes we may expect the Ekman drift to operate. However, it is also well known that there are motions present with scales ranging from a few meters to the width of the lake. It is therefore not surprising that the dispersion measurements of Murthy (1976) in Lake Ontario indicate a general increase in the horizontal mixing with increase in the size of the tracer cloud. In particular, Murthy (1976) showed that the diffusion coefficients in the direction x of the mean motion could be represented by

$$\varepsilon_x = 1.2 \times 10^{-2} \sigma_x^{1.07}, \tag{6.77}$$

where σ_x is the standard deviation in the x direction. The diffusivity transverse to the mean motion was generally smaller by a factor of ten.

By comparison, Csanady (1963) found the dispersion coefficient in Lake Huron asymptotic to a maximum of about 0.04 m^2/sec for cloud sizes up to 1 km. This is smaller by a factor of 100 than that predicted by Eq. (6.77) even though the winds were similar to those present in the Lake Ontario study. Furthermore, Csanady (1963) draws attention to the very irregular nature of the dispersion patterns, the effect of slicks and the strong influence of unsteady winds. Equations similar to Eq. (6.77) should thus be used with great caution when estimating the horizontal diffusion coefficient. Field experiments with dye tracers or floats should be carried out whenever possible, but especially on windless days when mixing conditions usually become most critical.

6.5.2   Horizontal Mixing in the Hypolimnion

The shear associated with the interleaving density structure is most probably the dominant source of horizontal diffusion in a stably stratified hypolimnion. We may use the work outlined in Chapter 4 to estimate an effective horizontal diffusion coefficient ε_x.

The intrusions are most probably not well correlated so that the model given in Section 4.5 may be usefully applied. A typical velocity gradient α would be $\Delta u/\delta$, where Δu is the velocity of a fluid lens or intrusion and δ the lens thickness. Substituting the vertical diffusion coefficient predicted by Eq. (6.76) into Eq. (4.68) with Saffman's coefficient [Eq. (4.67)], we get the required result

$$\varepsilon_x = 0.34\left(\frac{\Delta u}{\delta}\right)^{2/3}\left(\frac{CH^2}{T_m S}\right)^{1/3}\sigma_x^{4/3}. \qquad (6.78)$$

Two things are apparent from this equation. First, the intensity of mixing is once again a function of the degree of agitation of the lake. Second, for lakes which are elongated in shape and which have a through flow along the longitudinal axis, we may expect Δu to be larger in this direction and hence diffusion in the longitudinal direction may be expected to be greater than that found in the transverse direction.

If we now assume that the larger intrusions are inertia-buoyancy dominated, then as will be shown in Section 6.7.3 for inflows, the general intrusion thickness δ will be approximately given by

$$\delta = \Delta u/N. \qquad (6.79)$$

With $C = 0.048$, Eq. (6.78) then becomes

$$\varepsilon_x = 0.075\,\frac{N^{2/3}H^{2/3}}{T_m^{1/3}S^{1/3}}\,\sigma_x^{4/3}. \qquad (6.80)$$

Example 6.7.   Consider a typical lake situation in which $N = 10^{-2}/\mathrm{sec}$, $u^* = 10^{-2}$ m/sec, $H = 50$ m, $T_m = 100$ day, and $S \approx 1$. Calculate the horizontal diffusion coefficient.

Solution.   By substitution we get $\varepsilon_x = 2.3 \times 10^{-4}\sigma_x^{4/3}$ m^2/sec.   ∎

Murthy (1976) has reported on direct measurements in Lake Ontario. He calculated the diffusion coefficients from the growth of patches released in the hypolimnion. He found that in Lake Ontario under mildly variable conditions the data could be described by

$$\varepsilon_x = 1.8 \times 10^{-4} \sigma_x^{1.33}. \tag{6.81}$$

The scatter was approximately $\pm 80\%$ and the values of σ_x ranged from 10 m to ~ 1 Km. Unfortunately, there is little data given on the actual meteorological condition or the water structure at the times the dye experiments were carried out. However, the comparison between predictions from Eq. (6.80) (Example 6.7) and the measured value Eq. (6.81) are certainly encouraging. Until further experiments are conducted to fully verify all the implications of Eq. (6.80), it is advised that a great deal of caution be exercised in using it for predictive purposes. However, the plausibility of its formulation should make it useful for estimating the growth of individual clouds about their centers of gravity, and thus the peak value of the concentration. To estimate the exact location of a cloud after it is released, one must know the velocity structure. It is recommended that direct measurements be carried out to ascertain this.

6.6   OUTFLOW DYNAMICS

In the previous sections we have described the general mixing mechanisms and derived procedures to estimate the mixing in both the epilimnion and the hypolimnion. We showed how the wind often forms a very sharp transition zone at the thermocline which acts to protect the hypolimnetic waters from the surface induced mixing. Figure 6.2b clearly shows that the greater degree of mixing in the epilimnion allows adequate aeration of the water, whereas the water in the hypolimnion will often be depleted of oxygen. It is therefore of great importance to determine the origin of the water withdrawn at the offtakes, since this determines the quality of the outflowing water.

In the winter months when the reservoir waters are normally unstratified withdrawal from the reservoir will be described by classical potential theory which requires the water to flow radially towards the outlet, equally from all directions.

However, as we have seen, for much of the year the impounded waters of a storage reservoir are temperature stratified. This stratification is most pronounced at the thermocline, but there is also usually a weak, temperature gradient throughout the hypolimnion where the outlet structures of a reservoir are normally located. If water is withdrawn from such an outlet at small discharges, the vertical density gradient may produce buoyancy forces sufficiently strong to prohibit extensive vertical motions so that the water withdrawn comes from a thin horizontal layer at the level of the intake (see Fig. 6.14a). At somewhat larger discharges the withdrawal layer may intersect the thermocline (Fig. 6.14b), and at very large discharges (Fig. 6.14c) the effects of buoyancy may be completely overwhelmed and the flow returns to potential flow.

It is not the purpose here to describe the dynamics of the outflow since we are mainly concerned with mixing in reservoirs. A full treatment of selective withdrawal in an elongated reservoir, as illustrated in Fig. 6.14a, is given in Imberger *et al.* (1976) and the ratio of epilimnion to hypolimnion water withdrawn in large discharge situations (Fig. 6.14c) has been calculated by Wood and Binney (1976). We shall restrict ourselves to a brief description of the most essential formulae for steady flow necessary to determine the quality of the outflowing water.

Consider first withdrawals at small discharges from narrow elongated reservoirs. In most practical cases the outlet structure is a tower in the center of the lake with a gate at a predetermined elevation. The water must therefore flow towards a sink of quite small dimensions. However, since in the horizontal

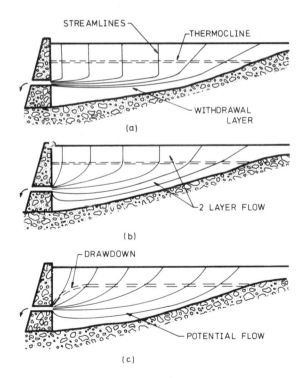

Figure 6.14 Withdrawal flow patterns. (a) Strong stratification in the hypolimnion and small discharges. (b) Weak stratification in the hypolimnion, but a strong temperature gradient across the thermocline; medium discharge. (c) Discharge strong enough to overcome buoyancy forces both in the hypolimnion and at the thermocline.

lateral direction there are no buoyancy forces the flow will be almost radial in the horizontal plane and the streamlines converge in about one reservoir width from a flow uniform over the width to a radial flow into the offtake. Upstream, where the flow is uniform across the width, we can define a discharge per unit width $Q_2 = Q_3/B$, where Q_3 is the total discharge and B the width of the reservoir at the elevation of the outlet. Suppose N_δ and $\varepsilon_{v,\delta}$ are the buoyancy frequency and the vertical diffusivity of momentum averaged over the withdrawal layer thickness 2δ. The withdrawal layer structure is determined by the value of the internal Froude number

$$F_2 = Q_2/N_\delta L^2, \tag{6.82}$$

the Grashof number

$$\text{Gr} = N_\delta^2 L^4/\varepsilon_{v,\delta}^2, \tag{6.83}$$

and the Prandtl number

$$\text{Pr} = \varepsilon_{v,\delta}/\varepsilon_{T,\delta}, \tag{6.84}$$

where $\varepsilon_{T,\delta}$ is the vertical average diffusivity of heat.
  Imberger *et al.* (1976) defined a parameter

$$R = F_2 \text{Gr}^{1/3} \tag{6.85}$$

and showed that a hierarchy of withdrawal layer structures exists which is completely determined by the relative magnitude of R compared to the value of Pr.

  However, it has been found that for general computations of steady two-dimensional withdrawal layers in a reservoir it is sufficient to assume the empirical horizontal velocity distribution

$$u = \frac{Q_2}{2\delta}\left(1 - \frac{x}{L}\right)\left(\cos\frac{\pi Z}{\delta} + 1\right), \tag{6.86}$$

where the origin of Z is at the offtake centerline and we distinguish between only two classes of layers:

$$R \geq 1; \delta = 2.0LF_2^{1/2} \tag{6.87}$$

and

$$R < 1; \delta = 3.5L \, \text{Gr}^{-1/6}, \qquad Z > 0 \tag{6.88a}$$

$$\delta = 2.0L \, \text{Gr}^{-1/6}, \qquad Z < 0. \tag{6.88b}$$

It will be noticed that Eqs. (6.87) and (6.88) depend on N_δ and $\varepsilon_{v,\delta}$, which in turn depend on δ. In practice a simple iteration scheme based on a first guess converges very rapidly.

Intake structures are often located in the center of open, bowl-shaped reservoirs where it is not possible to define a meaningful equivalent two di-mentional discharge. Little work has been carried out on axisymmetric selective withdrawal. At present we recommend that Eqs. (6.87) and (6.88) be applied, but with Q_2 replaced by $Q_R = Q_3/2\pi R$, where R is the average radius from the intake structure to the shoreline, at the level of the offtake. This procedure will produce reasonably reliable estimates of the withdrawal layer thickness, but model test should be considered.

The other important factor to be remembered when applying Eqs. (6.87) and (6.88) is that the time to reach steady state may be very long. Imberger *et al.* (1976) predict that the characteristic time T_δ of flow establishment again depends on the parameter R, such that

$$R \geq 1; \ T_\delta = N_\delta^{-1} F_2^{-1/2} \tag{6.89}$$

$$R < 1; \ T_\delta = N_\delta^{-1} \, \mathrm{Gr}^{1/6}. \tag{6.90}$$

If these times are much longer than the times over which the outflow is held steady then numerical or physical model studies must be used to estimate the outflow characteristics.

Lastly, the use of Eqs. (6.87) and (6.88) must be modified on days where the erosion due to wind mixing or penetrative convection has sharpened the upper edge of the thermocline so that a severe temperature step exist there. This leads to a density structure where the gradient in the hypolimnion is very much weaker than that at the thermocline and it may be assumed that the water behaves more like the two layer system analyzed by Wood and Binney (1976).

Let us consider this case in somewhat more detail. Suppose that the density structure in a lake at a particular time is as shown in Fig. 6.15a. There is now some ambiguity as to what is the equivalent uniform density of the hypolimnetic water appropriate for two layer calculations. We shall define a density jump $\alpha^2 \Delta^1 \rho$, where $\Delta^1 \rho$ is the difference in the density between the water at the offtake and that at the top of the hypolimnion. The coefficient α^2 is unknown. However, if the profile in the hypolimnion were linear and if we required equal work to be done against gravity for a particle in a two layer system as in the linearly stratified case then α^2 would be equal to 0.5.

The cone of depression of the thermocline associated with a drawdown will have a radius of the order of d, the depth of the offtake below the thermocline. Hence, since essentially all reservoirs have a width greater than d, we must model the flow as a point offtake and not an equivalent two-dimensional line slot as was required for the hypolimnion withdrawal layers.

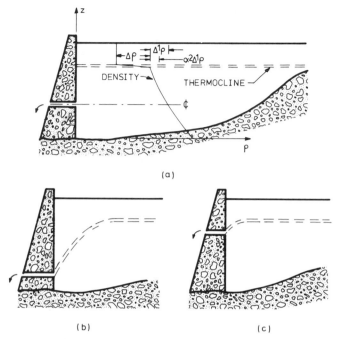

Figure 6.15 Thermocline drawdown configurations. (a) No drawdown. (b) Drawdown into a deep outlet; motion may be idealised by flow into a corner sink. (c) Drawdown into a surface outlet; motion may be modeled as flow into a plane wall.

For the radial outflow two extreme cases may be defined. First, as shown in Fig. 6.15b, the offtake is close to the bottom of the lake and second, the offtake may be very much closer to the thermocline (Fig. 6.15c). Wood and Binney (1976) defined a critical Froude number at which the upper layer begins to be drawn down into the outlet. Using the equivalent density difference $(\Delta\rho + \alpha^2\Delta^1\rho)$, their result becomes

$$F_{3C} = \frac{Q_{3C}}{\left[\left(\dfrac{\Delta\rho + \alpha^2\Delta^1\rho}{\rho_0}\right)gd^5\right]^{1/2}} = 1.02, \qquad d \approx (H - h) \qquad (6.91)$$

$$F_{3C} = 2.04, \qquad d \ll (H - h), \qquad\qquad\qquad (6.92)$$

where Q_{3C} is the critical discharge at which drawdown into the outlet is initiated.

On the other hand, Eq. (6.87) gives the thickness of the withdrawal layer close to the sink for all values of R, and may be used to estimate a discharge at which such an inertial-buoyancy layer reaches the thermocline:

$$Q_3|_{\delta=d} = BQ_2|_{\delta=d} = \frac{1}{4.0} Bd^2N_\delta. \qquad\qquad (6.93)$$

Using $N_\delta^2 \approx g\Delta^1\rho/\rho_0 d$ and some algebraic manipulation leads to a comparison of Eqs. (6.91) and (6.93):

$$\frac{Q_{3C}}{Q_3|_{\delta=d}} = 4.00\varepsilon F_{3C} \frac{d}{B} \left(1 + \frac{\Delta\rho}{\alpha^2\Delta^1\rho}\right)^{1/2}. \qquad (6.94)$$

If $Q_{3C} > Q_3|_{\delta=d}$, then the layer will thicken with increasing Q until it reaches the thermocline, after which its growth will be confined by the stability of the thermocline. At $Q = Q_{3C}$ the water from the epilimnion will be drawn into the outlet and the ratio of epilimnion water to hypolimnion water Q_E/Q_H will be approximately given by the analysis of Wood and Binney (1976). Their result is reproduced in Fig. 6.16.

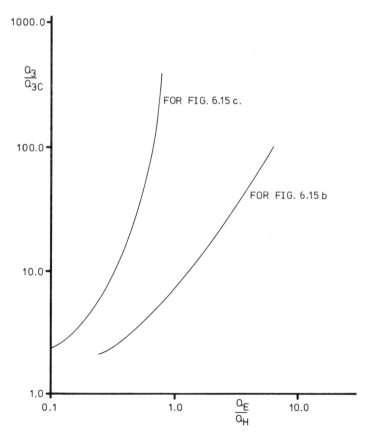

Figure 6.16   Ratio of outflow discharge Q_3 to the critical drawdown discharge Q_{3C} against the ratio of epilimnion to hypolimnion rates of withdrawal Q_E/Q_H. The two curves represent the two extreme geometries sketched in Figs. 6.15a and 6.15b. [After Wood and Binney (1976).]

On the other hand if $Q_3|_{\delta=d} \gg Q_{3C}$, then as Q_3 increases to Q_{3C}, the epilimnion water will be drawn into the offtake before the two-dimensional withdrawal layer has thickened to the thermocline. The flow may be likened to one induced by a sink distributed between the actual offtake and the thermocline, but no research results appear to exist for confirmation of this hypothesis. Once $Q_3 > Q_3|_{\delta=d}$, the details of the withdrawal layer become increasingly irrelevant and Fig. 6.16 will provide the most appropriate design guide line for estimating the quality of the outflowing water.

Example 6.8.   Figure 6.17 shows the average density structure from measurements taken on December 13 and 14, 1975 at the Wellington Dam. The outflow was constant at 4.5 m³/sec over these days. The width B is approximately 500 m, the length $L = 1.8 \times 10^4$ m, and the outlet is located on the side of the dam wall, 15.3 m above the base of the dam wall. Also shown in Fig. 6.17 is the vertical diffusivity of heat as computed for these days from Eq. (6.76). Investigate the withdrawal layer structure. Repeat calculations for a discharge of 20.2 m³/sec. Assume $\alpha^2 = 0.5$.

Solution.   As a first estimate let us assume a three segment density profile as shown by the dashed lines in Fig. 6.17.

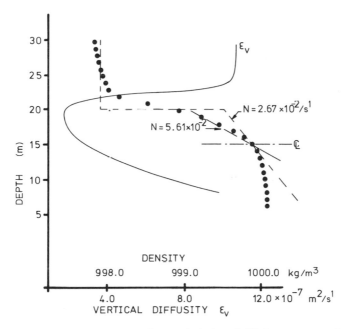

Figure 6.17 Example of a withdrawal layer calculation. Solid dots represent the measured density structure and the solid line shows the variation of the effective vertical mixing coefficient of heat. Two idealizations of the density structure are shown.

The buoyancy frequency then becomes

$$N_0 = \left(\frac{g}{\rho_0}\frac{d\rho}{dz}\right)^{1/2} = \left(\frac{9.8}{997}\frac{1}{13.8}\right)^{1/2} = 2.67 \times 10^{-2}/\text{sec}.$$

The average vertical eddy diffusivity for momentum

$$\varepsilon_{v,\delta} = \varepsilon_{T,\delta} \times 20 = 2.5 \times 10^{-7} \times 20 = 5 \times 10^{-6} \text{ m}^2/\text{sec}.$$

For the withdrawal layer calculations we require the parameters

$$F_2 = \frac{Q_2}{N_0 L^2} = \left(\frac{4.5}{500}\right)\frac{1}{2.67 \times 10^{-2} \times (1.8 \times 10^4)^2}$$

$$= 1.04 \times 10^{-9}$$

$$\text{Gr} = \frac{N_0^2 L^4}{\varepsilon_{\delta,v}^2} = \frac{(2.67 \times 10^{-2})^2 \times (1.8 \times 10^4)^4}{(5 \times 10^{-6})^2}$$

$$= 2.99 \times 10^{24},$$

$$R = F\,\text{Gr}^{1/3} = 1.04 \times 10^{-9} \times (2.99 \times 10^{24})^{1/3}$$
$$= 0.15.$$

This means that the withdrawal layer is governed by Eq. (6.88) with

$$\delta = 3.5L\,\text{Gr}^{-1/6} \quad (Z > 0)$$
$$= 3.5 \times (1.8 \times 10^4) \times (2.99 \times 10^{24})^{-1/6}$$
$$= 5.2 \text{ m}.$$

Hence, the layer just encroaches on the assumed two layer structure. The withdrawal layer thickness close to the outlet is

$$\delta = 2.0LF^{1/2} = 2.0(Q_2^{1/2}/N_0^{1/2}) = 1.2 \text{ m}.$$

We may repeat these calculations with a new value of $N = N_{\delta=5.2}$. We shall concentrate only on the layer thickness above the offtake. From Fig. 6.17 we see that $N_{5.2} = 5.61 \times 10^{-2}$, yielding a new value

$$\delta = 3.5L\,\text{Gr}^{-1/6}$$

$$= 3.5 \times 1.8 \times 10^4 \times \frac{(5.61 \times 10^{-2})^2 \times (1.8 \times 10^4)^{-16}}{(5 \times 10^{-6})^2}$$

$$= 4.1 \text{ m},$$

and so on. From Fig. 6.17, $(\Delta\rho + \alpha^2\Delta^1\rho) = 1.8$ kg/m^3 and $d = 5.0$ m, so that the critical drawdown discharge $Q_{3C} = 11.9$ m^3/sec. It must, however, be remembered that this discharge is for a plane wall. In the present example the offtake is located at one side of the dam wall, effectively restricting the flow to one-half of a plane wall flow, leading to a critical discharge $Q_{3C} = 6.0 \text{ m}^3/\text{sec}$. The flow $Q_3|_{\delta=d}$, required for the two-dimensional layer close to the offtake to possess a thickness of 5 m is given by Eq. (6.87) as 83.4 m³/sec, well above the critical drawdown value. Hence for a discharge of 4.5 m³/sec the withdrawal layer is completely contained by the stratification in the hypolimnion.

Raising the discharge to 20.2 m³/sec will cause the thermocline to be drawn down into the offtake even though the two-dimensional withdrawal layer will not have thickened to the base of the thermocline close to the offtake. No procedure appears to be available for the analysis of such a case, but an approximate answer may be obtained by assuming that the drawdown distributes the flow over the depth below the thermocline and the thermocline behavior is predicted by Wood and Binney (1976). From Fig. 6.16, with $Q_3/Q_{3C} = 3.4$, the ratio $Q_E/Q_H = 0.14$, so that 17.7 m³/sec flows from the hypolimnion into the offtake and the remaining 2.5 m³/sec comes from the epilimnion. The withdrawal layer in the hypnolimnion would have an approximate thickness of $d + \delta$.

■

6.7 MIXING OF INFLOWS

A river entering a reservoir will nearly always be at a different temperature, and thus density, than the surface water in the reservoir. Upon entering the reservoir it will thus push the stagnant lake water ahead of itself only until buoyancy forces, due to the density difference, have become sufficient to arrest the inflow. At that point the inflowing water will either flow over the surface of the lake if it is warmer or plunge and flow submerged down the old river channel if it is colder. In Fig. 6.18 we show how the inflow may be partitioned into three distinct mixing regimes.

First, there is mixing associated with the plunge line. Second, in the case of the underflowing situation the bottom roughness often leads to mixing, called entrainment, at the interface between the reservoir water and the inflow. Third,

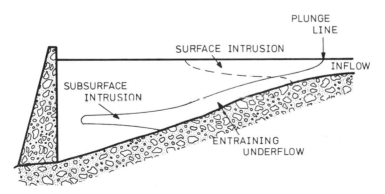

Figure 6.18 Possible river inflow patterns.

whenever the density of the inflowing water equals that opposite in the reservoir, then the inflowing water will leave the bottom and intrude horizontally into the reservoir. These intrusions may also occur along the surface if the density of the inflowing water is less than that of the surface water.

In Fig. 6.19 we show an example of a complete underflow in the Wellington Reservoir. Here the density differences were caused not by temperature differences, but rather variations in salinity. Particularly to be noted is the very well defined plunge line and the retention of the high salinity water in the depression of the river valley at the 10 km mark. Both indicate that the entrainment process proceeded at very low turbulence levels.

An example of a very different, more vigorous, underflow is given in Fig. 6.20 which shows a muddy river water intrusion into the epilimnion of the normally clear De Gray Lake. The inflow was only of short duration, so that the intruding slug of stream water quickly collapsed into a very thin tongue. In this case the river temperature dropped to about 19.0°C corresponding to the temperature in the lake at a depth of 8 m. Entrainment of warmer surface water at the plunge point however must have raised the overall temperature of the inflow since the intrusion actually entered the reservoir at a depth of only 5–6 m.

In both examples the reservoir was elongated with a small width so that the underflow was uniform across the cross section. However, there are many cases in which the river enters the lake over a very wide flat delta leading to a three dimensional under or overflow. For instance, a stream colder than the lake will enter the reservoir and immediately begin to spread laterally at the bottom. The plunge will then occur at a single point and not a line across the width, with the underflowing river water spreading in all forward directions. The shape of the incoming cold water body is similar to a symmetric spit protruding into the lake. Although this problem is of great importance little work has been carried out and essentially no definite recommendations may be made. The behavior of the overflowing hot water jet may provide some guidance, but the data are too sparse to be included here.

Lastly, we must again mention the effects of the earth's rotation. In Section 6.2 it was shown that internal waves with wavelengths that are short and vertical are strongly affected by Coriolis accelerations. Since inflows may be viewed analogously to such short internal waves, we may expect that for inflows into wide large lakes the inflows dynamics will be very strongly influenced by the earth's rotation and that the isotherms should be strongly tilted with the main inflow taking place on the right side of the reservoir in the northern hemisphere. A beautiful documentation is given by Serruya (1974) in Lake Kinneret. The river water was traced by its conductivity and the results are reproduced in Fig. 6.21.

Hamblin and Cormack (1978) have analyzed such flows and the reader should consult that paper when dealing with large nonchannelized lakes. In what follows we shall confine our attention to inflows into long elongated reservoirs with a narrow cross section.

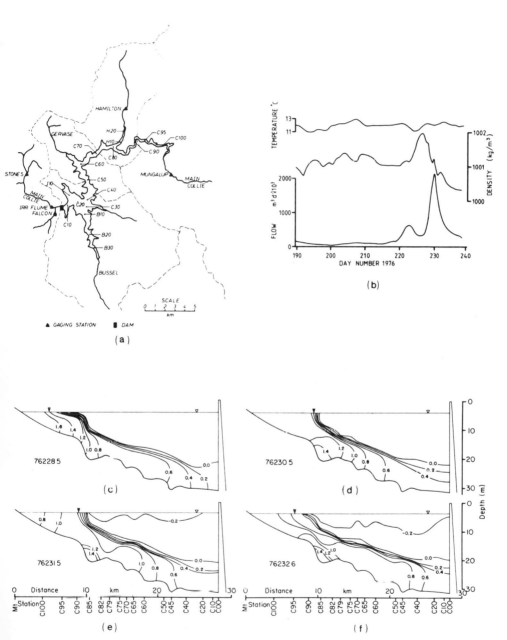

Figure 6.19 Inflow of the Collie River into the Wellington Reservoir. (a) Plan of reservoir showing the measuring stations. (b) Temperature, density and river discharge during the inflow flood. (a)–(f). Lines of constant density (kg/m³) along the central valley. The data illustrate the slow motion of the density surge down the drowned valley. [After Hebbert *et al.* (1979).]

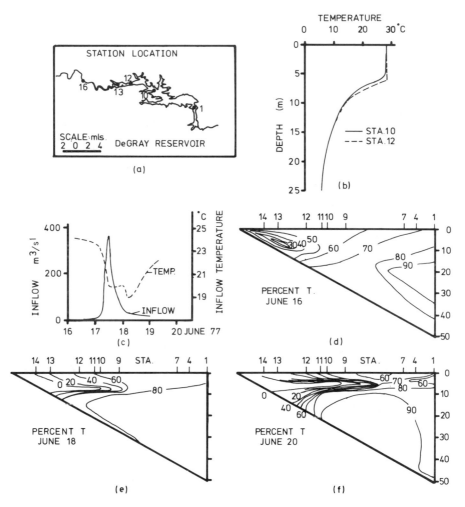

Figure 6.20 Muddy inflow into De Gray Lake during a storm on June 16, 1977. (a) Plan of De Gray Lake showing measuring stations. (b) Thermal structure at stations 10 and 12 just before the inflow. (c) Temperature and river discharge during flood event. (d)–(f). Lines of constant light transmittance defining the intruding water mass. The intrusion entered the reservoir just above the thermocline. [After Ford (1978).]

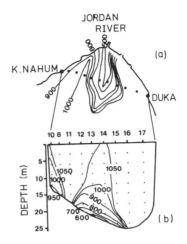

Figure 6.21 The mixing patterns of the Jordan River in Lake Kinnert on February 7, 1972 (μmho cm^{-1}). (a) The front of the Jordan flood progressing on the Kinnert bottom. (b) Kfar Nahum–Duka conductivity section. [After Serruya (1974).]

6.7.1 Entrainment and the Underflow

As shown in Fig. 6.19, when the heavy river water enters the reservoir it pushes the reservoir water ahead of itself, back into the reservoir, until at some distance the buoyancy forces due to the density difference at the interface between reservoir and river water become sufficient to stop the incoming water. At that station the river water plunges and flows underneath along the submerged river bed toward the dam wall. On its way it will entrain reservoir water by mechanisms which are identical to those described in Section 6.3.

For simplicity let us first consider a reservoir filled with water of uniform density ρ_0. The density of the inflowing water is designated by ρ_1. Furthermore, we shall assume that the slope of the river channel is uniform and the inflow is steady.

We shall analyze the downflow with an integral method similar to that given by Ellison and Turner (1959) for the two dimensional underflow and developed in detail for the jet problems in Chapter 9. For simplicity we shall assume both a uniform velocity u and a density ρ over cross-sectional area a of the under-flowing water.

Conservation of volume requires that

$$\frac{d}{dx}(ua) = EuT, \tag{6.95}$$

where E is the entrainment coefficient, T the top width of the underflowing water, and x is measured downstream along the river valley. In writing Eq. (6.95) we have made the assumption, discussed in Chapter 9, that the entrainment is proportional to the velocity u.

Since the reservoir is at a uniform density ρ_0, conservation of mass requires that

$$\Delta\rho au = \text{const.} \tag{6.96}$$

where $\Delta\rho = (\rho - \rho_0)$.

The remaining conservation law is that of momentum. This is most easily written down by imagining that the underflow is made up of water of density ρ_0 plus a small amount of added mass to bring it back to ρ. That part which has a density ρ_0 will be in static equilibrium and there will be no net pressure forces due to the density ρ_0. Assuming the pressure to be hydrostatic, conservation of momentum for the underflow is thus similar to the open channel formulation except that we now have an extra buoyancy force:

$$\frac{d}{dx}(au^2) = -C_D u^2 P - \frac{d}{dx}\{ag\Delta\rho(h - h_c)\cos\phi\} + \Delta\rho ga \sin\phi, \tag{6.97}$$

where C_D is the drag coefficient defined so that the stress on the perimeter is equal to $C_D \rho_0 u^2$, P the perimeter of the bottom boundary of the underflow, h the underflowing depth, h_c the depth of the centroid of the area a, and ϕ the slope of the valley.

In Section 6.3 we saw that the entrainment coefficient may be a function of the shear stress at the boundaries expressed by the drag coefficient C_D and the shear production parameterized by the internal Froude number F_i:

$$E = E(C_D, F_i), \tag{6.98}$$

where

$$F_i^2 = \rho_0 u^2 / \Delta\rho gd$$

and d is the hydraulic depth a/T of the underflow.

In general we must thus simultaneously solve Eqs. (6.95)–(6.98) in order to determine the dilution due to entrainment. However, Ellison and Turner (1959) noticed in the two-dimensional counterpart to our problem, that a special solution is given by

$$F_i = \text{const.} \tag{6.99}$$

The solution expressed by Eq. (6.99) is analogous to uniform flow in open channel hydraulics and it is possible to show that all underflows quickly asymptote to Eq. (6.99). Indeed, we shall now show that good results may be obtained by assuming that the flow is at all times locally uniform. In order to do this we must first determine the constant in Eq. (6.99).

Differentiating Eq. (6.99) and assuming the river section to be prismatic yields:

$$\frac{3}{u}\frac{du}{dx} + \frac{1}{T}\frac{dT}{dh}\frac{dh}{dx} = 0. \tag{6.100}$$

Equation (6.100) may be combined with the conservation of volume equation
(6.95) to determine the expression

$$\frac{dh}{dx} = \frac{-E}{\left(\dfrac{a}{3T^2}\dfrac{dT}{dh} - 1\right)}. \tag{6.101}$$

Substituting Eq. (6.101) back into Eq. (6.100) yields the rate of change of the
velocity u:

$$\frac{du}{dx} = u\left(\frac{1}{3T}\frac{dT}{dh}\right)\frac{E}{\left(\dfrac{a}{3T^2}\dfrac{dT}{dh} - 1\right)}. \tag{6.102}$$

The constraint on F_i, so that it is a solution, is now obtained by substituting the
expressions (6.101) and (6.102) into the momentum equation (6.97):

$$F_i{}^2 = \frac{\dfrac{TE}{P} + \left(\dfrac{(h-h_c)T^2E}{aP} + \dfrac{T}{P}\tan\phi\right)\left(\dfrac{a}{3T^2}\dfrac{dT}{dh} - 1\right)}{C_D\left(\dfrac{a}{3T^2}\dfrac{dT}{dh} - 1\right) + E\left(\dfrac{2}{3}\dfrac{a}{TP}\dfrac{dT}{dh} - 1\right)}. \tag{6.103}$$

Most river channels may be represented by a triangular section with a base
angle of 2α. For such a section we have

$$F_i{}^2 = \frac{2u^2\rho_0}{\Delta\rho gh} \tag{6.104}$$

and

$$F_i{}^2 = \frac{5\tan\phi - 8/3\,E}{4E + 5C_D/\sin\alpha}. \tag{6.105}$$

Equation (6.103) or Eq. (6.105), for a triangular channel, yields an implicit
equation for F_i which must be solved in conjunction with an entrainment law
[Eq. (6.98)] that is usually based on some type of entrainment model such as
Eq. (6.49).

Once F_i and E are known then we can integrate Eq. (6.101) to obtain the
relationship

$$h = \frac{E}{\left(1 - \dfrac{a}{3T^2}\dfrac{dT}{dh}\right)}x - h_0, \tag{6.106}$$

where h_0 is the initial flow depth and the denominator is assumed to be constant.

The dilution $\Delta Q/Q$ due to entrainment follows from Eq. (6.95) with $Q = ua$:

$$\frac{1}{Q}\frac{dQ}{dh} = \frac{T}{a} - \frac{1}{3T}\frac{dT}{dh}. \tag{6.107}$$

For the triangular cross section this may be simply integrated to yield:

$$\Delta Q = Q\left\{\left(\frac{h}{h_0}\right)^{5/3} - 1\right\}. \tag{6.108}$$

The dilution of the underflow in a stratified lake may also be computed using Eq. (6.108) provided account is taken of the variable density. The lake is divided into horizontal slabs in which the density is assumed to be uniform. Equations (6.99)–(6.108) are then applied within each slab using the density of the particular slab as ρ_0. Equation (6.108) then yields the approximate entrainment as the underflow moves through one slab. Transition between slabs is described by ensuring that Q and F_i are continuous. This procedure will lead to a step change in the depth h, but this will be quickly smoothed out provided the distance the flow requires to reach the normal state is small compared to the slab length.

The only problem remaining is the determination of h_0; this we shall take up in the next section.

Generalizations to unsteady flows or gradually varied flows do not require any additional fundamentals provided the variations are not so severe as to invalidate the assumed entrainment Eq. (6.98). Just as in open channel hydraulics integration of the conservation equations is then best done with the aid of a computer.

Example 6.9. Suppose a river is flowing into a reservoir that has essentially a two layer structure. Assume the epilimnion water is at 25°C and the hypolimnion water is at 20°C and the thermocline is located at 10 m depth. If the bed slope is 10^{-3}, the length of submerged river bed is 30 km, the bottom drag coefficient $C_D = 0.016$, and the initial depth $h_0 = 3$ m, calculate the entrained volume and the temperature by the time the water reaches the base of the dam. You are given that the initial river temperature is 5°C, the flow rate is 1.5 m³/sec, and the river valley can be approximated by a triangular shape with a base angle of 156°.

Solution. From the definition of C_D we have

$$u^* = C_D^{1/2}u. \tag{6.109}$$

Since the slope is very small we can expect that shear production is negligible so that Eq. (6.49) may be applied. Setting $u_f = 0$ in that equation yields the entrainment law:

$$E = \eta^3 C_k{}^f C_D^{3/2} F_i{}^2/2. \tag{6.110}$$

It is tempting to take the value of 0.23 for $\eta^3 C_k{}^f$ as found in the experiments by Wu (1973) (see Section 6.3.2), but experiments by Hebbert *et al.* (1979) and Elder and Wunderlich (1972), suggest a considerably higher efficiency of 3.2 for mixing in a downflow. Rather careful field experiments must in general be conducted therefore to estimate the appropriate value of $\eta^3 C_k$. Equally the contribution due to E in Eq. (6.105) may be expected to be small so that we can write

$$F_i^2 = \frac{\sin \alpha \tan \phi}{C_D} \left(\frac{1 - \frac{8}{15}E/\tan \phi}{1 + 4E \sin \phi/5C_D} \right)$$

$$\approx \frac{\sin \alpha \tan \phi}{C_D} \left(1 + \frac{8}{15} \frac{E}{\tan \phi} \right).$$

Substituting $F_i^2 = (\sin \alpha \tan \phi)/C_D$ into the right-hand side yields a consistent first approximation

$$F_i^2 = \frac{\sin \alpha \tan \phi}{C_D} (1 - 0.85 C_D^{1/2} \sin \alpha), \qquad (6.111)$$

where we have used $\eta^3 C_k = 3.2$ as suggested by field experiments.

Substituting for C_D, α, and ϕ yields $F_i = 0.23$ and a value of $E = 1.8 \times 10^{-4}$. For a triangular section, Eq. (6.106) reduced to

$$h = \tfrac{6}{5}Ex + h_0. \qquad (6.112)$$

For the downflow to the thermocline

$$h = \tfrac{6}{5} \times 1.8 \times 10^{-4} \times 10,000 + 3$$
$$= 5.16 \text{ m}$$
$$\Delta Q = 1.5[(5.2/3)^{5/3} - 1] = 2.25 \text{ m}^3/\text{sec}.$$

Hence the flow at the thermocline becomes 3.75 m³/sec. The temperature will be equal to that obtained by mixing the entrained water with the inflow. A volumetric mixing leads to a temperature

$$\frac{1}{3.75} (5 \times 1.5 + 25 \times 2.25) = 17°C.$$

The transition across the thermocline is accomplished by ensuring that Q and F_i are continuous. This is achieved if $(\Delta \rho h^5)$ is continuous. Hence the depth after the flow has passed underneath the thermocline will be equal to $(\tfrac{8}{3})^{1/5} \times 5.16 = 6.3$ m.

The flow under the hypolimnion is described in the same way. The final depth equals $\tfrac{6}{5} \times 1.8 \times 10^{-4} \times 20,000 + 6.3 = 10.6$ m and the final flow equals

$3.75[(10.6/6.3)^{5/3} - 1] + 3.75 = 8.97$ m^3/sec. The corresponding temperature will be

$$\frac{17 \times 3.75 + 20 \times 5.22}{8.97} = 18.8°C. \quad \blacksquare$$

6.7.2 Plunge Line Location and Mixing

At very low valley slopes, and thus small normal Froude numbers, the mixing at the plunge line is very small and we may obtain an approximate estimate for the depth at the plunge point by equating the normal internal Froude number F_{in} to the value based on the river inflow properties F_{io}. Specifically, if Q_0 is the flow in the river, $\Delta\rho_0$ is the density anomaly between the river water and reservoir surface water and the river cross-sectional area

$$a_0 = h_0{}^2 \tan \alpha, \tag{6.113}$$

then

$$h_0 = \left\{ \frac{2Q_0{}^2}{F_{in}^2 \Delta\rho_0 \tan^2 \alpha} \right\}^{1/5} \tag{6.114}$$

so that the location of the plunge point is where h_0 is equal to the depth of the reservoir. The excellent predictions obtained by this simple method are indicated in Fig. 6.19 where the plunge line position, as predicted by Eq. (6.114), is shown. For a more accurate analysis which takes into account the momentum balance at the plunge line the reader is referred to Imberger et al. (1978).

Figure 6.20 clearly indicates that as F_i increases, mixing at the plunge point becomes noticeable. The data from this underflow yields an entrance F_{io}^2 of approximately 0.4 and a total entrainment nearly equal to the flow itself. Unfortunately, little or no detailed work appears to have focused on this rather important problem, and we must recommend caution when applying Eq. (6.114) to flows with high entrance Froude numbers. Indeed, when h_0 becomes comparable to the total depth of the reservoir, then we may expect the whole character of the reservoir to change into one of violent mixing.

6.7.3 Inflow Intrusions

Figure 6.20 shows the case where the reservoir is stratified and the entrainment of water from the upper layers renders the mixed inflow water lighter than that in the deep parts of the hypolimnion. In such cases the flow will leave the river channel at some depth and penetrate horizontally into the reservoir.

Intrusion into a linearly stratified hypolimnion was analyzed by Imberger et al. (1976) by techniques similar to those established for the withdrawal problem. Again we assume that by the time the flow leaves the river channel and

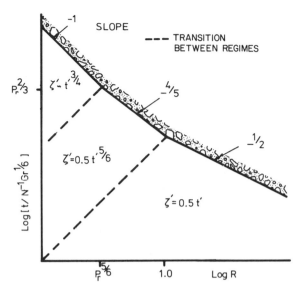

Figure 6.22 Regimes of penetration of density intrusions. The solid lines mark the time scale for the intrusion to reach the end wall. $R = F_2 G_r^{1/3}$, $\zeta' = \zeta/LR^{3/2}$ and $t' = t/N^{-1}G_r^{1/6}R$.

enters the stratified water column all turbulence has collapsed and the entering water assumes properties similar to that of the host water.

Under these circumstances we can form the dimensional parameter $R = F_2 \text{Gr}^{1/3}$, where F_2 is based on the inflowing discharge after entrainment and all the other variables have the same definition as in the withdrawal problem. The value of the parameter R determines the intrusion properties and we may write the following formulas for the intrusion length ζ,

$$t' \leq R, \qquad \zeta = 0.44LR^{1/2}t', \qquad (6.115)$$
$$R' < t' < \text{Pr}^{2/3}, \qquad \zeta = 0.57LR^{2/3}t'^{5/6}, \qquad (6.116)$$
$$\text{Pr}^{2/3} < t' < R^{-1}, \qquad \zeta = CLR^{3/4}t'^{3/4}, \qquad (6.117)$$

where the coefficients in Eqs. (6.115) and (6.116) were determined from experimental data and the nondimensional time $t' = t/N^{-1}\text{Gr}^{1/6}R$. The coefficient C in Eq. (6.117) is still unknown as there appears to be no experimental data for this regime. Figure 6.22 shows the transition zones including a boundary of when the intrusion will reach the end wall for a steady intrusion of a given initial value of R.

In an actual reservoir, the inflow is seldom steady for times comparable to the time the slug takes to reach the end wall, nor is the stratification linear. A particular case is the flow depicted in Fig. 6.20 which is "riding" on top of the thermocline and entering the unstratified epilimnion. Much still needs to be done to account for such flows.

6.8 USES OF A NUMERICAL MODEL: AN EXAMPLE

The Wellington Reservoir, situated approximately 160 km (100 miles) south of Perth in Western Australia (Fig. 6.1), is a medium-sized storage reservoir with a capacity of 185×10^6 m^3 (150×10^3 acre-feet) which is used to regulate the Collie River and to supply both the South Western Coastal Plains Irrigation System during the summer period and the small domestic requirement of several nearby inland towns throughout the year. In the past few decades inflow salinities to the reservoir have increased with a consequent deterioration in supply quality. The high variability of both the discharge and the salinity of the Collie River have compounded the problem. In wet years when the river flow is high, the average salinity of inflow reduces and the reservoir is flushed out so that the quality of the impounded water improves. Alternatively, a series of consecutive dry years causes a rapid deterioration of quality. We shall now describe the work of Patterson *et al.* (1978) who showed how a numerical model, based on the work outlined above, may be used to investigate management strategies based on inflow and withdrawal manipulations.

6.8.1   The Numerical Model

The numerical model DYRESM as developed by Imberger *et al.* (1978) is conceptually simple as illustrated in Fig. 6.23. It is based on the one-dimensional structure found so often in reservoirs and the scheme basically divides the body of the reservoir into horizontal slabs which retain their identity at all times. These Lagrangian slabs move vertically to accommodate volume changes which might occur below the level of the slab due to inflow or outflow. In this way no numerical diffusion is encountered.

The formation and deepening of the epilimnion dynamics is described by Eqs. (6.46) and (6.49). The mixing in the hypolimnion is modeled with a diffusion coefficient given by Eq. (6.76). The inflow dilution is assumed to be given by Eq. (6.108) and the thickness of the intrusion is calculated from intrusion length given by Eqs. (6.115) and (6.117). The withdrawal layer thickness is obtained from Eqs. (6.87) to (6.88) and the bookkeeping of the equations is carried out on a $\frac{1}{4}$ hour time step, as is the adjustment of the slab volumes necessary to conserve mass.

Spigel (1978) has replaced Eq. (6.49) and Eq. (6.50) and a simple parameterization of the momentum equation. This modified version of DYRESM was tested for the year 1975 for which the major characteristics are given in Fig. 6.2. The values of the coefficients used are shown in Table 6.8. Figure 6.24 shows a representative set of temperature and salinity profiles predicted and measured. One may conclude that the dynamical representation introduced by the fine vertical resolution and parameterization of the physical processes, made possible with

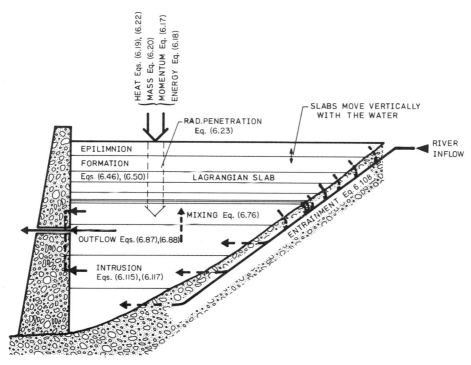

Figure 6.23 A schematic of the numerical model DYRESM. Slab volumes are kept between prescribed limits by slab partitioning and amalgamation.

TABLE 6.8
Values of the Coefficients Used in the Simulation Model DYRESM

Algorithm	Coefficient	Context	Value	Determination
Thermal heating	η	Eq. (6.23)	0.35	Thermal heating of bottom waters during summer heating
Mixing in the epilimnion	C_κ	Eq. (6.50)	0.13	Laboratory experimental evidence
	η		1.23	experimental evidence
	C_s		0.5	experimental evidence
	C_T		0.5	experimental evidence
Mixing in the hypolimnion	C	Eq. (6.76)	0.048	Reduction of high winter bottom salinities during spring
Inflow	C_D	Eq. (6.97)	0.016	Field experimental determination

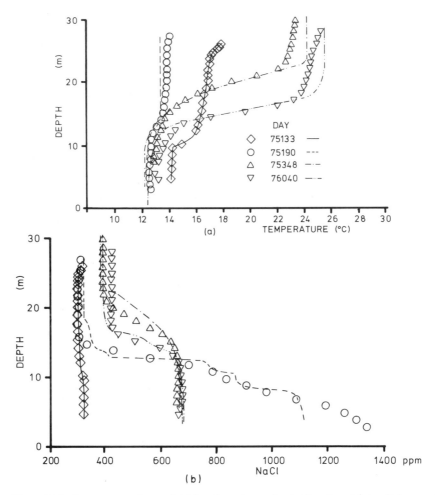

Figure 6.24 Comparison of mean field data (experimental points) and model simulation results (lines) using DYRESM with the coefficient values given in Table 6.8. (a) Temperature. (b) Salinity expressed as NaCL parts per million.

such a one dimensional scheme, far outweigh the added degrees of freedom gained with two or three dimensional schemes. To achieve the same order of physical representation with a three dimensional finite difference scheme is presently not possible.

6.8.2   An Example

The management strategies tested for the year 1975 include scouring of excess water from an offtake near the base of the reservoir wall, the use of base offtake supply for irrigation, and the diversion of the most saline inflow. As a result of these investigations, a number of general conclusions on aspects of reservoir management can be drawn, applicable to reservoirs with similar characteristics and problems.

The only controls available over the behavior of the reservoir are the release of water from the existing offtakes and the possible diversion of some of the inflow. Wellington Reservoir has two fixed level offtakes, a midlevel offtake located 15 m (49 ft) above the base of the wall, and a bottom offtake approximately 1 m (3.3 ft) above the base. The policies discussed below are based on the scouring of saline water to waste from the bottom offtake or the diversion of the most saline of the inflow to waste, prior to irrigation, followed by irrigation from either or both offtakes.

The quantities of water which can be scoured or diverted must, of course, be consistent with the projected future demand on the reservoir; the determination of these quantities is a complex statistical problem, depending in part on the future volume of inflow and the current storage volume. Long term monthly simulations indicate that there is considerable scope for the development of operating rules which minimize spillage and maximize the available water for scour or diversion. In this context it is sufficient to note that several techniques for determining such rules are available. For the 1975/76 cycle, it was determined that an additional 43×10^6 m^3 could be released without adversely affecting the reliability of supply. This additional release could be increased if required, but with some reduction in long term reliability.

A number of policies based on the release of 43×10^6 m^3 and two with larger releases have been tested, together with the diversion of the same amount of the most saline inflow. The variations of actual offtake quality with time from these policies and the existing operational policy are given in Figs. 25a (scour policies) and 25b (irrigation and diversion policies). The results of all policies are summarized in Table 6.9.

The scour and irrigation policies were designed to test the effects of firstly, the timing of the scour, secondly the rate of scour withdrawal, and thirdly, an increase in the scour volume either by increasing the rate or period of scour or the use of water from the bottom offtake for irrigation, possibly in mixture with midlevel offtake water. The results of these policies and of the diversion policy are compared with the no scour, midlevel offtake irrigation result, as the reservoir was actually managed in 1975/76; the percentage improvements listed in Table 6.9 are calculated on this basis.

The variation in midlevel offtake quality with time for the actual reservoir operation is shown for comparison in both Figs. 25a and 25b. A very rapid rise in offtake salinity is evident following the beginning of the winter inflows on day 75170 (Fig. 6.2). This was the result of the cold, salty underflow forcing the halocline above the offtake, which then drew exclusively from the lower, saltier regions with a consequently decrease in quality. A severe storm on day 75211 briefly depressed the halocline below the offtake and mixed some fresher

TABLE 6.9

Effect of Scour Policies in the Wellington Reservoir, Western Australia

Class	Policy	Scour rate (10⁶ m³/d)	Scour period (day)	Scour volume (10⁶ m³)	Average irrigation salinity (mg/liter TDS)	Percent improvement irrigation salinity	Final average storage salinity (mg/liter TDS)	Percent improvement storage salinity
First (top offtake irrigation)	No Scour	—	—	—	788	—	620	—
	Scour I	1.0	75170–75212	43	705	10.5	591	4.7
	Scour IIa	1.0	75210–75252	43	671	14.8	589	5.0
	Scour III	1.5	75190–75232	64.5	623	20.9	597	3.7
	Scour IV	1.5	75190–75252 75185–75214	94.5	598	24.1	597	3.7
	Scour V	1.0	75236–75242 75268–75273	43	681	13.6	575	7.3
	Scour VI	3.071	75210–75223	43	671	14.8	588	5.2
Second (bottom offtake irrigation	No Scour	—	—	—	859	−9.0	597	3.7
	Scour IIb	1.0	75210–75252	43	754	4.3	556	10.3
Third (inflow diversion)	Diverted inflow	—	—	43 (diverted)	515	34.6	440	29.0

water into the upper regions of the hypolimnion, improving the withdrawal quality slightly. This situation, with some minor excursions of the halocline as inflow peaks arrived, persisted until day 75265, when the final major inflow lifted the halocline well above the offtake. The subsequent irrigation water was taken exclusively from the hypolimnion until day 76035, when the draw down was sufficient to depress the halocline to the offtake level with a consequent decrease in offtake salinity. By day 76060, the withdrawal was taken exclusively from the epilimnion, and the salinity remained steady at around 575 mg/liter TDS for the remainder of the season. The average salinity of the reservoir rose from 378 mg/liter TDS to 620 mg/litre TDS during this period.

The first class of policy tested was based on a continuation of the practice of drawing the irrigation supply from the midlevel offtake through the hydro-electric plant, with trials to improve supply quality by scouring from the bottom offtake prior to irrigation. In this class, the first policy tested, Scour I, shown in Fig. 25a, scoured 1×10^6 m^3 daily from the bottom offtake over the period 75170–75212, coinciding with the first of the inflows. This policy delayed the rapid rise in salinity, reduced the salinity for the bulk of irrigation season with the average irrigation salinity falling from 788 mg/liter TDS to 705 mg/liter TDS, an improvement of 10.5%, and advanced the date of the depression of the halocline towards the end of the season. The average reservoir salinity at the end of the season improved by 4.7%.

Two factors were evident from the simulation of this scour: firstly, the salinity of the scoured water was consistently less than the salinity of the salt wedge, indicating that the scour was taking water from the fresher regions, and secondly, surface forcing events were penetrating the salt wedge and mixing it with the fresher water above, reducing the effectiveness of the scour. These effects were the result of initiating the scour before the salt wedge was fully inserted, thus preventing the development of a strong protective interface between the wedge and the fresher regions.

To confirm this, the scour was delayed until 75210, Scour IIa (Fig. 25a) at which time the inflow had reached the dam wall and a strong interface had built up. As expected, the scour was more efficient and the irrigation quality improvement became 14.5%.

When the scour rate was increased to 1.5×10 m^3/d over the period 75190–75232, Scour III (Fig. 25a), the average irrigation quality improved dramatically by 20.9%. This was further increased to 24.1% with an extension of the period to 75252 at the same rate (Scour IV, Fig. 25a). These scours, however, removed more than the recommended quantity, and the significant improvements were the result of scouring sufficient water to depress the halocline below the offtake for a significant part of the irrigation season, rather than a reduction in overall salinity, which improved in both cases by only 3.7%.

The results of Scours I and IIa indicated that it was desirable to delay the scour until the bulk of the first inflow had lodged in the base of the reservoir. To

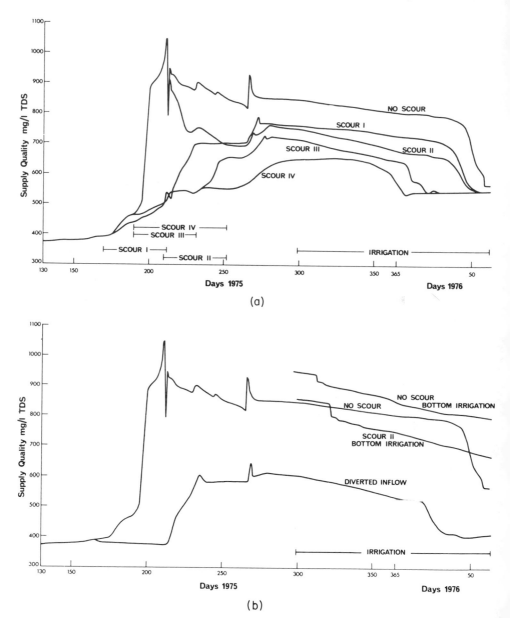

Figure 6.25   Water quality improvements resulting from different scour policies in the Wellington Reservoir, Western Australia. (a) Benefits from scour policies. (b) Benefits from irrigation policies. [After Patterson *et al.* (1978).]

determine if the subsequent timing or rate had any effect, two further scours were tested, Scour V, which scoured in three bursts after each of major inflow peaks had been fully inserted, and Scour VI, which scoured the same total quantity $(43 \times 10^6 \text{ m}^3)$ at a higher rate $(3.07 \times 10^6 \text{ m}^3/\text{d})$ for a shorter period. The offtake quality resulting from these policies was only marginally different during the irrigation period from the result of Scour IIa, and the results are not shown on Fig. 25a. Because of the more efficient removal of the saline slug by Scour V, however, the average reservoir salinity improvement rose to 7.3%. If the salt wedge is left in the base of the reservoir for extended periods, it will gradually be transported to the fresher regions by the internal mixing process. Thus the timing of the intervals of scour is important and should be initiated as soon as the inflow from each flood peak is fully inserted. The subsequent rate of scour however, has little effect.

It is clear from these results that the quality of the offtake supply may be improved in two ways:

(1)   By removal of some of the most saline of the inflow from the base of the reservoir before internal mixing distributes it over the remainder of the storage and raises the salinity of the region from which the supply is taken. Additionally, there will be some benefit with respect to the average reservoir salinity.

(2)   By manipulating the level of the halocline by controlled scouring in such a way that the irrigation draw from the midlevel offtake is taken from the epilimnion for the bulk of the irrigation season. Since the epilimnion is the best quality available, dramatic improvements in offtake quality can be expected. However, the effect on the average reservoir salinity will be only marginally beneficial, or even detrimental if the epilimnion salinity is significantly lower than the average.

Hence scouring from the bottom offtake followed by irrigation from the mid-level offtake may result in improvement to the average offtake quality but the effect on average reservoir salinity may be small.

The second class of policy tested involved the use of water from the bottom offtake for irrigation, at the expense of power generation, either solely, or if quality constraints so demand, in conjunction with water from the midlevel offtake. As such a policy amounts to an increase in the volume scoured to around $120 \times 10^6 \text{ m}^3$, an improvement in average reservoir salinity can be expected. However, as the irrigation supply is now taken from the lowest quality region, its salinity can be expected to increase.

Two trials were tested in this class. The first of these involved no scour but irrigation from the bottom offtake only. The resulting offtake quality is shown in Fig. 25b, commencing on the first day of the irrigation season. The actual mid-level offtake quality is also shown for comparison. The quality of supply was very much worse than that from the midlevel offtake, and the mixing occurring prior to the initiation of irrigation reduced the effectiveness of the removal of a large quantity of water from the base of the reservoir. Consequently, only a minimal improvement in average reservoir salinity was achieved.

An alternative which has not been tested would be to mix irrigation water from both offtakes to supply water with a particular salinity level. However, the result would lie between the qualities of Scours IIa and IIb.

The final class of policy tested was that of diverting some of the more saline inflow, by means of a barrier with a separate outlet, before it lodged in the body of the dam (Fig. 25b). Such a policy will have a major effect on the salinity distribution as little entrainment or mixing will occur, and the removed water will be of much higher salinity than any that could be scoured. The policy tested involved the removal of all inflow of salinity 1110 mg/liter TDS or greater, corresponding to a volume similar to that removed by the scour policies. The midlevel offtake quality improved markedly, the irrigation water being always below 600 mg/liter TDS. For this policy, the average salinity increased over the years by 70 mg/liter TDS to 440 mg/liter TDS, an improvement of approximately 30% on the actual value. A further improvement, with only a slight deterioration in offtake quality, would result were the irrigation to be taken from the bottom offtake.

Chapter 7

Mixing in Estuaries

7.1 INTRODUCTION AND CLASSIFICATION

An estuary is where a river meets the sea. Pritchard (1967) has given a more circumscribed definition that "an estuary is a semienclosed coastal body of water which has a free connection with the open sea and within which sea water is measureably diluted with fresh water derived from land drainage." Our purpose in this chapter, however, is to describe mixing in a class of water bodies that do not necessarily fit Pritchard's definition, or any other strict definition of an estuary. We are concerned, for example, with mixing in a tidal bay, like perhaps San Diego Bay, which has so little fresh water inflow that there is no measurable dilution. We are also concerned with mixing in the tidal portions of rivers upstream of the maximum extent of sea water intrusion. The studies we describe are even applicable, to some extent, to mixing on some continental shelves well outside what might be called a semienclosed body of water, but where the effect of river water is still evident. Thus our definition of the bodies of water we are concerned with must be an operational one. In Chapter 5 we discussed mixing in a flow driven by the slope of the water surface; in Chapter 6 we discussed mixing in flows driven primarily by wind stresses and by internal density variations. The flows discussed in this chapter are driven by all three; the primary flow is usually driven by the slope of the tidal wave, but wind stresses and internal density variations are often important. In addition we add a new complication, the flow oscillates. The result is the complex, unsteady and spatially varying flow which we customarily see in what are broadly referred to as "estuaries." Our purpose is to give an account of how the various aspects of this flow lead to different types of mixing, to synthesize the results as best we can, and then to describe some practical methods for studying the distribution of pollutants.

Part of the difficulty in describing estuaries is that the term covers such a diversity of sizes and shapes. A method for classifying estuaries into categories would be helpful, but no single scheme has been found sufficient. Bowden (1967a) and Pritchard (1967) distinguished three major hydrodynamic categories; sharply stratified estuaries such as fjords and salt-wedge estuaries; partially stratified estuaries, in which there is a significant vertical density gradient; and well mixed estuaries (see Fig. 7.1). A second method of classification is geomorphological: coastal-plain estuaries; fjord type estuaries; bar-built estuaries; and the rest. Fjord-type estuaries are generally formed by glacial action; they are usually very deep, narrow, and highly stratified. Coastal plain estuaries are generally formed by the gradual drowning of a river system, and are usually long and narrow with many branches. Bar-built estuaries are formed by the closing off of an embayment by a sand bar, and are generally found along coasts with large littoral drift. "The rest" includes such large closed bays as San

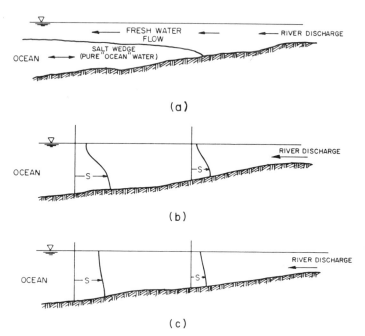

Figure 7.1   Salinity distributions along the axes of (a) a "salt wedge" estuary, (b) a "partially mixed" estuary, and (c) a well mixed estuary.

Francisco Bay, and estuaries formed by channels running across aluvial plains, such as the Columbia. A third possibility is to devise analytical methods of classification. Hansen and Rattray (1966) proposed a method based on the vertical variation of salinity and the strength of the internal density-driven circulation; some of their results are discussed in Section 7.2. A simple geometrical classification can be based on the ratios of length L, width W, and mean depth d. None of these schemes are able to express the unique characteristics of any estuary, however. Estuaries have individual personalities, made up of the distribution of sand bars, points of land, man made jetties and harbors, islands, deep channels, shallow bays, the characters of the tributary rivers, and the seasonal variation of the weather. Before doing a mixing study in an estuary it is best to get to know it.

This chapter begins with an enumeration of various causes of mixing in estuaries, and analytical treatments of them. We then discuss cross-sectional mixing and longitudinal dispersion in the same sequence as in the chapter on rivers, although with much less ability to give dependable results. We close with a discussion of the one-dimensional analysis as a practical tool. The discussion provides essential background to the descriptions of the use of numerical and physical estuary models in the next chapter.

7.2   THE CAUSES OF MIXING IN ESTUARIES

Mixing in estuaries results, as it does in rivers, from a combination of small-scale turbulent diffusion and a larger scale variation of the field of advective mean velocities. In rivers the combination is fairly simple, as explained in Chapter 5; the advective velocity field defines a set of approximately steady stream lines. The main role of turbulent diffusion is to transfer mass between stream lines, and longitudinal dispersion comes about mainly because the flow along different stream lines is going at different speeds. In estuaries we can also try to describe mixing in terms of advection by a mean flow along stream lines and turbulent diffusion between stream lines, but matters are nowhere near as simple as in rivers. The first problem is to differentiate diffusion from advection. If a current meter is held at a fixed point in an estuary and a long record is examined, spectral analysis can disclose fluctuations with a wide range of period. Fluctuations with a period of less than a few minutes can be identified as turbulence, and the transport resulting therefrom can be termed diffusive transport, just as we have done in rivers. The term "advection" can then be assigned to the remaining motion. The advective velocity is not constant, however, either in time, space, or direction. The velocity record obtained at a single point will contain semidiurnal and diurnal tidal variations, wind-induced variations of almost any period, an inertial frequency caused by the earth's rotation, and fluctuations of longer periods caused by the monthly and longer term variation of the tidal cycle and by seasonal variations of meteorological influences and tributary inflows. The direction of the velocity vector will often not be parallel to the channel axis, even if one can be defined. Often the flow is going in different directions at different depths; often the flow is one way near the shore and the opposite way in the center of the channel. Obviously, the analysis of mixing in terms of the interaction of advection and diffusion is much more complicated in estuaries than in rivers.

The proper way to begin seems to be to make things as simple as possible by considering different mechanisms in turn. Most of what is seen in an estuary can be related to one of three sources, the wind, the tide, and the river. Most of the analyses to be found in the engineering technical literature discuss the effect of only one or at most two sources, for example the current driven by the wind in a tideless bay or the circulation driven by the river inflow in a tideless estuary. Taking the literature as a guide, the next three sections discuss in turn the isolated effects of wind, tide, and river. In each case we will discuss qualitatively why and how the source causes mixing, and will quote what analytical results can be found. Later, in Section 7.4, we attempt an analytical synthesis based on a decomposition of the salinity and velocity profiles.

7.2.1 Mixing Caused by the Wind

Wind is usually the dominant source of energy in large lakes, the open ocean, and some coastal areas, but in estuaries it may or may not play a major role. Breaking waves, the most apparent result of wind, have little to do with large scale dispersion. In a long, narrow estuary the flow may be predominantly tidal, and the wind has little chance to generate much current. On the other hand, if the estuary is wide, or consists of a series of bays, wind stresses can generate currents of considerable importance. The wind exerts a drag on the water surface, and will pull floating objects, or even floating pools of oil or warm water, in the wind direction. Thus the dispersion of an oil spill is directly affected by the local wind. On the other hand, if a dissolved substance is distributed throughout the water column and vertical mixing is vigorous what matters is the mean motion of the water column. Thus the effect of wind depends primarily on the currents induced.

Suppose that a uniform wind blows over a wide, shallow basin containing water of constant density, and that the basin is deeper on one side than on the other. The resulting current is demonstrated by the laboratory photograph shown in Fig. 7.2; on the shallow side the current flows with the wind, and on the deep side it flows against the wind. By "current" we mean the vertically averaged mean flow; there is also a vertical velocity profile, not apparent in the photograph, such that the surface velocity is somewhat more in the direction

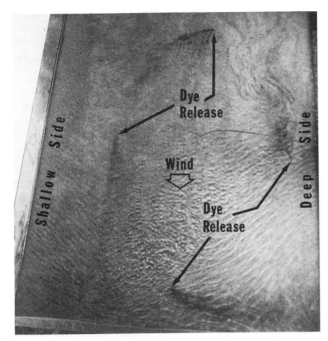

Figure 7.2 A rotational current caused by a uniform wind blowing over a basin of variable depth. The wind is blowing from the top of the picture towards the bottom at a velocity of 4.5 m/sec. The basin is 0.3 cm deep on the left side, sloping uniformly to 5 cm deep on the right side. Four dye plumes can be seen, each released from near the bottom. The plumes show that the flow is in the form of one nearly circular rotor extending throughout the basin. [Photo by R. Spigel, from Fischer (1976a).]

of the wind than the mean flow. The vertical velocity profile causes dispersion around the position of a particle traveling with the mean velocity, for the reasons explained in Chapter 4, and the circulatory current may be viewed as a larger scale mixing mechanism which will be additive to any other mixing mechanisms present because of other sources. For example, if an estuary has a deep channel alongside a shallow embayment, as illustrated in Fig. 7.3, the wind-driven steady circulation in the shallow bay will interact with the tidal current in the channel, and a complete analysis of mixing will have to account for both causes.

The simplest explanation for the current illustrated in Fig. 7.2 is as follows. The wind induces an approximately uniform stress everywhere on the water surface (estimates of the magnitude of the wind stress are given in Section 6.2.1). Therefore the line of action of the wind-induced force is through the centroid of the water surface. The center of mass of the water in the basin is displaced towards the deeper side, since there is more water there. Hence the line of action

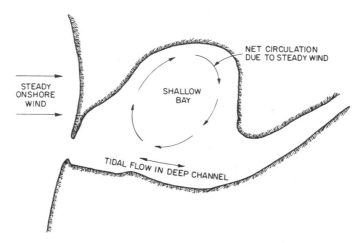

Figure 7.3   A sketch of a steady wind-driven circulation superposed onto the tidal flow in an estuary.

of the force passes on the shallow side of the center of mass of the water, and a torque is induced causing the water mass to rotate.

A more detailed prediction of wind-induced currents, and an examination of transient flows when the wind velocity changes, requires a solution of the equation of motion for the water. Usually the equations are averaged over the vertical, and the resulting depth-integrated two-dimensional equation of motion is solved numerically using any of a number of computer programs. A particularly well documented method is that of Leendertse (1967); some computations of wind-driven currents are illustrated in Fig. 8.1.

7.2.2   Mixing Caused by the Tide

The tide generates mixing in two ways. Friction of the tidal flow running over the channel bottom generates turbulence and leads to turbulent mixing, and the interaction of the tidal wave with the bathymetry generates larger scale currents. Efforts to quantify the rate of turbulent mixing are discussed in Section 7.3; here we discuss the effects of the larger scale currents. These include shear flow dispersion similar to that found in rivers, and in addition other circulations which we will classify by the terms "pumping" and "trapping" and discuss in detail below.

7.2.2.1   *The Shear Effect in Estuaries and Tidal Rivers*

The most obvious characteristic of tidal flow in most estuaries is that the flow is like a river, but goes back and forth. In Chapter 5 we showed how to apply shear flow dispersion theory to rivers, and in Section 4.3 we showed analytically the effect of oscillation on the longitudinal dispersion coefficient. Equations (4.55) and (4.56) combine to give

$$K = K_0 f(T'),\qquad(7.1)$$

where $f(T')$ is plotted in Fig. 4.7. $T' = T/T_c$ is the dimensionless time scale for cross-sectional mixing, T is the tidal period and T_c the cross-sectional mixing time. K_0 is the dispersion coefficient if the tidal period is much longer than T_c; if the cross section is relatively wide and shallow and density effects are absent we can make use of the result given for rivers in Eq. (5.17) that $K_0 = I\overline{u'^2}T_c$, in which $T_c = W^2/\varepsilon_t$ is the time scale for transverse mixing, and I is a coefficient whose value we found in Chapter 4 was generally approximately equal to 0.1. This result combines with Eq. (7.1) to give a prediction for the longitudinal dispersion coefficient in an estuary due to shear flow as

$$K = 0.1\overline{u'^2}T[(1/T')f(T')].\qquad(7.2)$$

The function $[(1/T')f(T')]$ is plotted in Fig. 7.4. It has a maximum of approximately 0.8 when T' is approximately one, and shows that the shear flow dispersion coefficient will be small if the estuary is very wide (T' small) or very narrow (T' large). Shear flow dispersion will have its maximum effect if the tidal period is similar to the time required for cross sectional mixing; even in that case Eq. (7.2) puts a limit on the magnitude of the coefficient. For example, if the tidal period is 12.5 hr, the tide has a mean velocity of 0.3 m²/sec, and if $\overline{u'^2} = 0.2\bar{u}^2$, as assumed in Chapter 5, the maximum possible value of the dispersion coefficient is approximately 60 m²/sec.

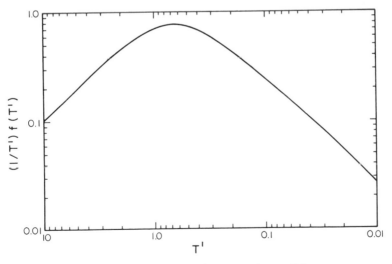

Figure 7.4   The quantity $T'^{-1}f(T')$ used in Eq. (7.2).

The limitations on the derivation of Eq. (7.2) should be recalled. The channel must be relatively uniform over a reach much longer than the channel is wide, must be much wider than it is deep, and the water must be of uniform density. Equation (4.55) was derived using a linear velocity profile with no phase shift across the channel. In real estuaries the flow often reverses near the side before it does in the deeper portions, and of course the velocity profile is never linear. Numerical studies have shown, however, that the phase shift has little effect and that the shape of the velocity profile only changes the constant of proportionality. Thus Eq. (7.2) may sometimes give a useful first estimate of the dispersion coefficient in constant-density portions of an estuary or in a tidal river upstream of the limit of salinity intrusion. It should also be emphasized that Eq. (7.2) gives only the effect of shear flow; all the mechanisms described elsewhere in this chapter can and often do increase the effective value of the dispersion coefficient by a large amount. Equation (7.2) should only be used if shear flow appears to be the dominant mechanism for dispersion.

An interesting aspect of Eq. (7.2) is that it predicts much smaller dispersion coefficients in tidal flows than in similar steady river flows, for the reasons described in Section 4.3 and illustrated in Fig. 4.8. The result is supported by observations of dye dispersion in the Potomac River by Hetling and O'Connell (1966). In their experiment it was intended that dye would be released continuously for a period of 13 days at a fixed point. Fortuitously for the measurement of dispersion, the dye injection equipment malfunctioned and large slugs of dye were released over a short period on two separate days. The resulting dye distribution was monitored for 34 days and the dispersion of the dye slugs was measured. The reach in which the study was conducted was just upstream from the maximum limit of salinity intrusion and the width of the channel varied from about 600 m at the upper end to 1000 m at the lower end. Dispersion coefficients computed from the data varied from about 6 m^2/sec at the upper end to 20 m^2/sec at the lower end. These coefficients are much smaller than would have been observed in a steady flow of the same size; compare, for instance, the dispersion coefficient measured by Yotsukura *et al.* (1970) in a 200 m wide reach of the Missouri River of 1500 m^2/sec.

Example 7.1: Dispersion in a tidal slough.   A slough connected to the open ocean is approximately 20 km long and 100 m wide. There is no fresh water inflow so the salinity is that of the ocean throughout the slough. The tidal range is approximately 1 m and the mean depth of water at mean tide is approximately 3 m. The tidal period is 12.5 hr. Estimate the value of the longitudinal dispersion coefficient near the ocean end.

Solution.   Since the tidal range is one-third of the mean depth, the tidal excursion will be approximately one-third of the length to the end of the slough, or approximately 7 km at the ocean end. The mean tidal velocity is $\bar{u} = 7$ km/6.25 hr or 0.31 m/sec. Since inadequate information is given about channel roughness or rates of transverse mixing we must make reasonable assumptions; assume that $u^* = 0.1\bar{u} = 0.031$ m/sec and $\varepsilon_t = 0.6du^* = 0.059$ m²/sec. Then $T_c = (100^2)/0.059 = 170,000$ sec and $T' = T/T_c = 45,000/170,000 = 0.27$. Using Fig. 7.4 an approximate estimate for the dispersion coefficient is

$$K = 0.1u'^2 T(0.4) \approx 0.1(0.2\bar{u}^2)(45,000)(0.4) = 35 \text{ m}^2/\text{sec} \quad \blacksquare$$

7.2.2.2   Tidal " Pumping"

A second important characteristic of most tidal flows, but one not at all obvious to the casual observer, is that superimposed on the back-and-forth flow is a net, steady circulation, often called the "residual circulation." The residual circulation is generally said to be the velocity field obtained by averaging the velocity at each point in the estuary over the tidal cycle. This definition has to be taken loosely, because no tidal cycle is identical to the one preceding or following it. We can conceptualize a residual velocity and discuss what causes it, and we can measure velocities over a 25 hr period, average, and claim to have measured the residual velocity, but we should be aware that the residual velocity field is itself a slowly varying, poorly defined quantity.

The concept of a residual velocity is well known to most navigators as well as researchers. For example, Fig. 7.5, a sketch by Bowden and Gilligan (1971), shows one flood and two ebb channels in the Mersey Estuary. "Flood," of course, means that the flood current is stronger than the ebb current, and "ebb" that the ebb current is stronger. In large estuaries one cause of the residual circulation is the earth's rotation, which deflects currents to the right in the Northern hemisphere and to the left in the Southern hemisphere. Therefore in the northern hemisphere flood tide currents are deflected towards the left bank (looking seaward) and ebb currents towards the right bank, resulting in a net counterclockwise circulation. As an example, this circulation is thought to explain why in Chesapeake Bay the salinity is on the average higher on the eastern shore (the left bank looking seaward) than on the western shore.

A second cause of residual circulation is interaction of the tidal flow with the irregular bathymetry found in most estuaries. Stommel and Farmer (1952) gave a simple analysis of one form of this circulation, as illustrated in Fig. 7.6. They envisaged an internally well mixed inlet of constant depth d with river inflow Q_f, mean salinity S, and a narrow mouth of width a. In such an inlet the flood flow enters as a confined jet, while the ebb flow comes from all around the mouth in the form of a potential flow to a sink; averaging over the tidal cycle yields an inward flow in the area of the jet and an outward flow elsewhere, i.e., a residual circulation.† Stommel and Farmer assumed that the jet would be confined to an area having the width of the mouth, and that the sink flow would come

---

† A large-scale example of this form of residual circulation is shown in Fig. 8.6a.

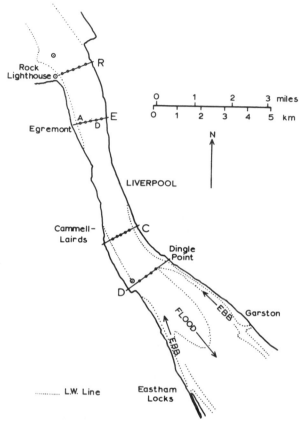

Figure 7.5 The narrows of the Mersey estuary, England. Dotted lines represent the low water line; arrows show the flood and ebb channels. [After Bowden and Gilligan (1971).]

uniformly from a semicircle. Setting the volume of ebb flow V equal to the volume of flood flow plus fresh water inflow gives

$$V = \tfrac{1}{2}\pi b^2 d = aLd + Q_f T,\tag{7.3}$$

where T is the duration of the tidal cycle and L and b are as defined in Fig. 7.6. Assuming that the ebb flow has salinity S and the flood flow ocean salinity S_0, and equating the transport of salt out to the transport in for steady state, gives

$$a(L - b)dS_0 = (\tfrac{1}{2}\pi b^2 d - abd)S.\tag{7.4}$$

Equations (7.3) and (7.4) combine to give the salinity in the inlet as

$$S/S_0 = 1 - Q_f T/(V - a\sqrt{2Vd/\pi}).\tag{7.5}$$

This result is of course limited to the conditions assumed; we have given its derivation to show how a simple residual circulation can be analyzed and to

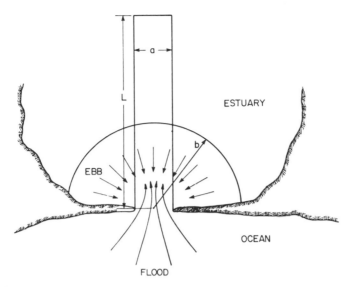

Figure 7.6   Stommel and Farmer's (1952) visualization of flood and ebb flows at an inlet.

introduce the concept of a "salt balance" which will be used later (see, for example, the derivation of Eq. (7.29) in Section 7.5.1, from which Eq. (7.5) can also be obtained).

A common example of a pumped circulation is the net flow found around islands or in braided channels. Prototype observations are few, because of the difficulty of averaging a tidal flow to determine the residual, but numerical models commonly compute net flows. For example, the net upstream flow in Montezuma Slough, a part of the Suisun Marsh, California (see Fig. 7.7), has been computed to be approximately 15 m^3/sec when the mean tidal elevation is the same at both ends. In that case the reason appears to be that the tidal curve at the upstream end lags the curve at the downstream end by approximately one hour. Even if the tidal curves are otherwise identical, the landward slope of water surface occurs when the tidal elevation is relatively high, and the seaward slope when the elevation is low. Therefore friction retards the seaward flow more than the landward flow, and the net flow is landward.

Combinations of channel geometry and separation at corners probably induce pumped gyres of various sorts in most large bays. Zimmerman (1978a) has shown analytically that an oscillatory tidal current flowing over an irregular bottom topography, such as a series of shoals, induces residual vortices. In the western Dutch Wadden Sea, a tidal area of about 130 km^2, Zimmerman (1978b) estimated that the magnitude of the longitudinal dispersion coefficient resulting from the residual vortices was approximately 800 m^2/sec. A similar mechanism may operate in South San Francisco Bay. Fig. 7.8 shows a counterclockwise

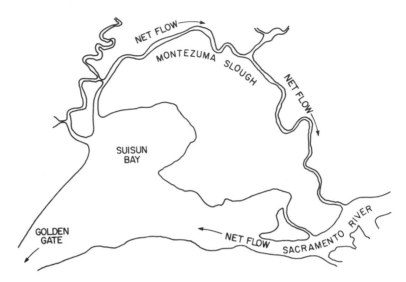

Figure 7.7 Suisun Bay and Montezuma Slough, part of the upper reaches of the San Francisco Bay estuary, showing the direction of the net flow in Montezuma Slough.

gyre predicted by a numerical model; however the currents shown have not actually been measured in the prototype, and whether the gyre exists in the real bay remains a subject of speculation.

It should be remembered that the residual circulation discussed in this section is additional to, and superimposed on, circulations driven by the wind

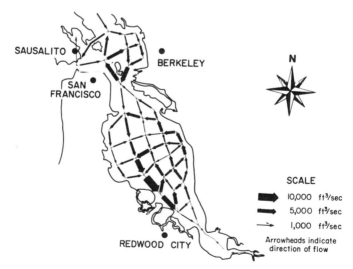

Figure 7.8 The distribution of residual currents in South San Francisco Bay as computed by a two-dimensional numerical program. [After Nelson and Lerseth (1972).]

and the river. We have referred to the circulation driven by the tide as tidal "pumping" to differentiate it from the others; as we have seen, the energy available in the tide is in part extracted to drive steady circulations whose net result is similar to what would happen if pumps and pipes were installed to move water about in circuits. This is an important mechanism for moving pollutants and transporting salinity upstream against a mean outflow of fresh water. We have not tried to quantify this mechanism in terms of a longitudinal dispersion coefficient, but in many estuaries it can be expected to be an important part of the flow distribution that produces longitudinal dispersion.

7.2.2.3   Tidal "Trapping"

"Trapping" is a term used by the writers to describe the effects of side embayments and small branching channels. In Section 5.2.2 we discussed the effect of "dead zones" in rivers; similar side embayments exist in estuaries, but their role is enhanced by tidal action in a way apparently first noticed by Schijf and Schonfeld (1953). These writers analyzed what they called a "storing basin" mechanism and concluded that it was responsible for all the diffusive salt flux in some Dutch estuaries. We prefer to call this mechanism "trapping," because it results from trapping of low velocity water along the sides of an estuary even if physical basins are not present.

The trapping mechanism can be explained as follows. The propagation of the tide in an estuary represents a balance between the inertia of the water mass, the pressure force due to the slope of the water surface (the shape of the tidal wave) and the retarding force of bottom friction. As a first example, consider the system shown in Fig. 7.9, a typical coastal plain estuary with one major channel and a number of side branches. In the main channel tidal elevations and velocities are usually not in phase; high water occurs before high slack tide and low water before low slack tide. This is because of the momentum of the flow in the main channel, which causes the current to continue to flow against an opposing pressure gradient. The side channel, in contrast, has less momentum, and the current direction changes when the water level begins to drop. Figure 7.9a shows a cloud of tracer particles being carried upstream by a flooding tide. Some of the particles go into the side channel and some continue upstream in the main channel (Fig. 7.9b). When the water surface begins to drop the particles in the side channel return into the main stream but now they are separated from their previous neighbors by unmarked water originally downstream of the whole cloud (Fig. 7.9c). The separation distance can be as much as the distance of travel in the main channel between high water and slack water, perhaps as much as a mile in a typical coastal plain estuary. Something resembling this effect probably occurs in almost all estuaries, and in many tidal rivers. In San Francisco Bay, for instance, the shoreline consists of irregularly shaped shallow basins, which probably play the same role as side channels. As another example, numerical studies of Jamaica Bay, New York, have shown that although in

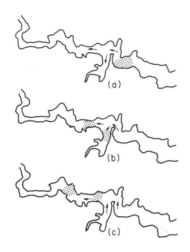

Figure 7.9 The phase effect in a branching channel. (a) A cloud of tracer being carried upstream on a flooding tide. (b) At high water some of the particles are trapped in the branch. (c) During the early stages of the receding tide the flow in the main channel is still upstream. The particles trapped in the branch reenter the main channel, but are separated from their previous neighbors.

overall plan the bay is almost round, trapping in the interior irregularities and inlets is an important dispersion mechanism.

Okubo (1973) has given an analysis which can be applied to the trapping mechanism. Okubo found that for a uniform velocity of flow in the main channel of velocity $u = u_0 \cos \sigma t$ and a uniform distribution of traps along the sides having a ratio of trap volume to channel volume of r, and a characteristic exchange time between traps and main flow of k^{-1}, the effective longitudinal diffusivity is given by

$$K = \frac{K'}{1 + r} + \frac{ru_0{}^2}{2k(1 + r)^2(1 + r + \sigma/k)}, \tag{7.6}$$

where K' is the longitudinal diffusivity in the main channel itself. Taking the Mersey as an example, $u_0 = 1.5$ m/sec and $\sigma = 1.4 \times 10^{-4}$ sec^{-1}; if we assume reasonable values of trap-volume ratio $r = 0.1$ and characteristic exchange time $k^{-1} = 10^4$ sec, $K = 0.9K' + 360$ m^2/sec. The second term is equal to the larger of two values of effective diffusivity reported by Bowden and Gilligan (1971). Thus it appears that the trapping mechanism alone can account for longitudinal dispersion in the Mersey.

7.2.3 Mixing Caused by the River

The river, or rivers if more than one enter the same estuary, delivers a discharge of fresh water Q_f. Analytical and laboratory studies usually assume that all the fresh water passing a given section comes from a single upstream source, and throughout this discussion we will do the same. The complications that arise when a number of rivers supply fresh water around the periphery do not change the qualitative description of how fresh water affects mixing, and we prefer to concentrate on what is known of the effect of a single source.

If a river discharges into an estuary connected to a nearly tideless sea, such as the Sea of Japan or the Mediterranean, the fresh water overrides the salt water and flows as a nearly undiluted layer into the sea. Salt water intrudes underneath the fresh water layer in the form of a wedge, as illustrated in Fig. 7.1. If there is some tide the wedge moves back and forth; the more the wedge motion the more kinetic energy is available to break down the interface and turbulently mix the fresh and saline layers. The river may be thought of as a source of deficit of potential energy, and the tide as a source of kinetic energy to overcome the deficit. More precisely, the river is a source of buoyancy, of amount $\Delta \rho g Q_f$, where $\Delta \rho$ is the difference in density between the river and ocean water. The dimensionless ratio

$$R = (\Delta\rho/\rho)gQ_f/WU_t^3,$$

where U_t is the rms tidal velocity and W the channel width, expresses the ratio of the input of buoyancy per unit width of channel to the mixing power available from the tide. R is a sort of Richardson Number; Fischer (1972a) called it the "Estuarine Richardson Number" in analogy with a "Pipe Richardson Number" defined by Ellison and Turner (1960). It is also equivalent to the ratio we used in Section 5.1.5 to express the likelihood that a buoyant discharge mixes vertically in a river flow. If R is very large we expect the estuary to be strongly stratified and the flow to be dominated by density currents. If R is very small we expect the estuary to be well mixed, and we might be able to neglect density effects. Observations of real estuaries suggest that, very approximately, the transition from a well mixed to a strongly stratified estuary occurs in the range $0.08 < R < 0.8$.

Figure 7.10 shows a vertical section along a typical partially stratified estuary. The isohalines (lines of constant salinity) slope upward towards the ocean. The natural tendency of isohalines is to become horizontal, because that is the condition of a stratified water body at rest; as illustrated in Fig. 6.4c sloping isohalines imply pressure gradients which, in the absence of other forces, will drive a current tending to bring the isohalines to the horizontal. In the case shown in the figure the necessary currents are a flow landward along the bottom and seaward at the surface. Such a flow is often referred to as the "classical estuarine circulation," or sometimes as "gravitational circulation." Internal flows driven by density variations are more properly called "baroclinic circulation," as distinguished from "barotropic circulation" which occurs in flows of constant density. The tide-driven flows discussed in the previous section are barotropic flows; the river-driven flows discussed here are baroclinic.

It should be noticed that if the estuary is perfectly well mixed vertically but has a horizontal density gradient the isohalines are vertical. Hence an internal baroclinic circulation is driven by a longitudinal density gradient even if the estuary is vertically well mixed.

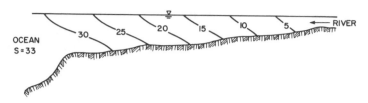

Figure 7.10   Isohalines in a typical partially stratified estuary.

Several attempts have been made to analyze density-driven circulation, but none are entirely satisfactory. Hansen and Rattray (1965, 1966) analyzed circulation in a vertical two-dimensional plane, assuming no variation across the channel. They visualized the steady flow illustrated in Fig. 7.11a, and assumed that the only tidal effect was to induce vertical and longitudinal turbulent mixing. Graphs were obtained showing the vertical distribution of velocity and salinity as a function of the river inflow, depth, and width of the channel, and vertical and longitudinal mixing coefficients. Going further, they used data from six real estuaries to relate the effect of the interior mixing to bulk channel parameters and eliminate the need to specify the mixing coefficients. The main

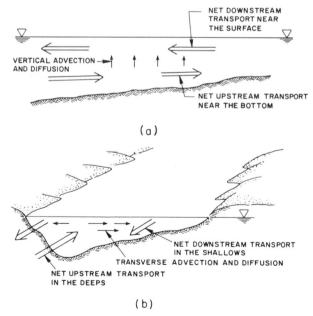

Figure 7.11   The internal circulation driven by the river discharge in a partially stratified estuary. Currents shown are after averaging over the tidal cycle, and are superimposed on the back-and-forth tidal flow. (a) A vertical section along the deepest part of the channel axis. (b) A transverse section showing the transverse distribution of net currents. [After Fischer (1976a).]

result is shown in Fig. 7.12. F_m is a bulk densimetric Froude number defined by

$$F_m = Q_f/A[(\Delta\rho/\rho)gd]^{1/2}, \tag{7.7}$$

where A is the cross-sectional area and d the mean depth. R is the previously defined estuarine Richardson number. According to the theory the values of R and F_m determine the magnitudes of the salinity and velocity deviations, $\delta S/\bar{S}$ and U_s/U_f. $\delta S/\bar{S}$ represents the difference in salinity between surface and bottom divided by the mean salinity, U_s is the residual velocity at the surface and $U_f = Q_f/A$ the fresh water discharge velocity. The parameter v shown in the figure comes from Hansen and Rattray's solution; according to their study it represents the fraction of landward transport of salinity caused by turbulent diffusion, the remainder being by the density-driven circulation. Fischer (1972a) suggested that a better interpretation of Hansen and Rattray's analysis would be that v is the fraction of landward transport of salinity caused by all dispersion mechanisms other than the density-driven circulation.

As an example of the use of Fig. 7.12, let us compute the values of the parameters for the case of northern San Francisco Bay, using the approximate description of flow and geometry given in the worked example on page 276. These are $Q_f = 100$ m^3/sec, $d = 8$ m, $W = 3,125$ m, $\Delta\rho/\rho = 0.025$ (the density difference used here is that between Sacramento River water and ocean water off the Golden Gate), and $U_t = 0.75$ m/sec. These values yield $R = 0.019$ and $F_m = 0.0029$. Figure 7.12 gives $\delta S/\bar{S} = 0.08$, which is approximately the observed value corresponding to a salinity difference of 2 ppt and a mean of 25 ppt

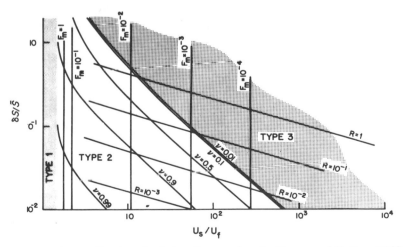

Figure 7.12   The stratification–circulation diagram given by Hansen and Rattray (1966) with lines of constant R added by Fischer (1976a).

as given in the worked example. The value of v is given as approximately 0.7, meaning that according to Hansen and Rattray's theory approximately 70% of the salt balance in this bay is maintained by mechanisms other than the density-driven circulation. Experiments have supported this latter conclusion; Fischer and Dudley (1975) reported the comparison of two experiments in the hydraulic model of San Francisco Bay, one with normal density gradients and one with greatly reduced density gradients, which gave similar salinity intrusion (see Fig. 8.16). These writers concluded that trapping was probably the main cause of salinity intrusion in San Francisco Bay, although pumping and wind-driven gyres might also be important.

In spite of the apparent success of Hansen and Rattray's analysis in the complex geometry of San Francisco Bay, as well as in a number of other estuaries, the result is strongly limited by the assumption of two dimensionality. Fischer (1972a) pointed out that the actual baroclinic circulation is complicated by the variation of depth across the channel. Figure 7.11b shows how the upstream flow is expected to be concentrated in the deeper portions of the channel, because the upstream pressure gradient increases linearly with depth below the water surface. The return current is spread over the cross section, and a net transport from the deeper to the shallower section is required to complete the circulation. Thus the channel geometry turns what would otherwise be a vertically two-dimensional circulation cell into a horizontal circulation cell. We have already seen that horizontal velocity gradients generally lead to much larger dispersion coefficients than do vertical ones, because in real channels the widths are so much greater than the depths. The baroclinic residual circulation is effectively a steady shear flow, for which Eq. (4.47) gives $K \propto l^2/\varepsilon$, where l is a characteristic dimension of the cross section and ε the mixing coefficient for transport across that dimension. Hence the ratio of the longitudinal dispersion coefficient caused by the transverse gradient to that caused by the vertical gradient will be approximately $(W^2/\varepsilon_t)/(d^2/\varepsilon_v)$, and in all but strongly stratified estuaries this number will be large. (The magnitudes of ε_t and ε_v are discussed in Section 7.3.) Following this reasoning, Fischer analyzed the transverse gravitational circulation in a triangular channel of depth d and width W, and obtained for the longitudinal dispersion coefficient

$$K = 1.9 \times 10^{-5} \left(\frac{g}{\rho}\frac{\partial\rho}{\partial x}\right)^2 \frac{d^6 W^2}{E_0^2 \varepsilon_t}, \tag{7.8}$$

in which E_0 is the vertical mixing coefficient for momentum. Fischer found that Eq. (7.8) gives approximately $K = 360$ m^2/sec in the Mersey estuary, the same value predicted by Okubo's trapping formula Eq. (7.6) and measured by Bowden and Gilligan. We are not sure what to conclude, except that the reader should beware of applying any of the formulas in this chapter by themselves.

There have also been several laboratory studies of baroclinic circulation.

The most extensive was conducted in the Delft Hydraulics Laboratory, The Netherlands, over a several year period in the late 1960s (Rigter, 1973). The results supplement an earlier series of tests, in the period 1959–61, conducted by Ippen and Harleman (1961) in the Waterways Experiment Station in Vicksburg, Mississippi. In both cases the test flume had a cross section of approximately 1 ft depth and 0.75–2.2 ft width. Both flumes were approximately 300 ft long, but the Delft flume had a pumping arrangement at the upstream end so that it could simulate a flume of greater length. More recently some tests have been made in a flume with an 11 ft wide test channel approximately 600 ft long, at the University of California, Berkeley. Some of the tests in the Berkeley flume used a nonrectangular channel. In both the Delft and Vicksburg tests the channel was rectangular, and vertical strips were placed along the sides of the channel to induce vertical mixing. The vertical strips were found to be essential to simulate a reasonable vertical salinity gradient; Abraham et al. (1975) show some results of flume tests with roughness on the channel bottom instead of the sides; they obtain a nearly two-layer flow with a sharp middepth halocline, instead of the distributed halocline caused by vertical strips. It appears that the vertical strips in the small flume tests produce vertical mixing similar to mixing in the prototype by mechanisms much more complicated and not fully understood. Thus one has to exercise considerable caution in applying the flume results to real estuaries. Vertical strips were not used in the larger flume at Berkeley, but the salinity gradients observed in the Berkeley flume were similar to those in the smaller flumes with strips.

One way to summarize the results of all the laboratory studies is to relate the observed intrusion of salinity to the estuarine Richardson number. Some judgment has to be used, since the different experiments are reported in different ways. Figure 7.13 shows a reasonable comparison of the several results, and a reasonable agreement on the trend. The abscissa in the figure is a modified estuarine Richardson number

$$R' = \frac{\Delta \rho}{\rho} \frac{gQ_f}{Wu^{*3}}, \tag{7.9}$$

in which the shear velocity is used in place of the mean velocity to include the effect of varying bottom friction. The ordinate in the figure is proportional to K/du^*, as suggested by Elder's result (see Eq. 4.46). K has been replaced by $U_f L_i$, where L_i is the length of salinity intrusion, in order to use the results of the Delft experiments. It should be remembered, however, that Fig. 7.13 shows mostly results in narrow rectangular laboratory flumes, where tide- and wind-driven mixing mechanisms were absent. The figure might be used to predict a minimum length of salinity intrusion in a real estuary, but it certainly should not be used to predict a maximum because of the likelihood of intrusion caused by the other mechanisms.

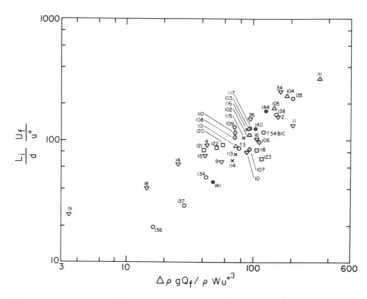

Figure 7.13 A correlation of observed salinity intrusions in laboratory flume experiments. Numbers beside points are run numbers given in the references. ▽ Ippen and Harleman (1961); other data from Rigter (1973) using one base condition and varying one parameter at a time identified as follows: ○ tidal amplitude, × channel roughness, △ channel length, □ fresh water discharge, ◇ mean depth, ● ocean salinity.

7.2.4   Synthesis and Summary

We have discussed three main causes of mixing and several effects resulting from each cause. In real estuaries motions resulting from all three causes are superposed, although one or two causes may dominate. Moreover, the main cause may change from season to season, or even from week to week. Many estuaries change from partially stratified or salt wedge in the wet season to well mixed in the dry season. A flood of a week or so may stratify a previously well-mixed estuary and the stratification may persist or be slowly eroded away over perhaps the following month. On the other hand, a stratified estuary may be suddenly well mixed by a passing hurricane, or a storm at sea may raise the mean tidal level at the coastline and force in a large quantity of ocean water. All of these seasonal or catastrophic events may be categorized as coming from the wind, tide, or river, but most of them are not adequately described by the simple steady-state analyses reviewed in the previous sections. Therefore, the engineer must beware that even if the main causes of mixing can be analyzed for a particular site and set of conditions, most estuaries are hardly ever in steady state and an analysis that is suitable for one season may not be suitable in another season or to describe the sudden effects of a massive change like a storm or flood.

Nevertheless, it is helpful and sometimes essential to at least conceptualize the different mechanisms. There are only a few ways to make practical engineering studies of estuaries: they include physical modeling, numerical modeling, and one-dimensional analytical modeling. Physical and numerical models inevitably neglect certain mechanisms; it is essential to know what mechanisms are important in the estuary at hand, and which are neglected in the model, to make sure that the two don't match. One-dimensional analytical models generally lump all of the mixing mechanisms into a single longitudinal dispersion coefficient; sometimes the magnitude of the coefficient can be determined from natural tracers, but often it must be estimated, or an estimate must be made of what will happen if natural conditions change. We have given several formulas for estimating the value of the longitudinal dispersion coefficient, but each one has been based on an analysis of one mechanism at the neglect of others. In the following chapter we will discuss numerical and physical models. Throughout the discussion it will be helpful to bear in mind that almost all practical methods now in use are based on recent and possibly incomplete research results, and that what really happens in estuaries may be much more complex than our present ability at description.

7.3   CROSS-SECTIONAL MIXING IN ESTUARIES

Some estuaries are so wide and irregular that they cannot be said to have an identifiable cross section; for them the concept of cross-sectional mixing in the sense discussed in Section 5.1 for rivers has no meaning. Other estuaries, however, are relatively long, narrow, and uniform; for these we may define vertical and transverse mixing coefficients ε_v and ε_t for use in formulas similar to those already given for rivers. As compared to rivers, however, the magnitudes of the coefficients in estuaries are much more difficult to establish, and even the correct manner of definition and usage is not always clear. In this section we discuss efforts to measure cross-sectional mixing coefficients, give some approximate values, and discuss why the values we give are so inexact.

7.3.1   Vertical Mixing

The simplest case is a constant-density tidal flow, say the flow in a tidal river upstream of the limit of salinity intrusion or in a fully marine bay. In this case vertical mixing is caused predominantly by turbulence generated by bottom shear stress, and we would expect Eq. (5.3), $\varepsilon_v = 0.067du^*$, to be adequate. The only complication is that u^* varies from nearly zero at slack tide to a maximum at the time of highest velocity. It seems likely that ε_v varies in the same way, although we know of no experimental verification. For engineering studies it is generally thought adequate to use the average value of u^*. Since shear is difficult to measure in an unsteady flow, it is usual to assume a relation between the shear velocity and the mean velocity; for example, Bowden (1967b) suggests that at middepth

$$\varepsilon_v = 0.0025 d U_a \tag{7.10}$$

for constant density conditions, where U_a is the depth mean amplitude of the current.

If the water column is stably stratified, turbulent mixing requires that some of the tidal energy be used to raise the potential energy of the water column. In Chapter 6 we discussed how to predict the deepening of the surface layer in a reservoir from the input of energy at the water surface. The same concepts apply to estuaries, but most of the energy for mixing is extracted from the bottom and internal shear. There have been so few adequate studies of vertical mixing in stratified estuaries that we hesitate to quote values. Two frequently quoted formulas are those of Munk and Anderson (1948),

$$\varepsilon_v = \varepsilon_0 (1 + 3.33 Ri)^{-3/2}, \tag{7.11}$$

in which ε_0 is the value of ε_v for neutral stability and $Ri = g(\partial \rho / \partial z) / \rho (\partial u / \partial z)^2$ is a gradient Richardson number, and Pritchard (1960) that in the absence of surface waves

$$\varepsilon_v = 8.59 \times 10^{-3} U_t [z^2 (d - z)^2 / d^3][1 + 0.276 Ri]^{-2}. \tag{7.12}$$

Bowden (1963) found values of ε_v in the Mersey ranging from 5 cm²/sec at the surface to up to 71 cm²/sec at middepth which fit Munk and Anderson's formula reasonably well; however, Pritchard (1971) also noted that the Mersey values fit his formula if he used $Ri = 0$. In the Mersey studies Ri ranged from approximately 0.1 to 1.0, and the neutral stability value given by Eq. (7.10) was approximately 500 cm²/sec. Bowden (1967b) reported values of ε_v, $\varepsilon_v / \varepsilon_0$, and Ri in the Mersey estuary, off Cumberland, and at three stations in Liverpool Bay. The ratio $\varepsilon_v / \varepsilon_0$ decreased with increasing Ri, but the scatter was large. Bowden thought that "the average value of $\varepsilon_v / \varepsilon_0 = 0.064$ at $Ri = 0.5$ approximately in the Mersey estuary is probably the most reliable." More recently Partch and Smith (1978) have reported time-dependent values of the vertical mixing coefficient in the Duwamish Waterway, near Seattle, Washington. The Duwamish is a saltwedge estuary with a strongly stratified surface layer. Partch and Smith used an array of current and salinity sensors to measure the variation of the vertical mixing rate during the tidal cycle. They found a value of ε_v of approximately 0.5 cm²/sec during most of the cycle, increasing to approximately 5 cm²/sec during the period of maximum ebb flow. Equation (7.10) gave a value of 55 cm²/sec. Thus Partch and Smith's results are similar to Bowden's, even though the conditions at the site of the experiment were quite different, in that ε_v ranged between $\frac{1}{10}$ and $\frac{1}{100}$ of the value in the absence of stratification. Even this vague guide may not apply to other estuaries, however; Partch and Smith suggested that the higher rate of mixing they measured during the ebb was caused by an internal hydraulic jump, a site specific mechanism that may occur in many estuaries but whose result would not be represented adequately by any of the formulas given here. Some further evidence has been given by Blumberg (1975) in connection with numerical modeling studies. Blumberg found it necessary to use a relationship between ε_v and Ri quite different in form from either Eq. (7.11) or (7.12), as illustrated and discussed by Bowden (1978). It seems clear, as Bowden concluded, that the properties of turbulent flow in the presence of stable density gradients are still poorly understood and that further observations are needed.

7.3.2   Transverse Mixing

For the case of rivers we have already seen (Section 5.1) that bottom-generated turbulence gives a transverse mixing coefficient of approximately $\varepsilon_t = 0.15du^*$ in a straight, rectangular section, but that much larger values are induced by sidewall irregularities and channel curvature. Estuaries are even more complex because the mixing mechanisms reviewed in Section 7.2 are in part mechanisms for transverse mixing. The flow into and out of "traps" (Section 7.2.2.3) is a transverse flow, and the "pumped" circulation in bays (Section 7.2.2.2) has a transverse component. Figure 7.3, the sketch of a wind-driven gyre, shows the depth-averaged velocity vector, but the profile of the velocity over the vertical includes transverse components similar to those sketched in Fig. 4.8. The sketch of the transverse baroclinic circulation shown in Figure 7.11b has a transverse component. When we define a transverse mixing coefficient at a point, we are attempting to express the effect of both small-scale turbulent fluctuations and transverse shear flow dispersion due to whatever transverse velocity profile is caused by the superposition of all the mechanisms. In an estuary it is not unusual to find the current going towards one bank at the surface and towards the other at the bottom, especially if the water column is stratified. Thus, transverse shear flow dispersion is probably more important than turbulent diffusion in most estuaries, and it should not be surprising to find a large range of measured transverse mixing coefficients. There is, in addition, the problem of spatial inhomogeneity of the currents. The flow into a side embayment is an example of a large-scale mixing mechanism, for which the Lagrangian length scale [Eq. (3.35)] is on the order of the width of the estuary. The discussion in Sections 3.3 and 3.4 suggests that the variance of a cloud of tracer released into an estuary should grow at a rate increasing with the size of the cloud, or in other words that it is not possible to define a Fickian transverse mixing coefficient. On the other hand, numerical models sometimes attempt to express medium scale currents, such as the flow into an embayment, as advection, and to include in a Fickian mixing term only those velocity fluctuations whose scale is smaller than the grid size of the model. In that case it may be reasonable to speak of a transverse mixing coefficient whose magnitude depends primarily on the vertical variation of the transverse component of the velocity vector. If the vertical mixing time is also sufficiently short, the shear flow dispersion tensor analysis given in Section 4.4 can be used to compute a transverse mixing coefficient from an observed velocity profile. This is the analytical concept used by Fischer (1969) to make the prediction of the transverse mixing rate in a curving flow given as Eq. (5.5), but to date it has generally not been used in estuaries because of the lack of observed velocity profiles.

There have been a few attempts to measure transverse mixing coefficient in well-mixed estuary flows. Ward (1974) extracted mixing coefficients from some previously published field data. His results for San Francisco Bay, Cordova Bay, British Columbia, and the Gironde estuary, France, were, respectively, $\varepsilon_t/du^* = 1.00$, 0.42, and 1.03. Fischer (1974) obtained $\varepsilon_t/du^* = 1.2$ from a dye release in the Delaware estuary. Ward (1976) reported observations of the growth of two dye clouds in the Fraser estuary, British Columbia, giving $\varepsilon_t/du^* = 0.44$ and 1.61. Both values were obtained from experiments lasting four to five hours; the smaller value was during a period including a slack tide, and the larger entirely during an ebb tide. All of the values mentioned here are thought to represent approximately constant density reaches of the estuaries.

Experiments described by Sumer and Fischer (1977) suggest that density stratification affects transverse mixing even more than vertical mixing. These writers conducted two sets of stratified-flow experiments, using a 3.5 m wide laboratory channel with a trapezoidal cross section; in one set of experiments the channel was uniform, and in the other one side was made wavy by varying the transverse slope of the bottom with a longitudinal period of approximately 3.5 m. The sidewall waviness generated vertical mixing on the wavy side, which in turn established transverse density gradients which drove a transverse baroclinic circulation similar to that sketched in Fig. 7.14. The result was a greatly

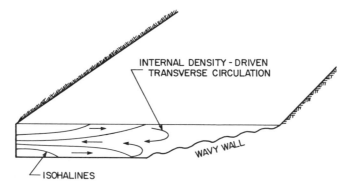

Figure 7.14 A typical distribution of isohalines and the inferred internal density-driven circulation in a channel with one wavy side.

enhanced rate of transverse mixing in the case of the wavy wall, as compared to results in the uniform channel. In real estuaries points of land, jetties, sand bars, and the like almost always create the equivalent of the waviness in Sumer and Fischer's laboratory channel; therefore, the suggestion from the laboratory experiment is that stratification in estuaries may greatly enhance the rate of transverse mixing by driving transverse circulations. Unfortunately, no proto-type confirmation is yet available, nor have the details of this sort of circulation received much study of any sort.

7.4 LONGITUDINAL DISPERSION AND SALINITY INTRUSION

Now suppose that we are dealing with a long, relatively narrow estuary with an identifiable channel axis. The Delaware in the United States and the Thames in Great Britain are examples; the reader may wish to refer to the detailed presen-tation of physical aspects of the Thames by Inglis and Allen (1957) as a way of becoming familiar with the physical oceanography of a particular estuary. Another comprehensive description of an estuary is given in the report on the Mersey, near Liverpool, by Price and Kendrick (1963). Similar detailed descrip-tions of United States estuaries, assembled in one publication, are difficult to find, but the reader will find a wealth of detail on the Delaware, San Francisco Bay, and some others contained in the references scattered throughout this chapter and the next.

Previously in this chapter we have discussed the various mechanisms that cause mixing. Now we wish to limit our attention to longitudinal dispersion of pollutants along the channel axis and intrusion of ocean-derived salinity up the channel axis by dispersive mechanisms. In either case, it is customary to combine the result of all mechanisms into a single dispersion coefficient K. The "salt balance" in an estuary in steady state is expressed by the equation

$$U_f S = K(\partial S/\partial x), \tag{7.13}$$

where $U_f = Q_f/A$ is the net downstream velocity caused by the freshwater discharge. Equation (7.13) states that the downstream advection of salt by the mean flow, $U_f S$, is in balance with upstream transport by all other mechanisms. In the next section we will divide the dispersive term into components, and discuss the space and time averaging implied in the equation.

The magnitude of K is usually determined by observation of gradients of in situ tracers, in particular the salinity itself. Arons and Stommel (1951) seem to have been the first to provide a formula to predict the value of K, and their method was successful in a small estuary in New Jersey. Stommel (1953) found that the method did not work for the Estuary of the Severn, in western England, and he began his 1953 paper with the remark, "In short, it does not appear likely that any good purpose can be served at present by making a priori suppositions about the turbulent mixing process." Therefore he went on, in the 1953 paper, to discuss how to measure K from the salinity distribution and to use the result to predict the distribution of pollutants. We will return to Stommel's results in the next section; his pessimism concerning predicting K was well taken, for at this writing 25 years later there is still no predictive formula that works in general. Nevertheless, the studies that have occurred in the interim have shed some light on how the mechanisms can be sorted out and their relative magnitudes assessed, as we shall now see.

7.4.1 Decomposition of the Salinity and Velocity Profiles

The technique for identifying mechanisms is to divide the observed fluctuations in velocity and salinity into components. The components are of two types; timewise variations and spacewise variations. The timewise variations are predominantly the periodic variations at the frequency of the tidal cycle. Longer period storm and seasonal fluctuations undoubtedly play a role in dispersion, but how much is not really known. Spacewise variations are caused by the variation of depth across a cross section, and by the variation of cross sectional shape along the axis of the estuary. A quantitative decomposition of the components is given by Fischer (1972a). The tidal rise and fall of the water surface is neglected to obtain a constant cross section throughout the tidal cycle,† and variations with periods longer than a tidal cycle are neglected. The velocity and salinity observed at any point can then be divided into four components

$$u(x, y, z, t) = u_a + u_c(x, t) + u_s(x, y, z) + u'(x, y, z, t) \qquad (7.14)$$

$$S(x, y, z, t) = S_a + S_c(x, t) + S_s(x, y, z) + S'(x, y, z, t), \qquad (7.15)$$

where u_a and S_a are what is obtained by averaging over the cross section and the tidal cycle. In this chapter we will use angle brackets $\langle \ \rangle$ to denote a tidal cycle average of any quantity, and an overbar to denote a cross-sectional average.‡ Then $u_a = \langle \bar{u} \rangle$ and $S_a = \langle \bar{S} \rangle$ are averages of u and S over the cross section and tidal cycle (formally, $\langle \bar{u} \rangle = (1/T)\int_0^T (1/A)\int_0^W \int_0^d u \, dy \, dz \, dt$, etc.). u_c and S_c are defined as the cross-sectional averages at any time during the tidal cycle, minus the tidal cycle averages, i.e.,

$$u_c = \bar{u} - u_a, \qquad S_c = \bar{S} - S_a. \qquad (7.16)$$

u_s and S_s are defined as the tidal cycle averages at any point, minus the cross sectionally averaged tidal cycle averages, i.e.,

$$u_s = \langle u \rangle - u_a, \qquad S_s = \langle S \rangle - S_a. \qquad (7.17)$$

---

† Dyer (1974) considers the tidal variation of the cross section and obtains an expansion of the product uS into 11 terms, but his data show that all the terms we omit are of small magnitude.

‡ Note the change in usage from Chapter 3, in which $\langle \ \rangle$ meant an ensemble average.

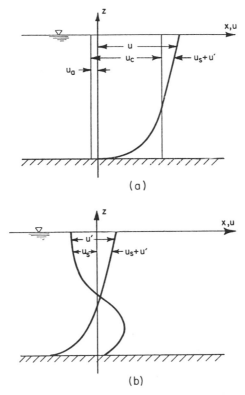

Figure 7.15   (a) Decomposition of a two-dimensional velocity profile $u(z, t)$ into components u_a, u_c, and the deviation from the cross-sectional mean $u_s + u'$. (b) Further decomposition of the deviation from the cross-sectional mean into components u_s and u'.

u' and S' are the remainders, that is what is left over when the various averages are subtracted from the observed velocity. Figures 7.15 and 7.16 illustrate how typical velocity and salinity profiles are decomposed.

The total transport of salinity through a cross section during a tidal cycle is given by

$$\dot{M} = A\langle \overline{uS} \rangle = Q_f S_a + A\{\langle u_c S_c \rangle + \overline{u_s S_s} + \langle \overline{u'S'} \rangle\}, \qquad (7.18)$$

where Q_f is the tributary discharge.† Note that all of the cross product terms (terms like $\overline{u_c S_s}$, for instance) average to zero. The four terms on the right side of the equation can be identified as the mean advection and three mechanisms for dispersion, and the three dispersion terms can be analyzed separately. If the

---

† The net downstream discharge is $Q_f = \langle \bar{u}A \rangle = A_a u_a + \langle A_c u_c \rangle$ where $A = A_a + A_c$ and $A_a = \langle A \rangle$. $A_c u_c$ is often a substantial fraction of Q_f; in theory Eq. (7.18) should contain other terms involving A_c, such as $u_0 \langle A_c S_c \rangle$, but these terms are generally negligible.

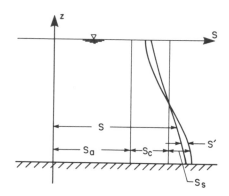

Figure 7.16 Decomposition of a salinity profile into the four components given by Eq. (7.15).

estuary is in steady state, i.e., the salinity distribution is the same at the beginning and end of the tidal cycle, the four terms sum to zero, giving Eq. (7.13) with

$$-K\frac{\partial S}{\partial x} = \langle u_c S_c \rangle + \overline{u_s S_s} + \langle \overline{u'S'} \rangle. \qquad (7.19)$$

More commonly the salinity distribution is slowly varying. It would be possible to introduce a fifth term expressing the long term correlation of the daily averages of velocity and salinity; this correlation would express the dispersive effect of storm and seasonal variations, but since we could go no further with its analysis, and since what remains is sufficiently complex already, we do not show the long term correlation explicitly.

Each of the terms in Eq. (7.19) can be identified with one or more of the mechanisms discussed in Section 7.2. The term $\langle u_c S_c \rangle$ is the tidal cycle correlation of the cross-sectional averages. It is not obvious why this term is important, and some researchers have assumed it was zero. We might expect the main effect of the tidal current to be advection of the salinity gradient back and forth with little change, so that if, say, $u_c = u_{cm} \sin(\omega t)$ we would find $S_c = S_{cm} \cos(\omega t)$, where $\omega = 2\pi/T$, and $\langle u_c S_c \rangle = u_{cm} S_{cm} \int_0^T \sin(\omega t) \cos(\omega t)\, dt = 0$. In real estuaries, however, we sometimes find that the peak salinity occurs before high slack water and the minimum salinity before low slack water, meaning that the salinity is on the average higher during the flood and that the correlation gives a net landward transport of salt. One explanation of this phenomenon is the trapping mechanism illustrated in Fig. 7.9; during the flood the traps will generally contain lower salinity water than the main channel, and if the traps partially discharge into the channel before high slack water the main channel salinity may be reduced.

The term $\overline{u_s S_s}$ is the residual circulation. Probably more than half of the literature on circulation in estuaries is concerned with the residual circulation, so it may be surprising that we still have no adequate way of predicting its effects in general. The difficulty is that wind-driven gyres, bathymetric tidal pumping, and density-driven currents all contribute, and a set of observations in an estuary may result from any combination thereof. For example, Fig. 7.11b illustrates that the residual baroclinic circulation will be expected to be seaward in the shallow portions of a channel, but tidal pumping of the sort illustrated in Fig. 7.8 or a maintained wind may reverse the flow and set up an opposite circulation. In prototype studies it is usually difficult or impossible to relate the observed residual currents to their cause, even though in the next section we give some field data which permit some speculative assessments.

The term $\langle \overline{u'S'} \rangle$ expresses the result of an oscillatory shear flow, as well as any random motions on time scales shorter than the tidal cycle, especially short term variations in the wind. The trapping mechanism may also effect this term, since dead zones along the side of a channel can be viewed as a nonmoving part of the velocity profile and included in a shear flow analysis.

7.4.2 The Relative Magnitudes of the Terms; Some Observations in Real Estuaries

The relative magnitudes of the terms in Eq. (7.19) can be measured by establishing an observation transect and measuring the velocity at all points in the cross section throughout a typical tidal cycle. The task is not easy and there have not been many adequate studies. The data used here are reported by Dyer (1974) and Murray et al. (1975). Dyer's data are barely adequate; they were taken from earlier studies of the Vellar estuary in India and Southampton Water and the Mersey estuary in England. In the Vellar and the Mersey there were only three stations for vertical profiles, and in Southampton Water only four, so the lateral distributions are based on few points. Murray et al. obtained six vertical profiles across one cross section in the Guayas estuary, Equador. The profiles were obtained by two ships, working one profile per tidal cycle each, so three days were required; however, the authors claim that the tidal cycles were sufficiently alike to draw conclusions as though the data were taken at the same time.

Both authors followed a suggestion by Fischer (1972a) that u_s and u' should be separated into transverse and vertical variations. Let

$$u_s = u_{st} + u_{sv}, \qquad u' = u_t' + u_v' \qquad (7.20)$$

where u_{st} is the transverse variation of the vertical mean and u_{sv} is the vertical deviation from the vertical mean. u_t' and u_v' are defined similarly. The separate contributions of the transverse and vertical circulations and shears are shown by expanding the last two terms in Eq. (7.19) to

$$\overline{u_s S_s} + \overline{u's'} = \overline{u_{st} S_{st}} + \overline{u_{sv} S_{sv}} + \langle \overline{u_t' S_t'} \rangle + \langle \overline{u_v' S_v'} \rangle. \qquad (7.21)$$

Table 7.1 gives Dyer's data for one section in the Vellar, two in Southampton Water, and one in the Mersey. Dyer noted that there might be considerable

TABLE 7.1

Relative Magnitudes of Terms in Eq. (7.19)[a]

	$A_a\langle u_c S_c\rangle$	$A_a\overline{u_{st} S_{st}}$	$A_a\overline{u_{sv} S_{sv}}$	$A_a\langle u_t'S_t'\rangle$	$A_a\langle u_v'S_v'\rangle$
Vellar estuary, 9/2/77	-75	-14	-105	-4	-42
Southampton Water					
Transect A	-25	250	102	-20	-50
Transect B	0	-220	-200	-20	-12
Mersey estuary	2200	-300	-350	-220	-280

[a] Data from Dyer (1974); positive indicates downstream salt flux; values in kg/sec.

error in measurements of u_a and u_c, but thought his data sufficient to compare the magnitudes of the terms. The Vellar is a strongly stratified estuary and as expected the vertical residual circulation dominates transverse or shear effects. It is notable, though, that even in the strongly stratified Vellar the trapping term is important. The measurements in Southampton Water, a partially stratified estuary, suggest that transverse and vertical residual circulations dominate, but are of equal importance (the positive flux at transect A is not explained). The large downstream flux by trapping in the Mersey is unreasonable and probably incorrect; the remaining terms are of roughly equal magnitude.

The data reported by Murray *et al.* in the Guayas estuary, Equador, have yielded among the most complete synoptic cross-sectional observations known to the writers. Figure 7.17 shows a plan of the estuary. The observation transect was in the straight reach of the Rio Guayas 16.5 km south of Guayaquil, approximately where the longitudinal salinity gradient is at its steepest. Cross sections of salinity and velocity at two hour intervals are shown in Fig. 7.18. We quote at length from Murray *et al.* because their description of their observations exemplifies several of the mechanisms we have previously discussed in general terms:

> Lunar hour zero in this series of figures (Figs. 7.18a–f) was chosen as the time at this section of slack current at the 2-m depth after ebb flow, or 20 minutes before predicted low tide at Guayaquil. The view is upstream, toward the north, and so references to the right and left sides of the channel are reversed from the standard usage.
>
> As seen in Figure 7.18a, only the middle section of the channel is at slack; ebb (positive) currents are still prevailing near the surface, and flood (negative) currents are already appearing near the bottom. This could be considered evidence for a weak density-driven flow, but it is equally likely that this advancement of the stage of tide closer to the bottom (priming) results from the role of the bottom frictional forces, as discussed by Proudman (1953) and Cannon (1969). Inasmuch as this is slack after ebb current, salinity is at its minimum for this location. Note the distinct stratification in the central, deep part of the channel but the well-mixed aspect along the channel walls. Within one lunar hour, flood speeds (negative) are well in excess of 50 cm/sec across the channel, and a pronounced subsurface jet occupies the east side of the channel, while the salinity is relatively unchanged.

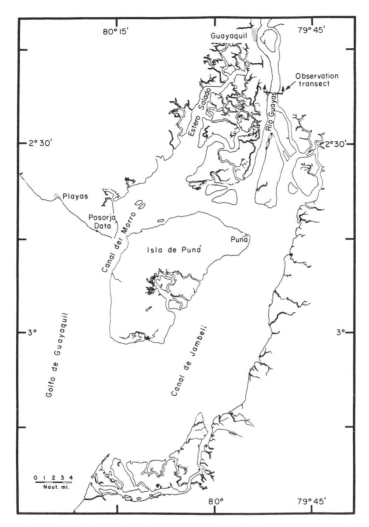

Figure 7.17 Plan of the Guayas estuary. [After Murray *et al.* (1975).]

At LH 2 (Fig. 7.18b), the flood velocity field is nearly at its maximum development; high velocities in excess of 100 cm/sec still occupy the east side of the channel, and the center channel pool of high-salinity water has expanded considerably as a result of the upstream movement of the densest water in the deepest part of the channel. Figure 7.18c (LH 4) shows that, even as the flood decelerates, maximum velocities occupy the east side of the channel; salinities have increased markedly (~ 5 ppt) throughout the channel in the previous two hours, maintaining a distinct stratification despite the considerable current speeds and shear. High water occurs at LH 5, while the

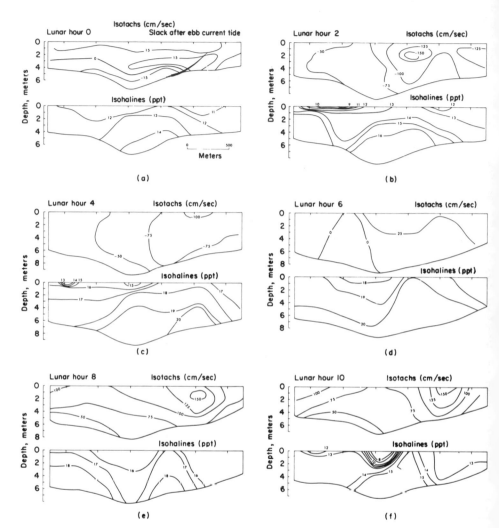

Figure 7.18 Velocity and salinity distributions in the Guayas estuary. [After Murray *et al.* (1975).]

decelerating current still exceeds 50 cm/sec upstream in more than half the channel; at this time, salinities are close to maximum, water of more than 21 ppt being present along the bottom in the east side of the channel.

The distributions at LH 6 (Fig. 7.18d) are very interesting in that currents have begun to ebb on the west side, while flood is still in progress on the east side. Note that the salinity distribution mirrors the velocity distribution inasmuch as the 20-ppt isohaline has bulged up to enclose most of the eastern half of the channel. An hour later (LH 7), the ebb is already well established, velocities of more than 75 cm/sec occurring in the eastern half of the channel. By LH 8 (Fig. 7.18e), the ebb is fully established; the speed distribution, except for directional reversal, is similar to that of the flood in that there is a subsurface jet again on the eastern side of the channel and speeds drop off slowly laterally across to the western bank. Clearly seen in Fig. 7.18e and persistent, but more pronounced, throughout the remainder of the ebb cycle is a depressed lens of fresher water occupying the west-central part of the channel adjacent to the dome of saltier water in the east-central channel noted during the flood cycle. At LH 9 (not shown) and LH 10 (Fig. 7.18f), the velocity distributions differ only in minor details from those at LH 8, but the fresher water lens in the west-central channel has both deepened and strengthened considerably.

It appears from the hourly salinity maps that a major zone of fresh-water advection down the estuary in the shape of a tongue lies in the surface layer of the west-central channel. In contrast, a dome of high-salinity water persistently occupies the east-central channel bottom. Coriolis forces in the southern hemisphere should deflect the fresher waters to the eastern side of the channel and the denser saltier waters to the west side of the channel. Hansen (1965) noted similar discrepancies in the Columbia River estuary and attributed them to accelerations related to channel shape. It is open to question whether a large bar growing out from the west channel wall about 750 m upstream of our section is deflecting the flow in such a manner as to cause these observed concentrations of high- and low-salinity waters. Waters near the western and eastern channel walls (roughly one-third the channel width on each side) are usually the best mixed in the channel. The role of these regions as mixing zones owing to channel-wall turbulence also can only be the subject of conjecture at this point.

Murray and Siripong (1978) analyzed the Guayas data to determine the relative contributions of transverse versus vertical variations. The decomposition is formally slightly different from that given in Eq. (7.19), but the results are essentially the same. They found that approximately 53% of the upstream advective transport was by transverse variations, and 35% by vertical variations; the remainder is by cross-product terms arising in their decompositions. Their analysis did not separate residual from fluctuating components, although the residual distributions are shown in Fig. 7.19. They also did not separate the term $Q_f S_a$ from $A\langle u_c S_c \rangle$ in Eq. (7.18) so the contribution of $\langle u_c S_c \rangle$ is not known.

The data of Dyer and Murray et al. are not conclusive. They demonstrate as much as anything the difficulty of obtaining field data. Nevertheless we can reach a few general conclusions. In a strongly stratified estuary the vertical residual circulation dominates the other cross-sectional variations, although the trapping mechanism can be important. In less well stratified estuaries transverse variations become progressively more important as stratification decreases, but trapping remains an important mechanism in most estuaries. It is not possible to differentiate density-driven from shape-induced residual circulations by the existing field data, but we suspect that the vertical residual circulations are primarily gravitational, while the transverse residual circulations are primarily shape induced.

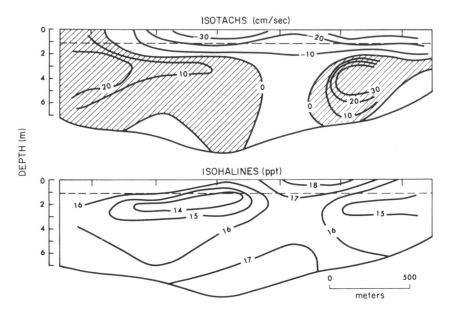

Figure 7.19   Residual velocity and salinity distributions in the Guayas estuary. [After Murray *et al.* (1975).]

7.4.3   Observed Values of the Longitudinal Dispersion Coefficient

The engineering literature contains many reports of observations and uses of longitudinal dispersion coefficients. Many are obtained by observing a longitudinal salinity gradient and a corresponding fresh water outflow and setting $K = U_f S/(\partial S/\partial x)$ as per Eq. (7.13). The result depends on whether S is observed at high slack water, low slack water, or is an average over the tidal cycle, and also on whether the salinity distribution is in steady state as implied by the omission of the time derivative in the equation. Therefore coefficients observed in this way are subject to some interpretation. Nevertheless engineering studies sometimes require an estimate of the magnitude of the coefficient, and engineers like to quote historical precedent for the numbers they use; Table 7.2 is offered in that spirit. Many of the values are approximately in the range of 100–300 m^2/sec, which is notably smaller than the values observed in moderately sized rivers such as the value of 1500 m^2/sec observed in a 200 m wide reach of the Missouri (Table 5.3). The reason is that the shear flow mechanism in rivers is limited in estuaries, for the reason discussed in Section 7.2.2.1. The low values of K listed in the table in the range of 10–50 m^2/sec are generally found in the constant density portions of estuaries and are consistent with the limit mentioned in Section 7.2.2.1 of approximately 60 m^2/sec resulting from shear flow dispersion alone.

TABLE 7.2

Some Observed Longitudinal Dispersion Coefficients in Estuaries

Estuary	Characteristic value or range of observed values of dispersion coeff. (m^2/sec)	Source	Comments
Hudson	160		Values given are K in Thatcher
Rotterdam	280	Thatcher and	and Harleman's model. Their
Waterway		Harleman	K differs from the one defined
Potomac	55	(1972)	by Eq. (7.13) by a factor usu-
Delaware	500–1500		ally not greater than three.
San Francisco Bay	200	Glenne and Selleck (1969)	
San Francisco Bay	200	Cox and Macola (1967)	Approx. value used in numerical model
Severn	10–100	Stommel (1953)	
Potomac	20–100	Hetling and O'Connell (1966)	From dye experiment
Delaware	100	Paulson (1969)	
Mersey	160–360	Bowden (1963)	
Rio Quayas, Equador	760		Computed from data given by Murray *et al.* (1975)
Severn (summer)	54–122	Bowden (1963)	
Severn (winter)	124–535	Bowden (1963)	
Thames (low river flow)	53–84	Bowden (1963)	
Thames (high river flow)	338	Bowden (1963)	

7.5   ONE-DIMENSIONAL ANALYSIS OF DISPERSION OF WASTES

Some estuaries are long enough, narrow enough, and sufficiently unstratified to be analyzed as though they were one dimensional. The flow velocity, salinity, and concentration of any dissolved substance are assumed to depend only on distance from the mouth; the effect of cross-sectional variations and all the mixing mechanisms we have discussed earlier in this chapter are lumped into the longitudinal dispersion coefficient K. Obviously the one-dimensional analysis is not suitable for all estuaries; Galveston Bay, for example, is wider than it is long, and a one-dimensional assumption would not give even a first approximation for an analysis. The Delaware, on the other hand, is very wide near its mouth but extends a long distance inland and has been analyzed using one-dimensional models. So has the San Francisco Bay system, a very irregular estuary with narrow straits between several wide bays, culminating in a braided delta. The one-dimensional analysis is a firmly established engineering tool because it is convenient, relatively simple, and capable of giving practical answers. We begin this section by discussing the tidal exchange ratio, an essentially one-dimensional concept. Then we give some one-dimensional methods for predicting the extent of conservative and decaying tracers, and finally we discuss limitations on the proper use of the one-dimensional analysis.

7.5.1   Tidal Exchange at the Mouth

Part of the volume of water that enters an estuary during the flood tide is made up of water that left the estuary on the previous ebbs. The remainder is water that we may think of as "new" ocean water, and since this portion is what is available for dilution of pollutants inside the estuary an estimate of its amount is an important part of a one-dimensional analysis. We define the tidal exchange ratio R, to be the ratio of new ocean water to total volume of water that enters the estuary during a flood tide. Usually it is not possible to predict the tidal exchange ratio from theory. As sketched in Fig. 7.20, the most important determinate is the longshore current. The longshore current deflects the ebb flow downcoast and delivers the supply of new ocean water for the flood. Without a longshore current all of the exchange would be by relatively inefficient local mixing processes in the coastal zone.

The tidal exchange ratio may be measured empirically as follows. Let

V_f be the total water volume entering the estuary on the flood tide,

V_{fe} be that part of V_f which flowed out of the estuary on the previous ebb,

V_0 be the volume of new ocean water entering the estuary during the flood tide,

V_e be the volume of water leaving the estuary on the ebb tide,

$V_Q = Q_f T$ be the volume of fresh river water entering the estuary during the tidal cycle,

S_f be average salinity of water entering the estuary on the flood tide,

S_e be average salinity of water leaving the estuary on the ebb tide,

S_0 be salinity of the ocean water.

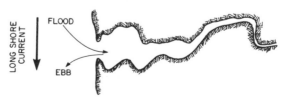

Figure 7.20   How the longshore current affects the tidal exchange ratio.

If the total salt and water contents of the estuary are to remain constant we have

$$S_f V_f = S_e V_e \tag{7.22}$$

and

$$V_f + V_Q = V_e. \tag{7.23}$$

Also

$$S_f V_f = S_e V_{fe} + S_0 V_0 \tag{7.24}$$

and

$$V_f = V_{fe} + V_0. \tag{7.25}$$

The tidal exchange ratio can be defined as

$$R = V_0/V_f. \tag{7.26}$$

Combining the above equations it is easy to show that

$$R = (S_f - S_e)/(S_0 - S_e). \tag{7.27}$$

Thus the tidal exchange ratio can be computed if enough measurements are available to determine the average salinity of the flood and ebb flows and the ocean salinity. It should be noted, however, that accurate determination of average flood and ebb salinities requires complete cross-sectional measurements of both salinity and velocity throughout the tidal cycle; for example, accurate measurement of the ebb salinity requires enough observations to compute the values of the integrals in the definition

$$S_e = \int_0^{T_e} \int_A Su \, dA \, dt \Big/ \int_0^{T_e} \int_A u \, dA \, dt, \tag{7.28}$$

where S and u are point values of salinity and velocity, T_e is the duration of the ebb flow and A the time-varying cross-sectional area.

Equation (7.22) can be used to eliminate S_f from Eq. (7.27) and obtain

$$R = [S_e/(S_0 - S_e)](V_Q/V_f), \tag{7.29}$$

a useful form of the result if the freshwater discharge is known. The reader may show as an exercise that Eq. (7.5) follows from Eq. (7.29) by setting $R = 1$.

Nelson and Lerseth (1972) describe some measurements of tidal exchange at the entrance to San Francisco Bay. Salinity and velocity were measured throughout the tidal cycle at a number of points on a transect at the Golden Gate. Unfortunately the number of measuring points is not stated. Measurements were made on two separate occasions with different tides. Values of R shown in Fig. 7.21 were computed by Eq. (7.27). Nelson and Lerseth also computed pollutant concentrations inside the bay, using a numerical model. They

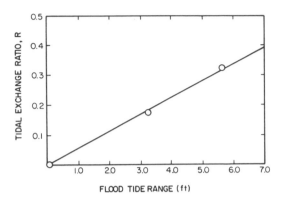

Figure 7.21 Tidal exchange ratio at the mouth of San Francisco Bay versus flood tide range. [After Nelson and Lerseth (1972).]

found that increasing the model value of R from 0.20 to 0.30 decreased the concentrations near the Golden Gate by 30%.

7.5.2   Tidal Exchange within the Estuary; the "Dilution Discharge"

Perhaps the most often posed problem in estuarine analysis is of the sort sketched in Fig. 7.22. A given loading of effluent is to be discharged at a given point, and a prediction is needed for the concentration at other points up and downstream. A complete answer requires a complete understanding of the three-dimensional flow and turbulent exchange structure of the estuary, but since this is never available it is common to use the one-dimensional analysis and to use the distribution of ambient salinity as a guide.

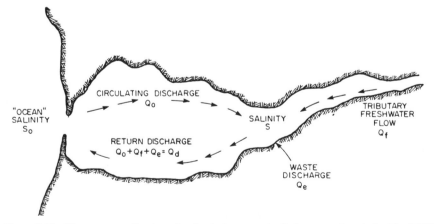

Figure 7.22   The conceptualization of a circulating oceanic discharge used to derive Eq. (7.30).

For the sake of visualization we can assume that the salinity at any point results from the mixing of two flows, the tributary flow which enters the estuary at its upstream end and is essentially fresh water, and a flow of ocean water which circulates in and out from the ocean. In actual fact the oceanic water is progressively diluted as it moves upstream from the mouth of the estuary, but for purposes of analysis one can assume, as sketched in Fig. 7.22, a flow of pure ocean water coming from the ocean to the point where the effluent is discharged, mixing with the effluent, and tributary discharges, and returning to the ocean.

Let

Q_0 be the circulating flow of ocean water,
Q_f be the tributary discharge from all tributaries upstream of the effluent discharge point,
Q_e be the effluent flow.

The salt balance in the estuary requires that

$$Q_0 S_0 = (Q_0 + Q_e + Q_f)S, \tag{7.30}$$

where S_0 is the ocean salinity. Solving for Q_0 gives

$$Q_0 = (Q_e + Q_f)S/(S_0 - S). \tag{7.31}$$

The total flow available for diluting the effluent is

$$Q_d = Q_0 + Q_e + Q_f = (Q_e + Q_f)S_0/(S_0 - S). \tag{7.32}$$

The mean concentration of effluent near the point of discharge can be estimated from

$$C_d = \dot{M}/Q_d, \tag{7.33}$$

where $\dot{M}$ is the discharge rate of material in units of mass per unit time.

In Section 7.5.1, Eq. (7.26) defined a tidal exchange ratio by the equivalent of $Q_0 T = RV_f$, where T is the duration of the tidal period. Introducing the result given in Eq. (7.29) gives

$$Q_0 = RV_f/T = \frac{S_e}{(S_0 - S_e)} \frac{V_Q}{T} = \frac{S_e}{S_0 - S_e}(Q_e + Q_f), \tag{7.34}$$

which is the same as Eq. (7.31) if S is taken to be the average ebb flow salinity. Thus the tidal exchange analysis of the previous section can be applied at any cross section within the estuary and leads to the same result as the salt balance analysis illustrated in Fig. 7.22. In bays with low tributary inflow the salinity may be too near oceanic to permit computation of Q_0 by Eq. (7.31). It may be possible to estimate a value of R, however, and to compute Q_0 by the relationship,

$$Q_0 = RP/T, \tag{7.35}$$

where P is the tidal prism and is approximately equal to V_f if the freshwater and effluent inflows are much smaller than the tidal flow. The value of R can be expected to decrease rapidly as one moves upstream from the mouth of the estuary. For example, in South San Francisco Bay a dye study in the physical model gave values of R varying from 0.076 at the Oakland–San Francisco Bay Bridge to 0.039 at the Hayward-San Mateo Bridge and 0.031 at the Dunbarton Bridge (Fischer and Kirkland, 1978). These values may or may not be typical of similar water bodies; only a prototype study can tell for sure.

Example 7.2: Discharge to an estuary with tributary inflow.   An industrial plant is to discharge 100 cfs of effluent containing ten parts per thousand (ppt) of toxic material into an estuary. The minimum tributary inflow upstream of the discharge point is 1000 cfs. The salinity observed at the discharge point is 19 ppt and the ocean salinity 33 ppt. Estimate the average concentration of the toxic material in the estuary in the vicinity of the discharge point.

Solution.   From Eq. (7.32)

$$Q_d = (1000 + 100) \times 33/(33 - 19) = 2596 \text{ cfs.}$$

From Eq. (7.33)

$$C = 10 \text{ ppt} \times 100 \text{ cfs}/2596 \text{ cfs} = 0.385 \text{ ppt.} \quad \blacksquare$$

Example 7.3: Discharge to an estuary with no inflow; use of a dye study.   The same effluent as in the previous example is to be discharged in the center of a bay which has effectively no tributary discharge. In this bay 33 cc/min. of a dye with an initial concentration of 200,000,000 ppb (parts per billion) are released continuously for a period of 15 tidal cycles at a point in midbay. The concentration near the injection point at the end of the injection is 8.5 ppb. Assuming no dye decay and that 15 tidal cycles is sufficient to reach equilibrium, estimate the concentration in the bay.

Solution.   The dilution discharge is found by solving for Q_d in Eq. (7.33). The result is

$$Q_d = \frac{33 \text{ cc/min} \times 2 \times 10^8 \text{ ppb}}{8.5 \text{ ppb}} = 7.76 \times 10^8 \text{ cc/min} = 459 \text{ cfs,}$$

which is essentially equal to Q_0. The concentration resulting from the discharge is

$$C = \frac{M}{Q_0 + Q_e} = \frac{10 \text{ ppt} \times 100 \text{ cfs}}{459 \text{ cfs} + 100 \text{ cfs}} = 1.79 \text{ ppt.} \quad \blacksquare$$

CAUTION: It should be recalled that average concentrations computed by these formulas are just that—averages. Peak concentrations in the effluent plume near the discharge point may be much higher—refer to the discussions of jets, plumes, and initial mixing elsewhere.

If the effluent material is conservative (i.e., does not decay with time), the concentrations up and downstream of the outfall are easily computed. Downstream the tracer continues to be diluted as it approaches the ocean, just as fresh water is diluted. If C_d represents the concentration near the outfall, then at any point further downstream

$$C_x = C_d(S_0 - S_x)/(S_0 - S_d), \tag{7.36}$$

where the subscript x represents any point in the estuary and the subscript d represents the discharge point.

Upstream of the discharge point concentration of the effluent is reduced in the same way that salinity is reduced by mixing with the tributary fresh water. Thus

$$C_x = C_d(S_x/S_d). \tag{7.37}$$

Example 7.4:   Distribution along the estuary.   The same conditions as for worked example 7.2. Find the concentrations downstream of the outfall at a point where the salinity is 24 ppt (parts per thousand) and upstream of the outfall at a point where the salinity is 5 ppt.

Solution.   Using Eq. (7.36), the downstream concentration is

$$C = 0.385 \text{ ppt} \times (33 - 24)/(33 - 19) = 0.248 \text{ ppt}.$$

Using Eq. (7.37) the upstream concentration is

$$C = 0.385 \text{ ppt} \times \tfrac{5}{19} = 0.101 \text{ ppt}. \quad \blacksquare$$

It is worth recapitulating that the results given in these examples are based on the following simplifications:

(i)   It is assumed that the salinity observed at some instant is representative of steady-state conditions. The assumption is usually not correct, and can be grossly misleading. However, a safe design procedure might be to use the maximum salinity observed in some given frequency, say once every five years, as a basis for a conservative design.

(ii)   The results are valid only for conservative substances.

(iii)   The results give cross-sectional averages, not peak concentrations. Near the source the peak concentration will greatly exceed the cross-sectional average. In a stratified estuary the discharge may be confined to one layer, and concentrations within the layer will always be higher than the cross-sectional average.

7.5.3   Dispersion of Decaying Substances

A one-dimensional analysis for the dispersion of pollutants can be based on the time and space averaged equation

$$A \frac{\partial C}{\partial t} + Q_f \frac{\partial C}{\partial x} = \frac{\partial}{\partial x}\left(KA \frac{\partial C}{\partial x} \right) + \text{(source or sink terms)}, \qquad (7.38)$$

in which the time derivative means the change per tidal cycle, K expresses the result of all the mixing processes that occur within the tidal cycle, and A may be taken to be the cross-sectional area at mean tide. Efforts have been made to derive Eq. (7.38) by averaging the advective diffusion equation over the cross section and over the tidal cycle, but they are not wholly satisfactory because of assumptions required during the averaging. It seems better to look on Eq. (7.38) as a postulated "model," subject to verification. The model says that a pollutant is transported downstream by the fresh water discharge and upstream by a Fickian exchange process. The best argument that the upstream exchange is Fickian may be that the distance traveled by a water particle during a tidal cycle, when observed in a coordinate frame moving with the mean motion, is approximately a stationary random function leading to the random walk described in Section 2.2.1. The same reasoning argues against a Fickian co-efficient, however, because as a water particle travels down an estuary it en-counters a constantly varying cross section and velocity distribution and its Lagrangian time scale (see Section 3.3) may be infinite. Since these theoretical questions have not been resolved, it is better to look on Eq. (7.38) as an em-pirical model.

In most waste discharge studies Eq. (7.38) is solved numerically as discussed in the next chapter. A numerical solution allows the freshwater discharge to be varied on a day-to-day basis, and some models vary the dispersion coefficient depending on the computed salinity gradient or the fresh water inflow. An example is the "seven-segment" model of northern San Francisco Bay (Cox and Macola, 1967), which models an extremely irregular channel approximately 100 km long by the seven segments shown in Fig. 7.23. The model amounts to a numerical solution of Eq. (7.38) using seven grid points. Historical salinity records have been used to devise a correlation between the value of K necessary to make the model work and the freshwater discharge, and the derived formula-tion $K = f(Q_f)$ is used to make predictions of salinity intrusion under such possible future conditions as construction of a new diversion canal to route water around the Sacramento-San Joaquin Delta. Experience with the model seems to support the view that for long-term planning purposes it is as depend-able as more sophisticated numerical models, in spite of its drastic simplification of the physics.

Equation (7.38) can also be solved analytically for the steady-state distribution

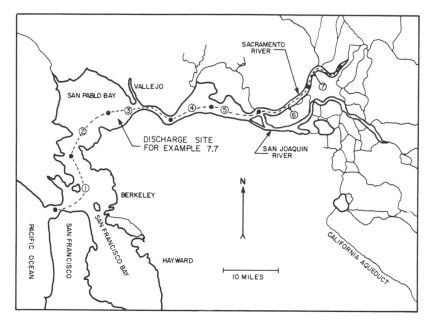

Figure 7.23   Plan of San Francisco Bay showing Cox and Macola's (1967) representation by seven segments.

of a tracer undergoing first order decay in a channel with constant cross section and dispersion coefficient, for which the equation becomes

$$U_f \frac{\partial C}{\partial x} = K \frac{\partial^2 C}{\partial x^2} - kC, \qquad (7.39)$$

where k is the rate coefficient defined in Section 5.5. Even though this case is of limited direct application the solution is of interest in showing the approximate forms of solutions to be expected. The equation has the two solutions (one of which has already been given in Section 5.5):

$$C = C_0 \exp[x'(1 \pm \sqrt{1 + \alpha})],$$

in which $x' = U_f x / 2K$ and $\alpha = 4Kk/U_f^2$. Suppose that a pollutant is introduced into an estuary at a rate $\dot{M}$ units of mass per unit time at a point $x = L(x' = L' = U_f L/2K)$, where $x = 0$ is the mouth of the estuary and x positive landward. Note that since U_f is a seaward velocity it has a negative value and x' is negative everywhere in the estuary. As a first case, suppose that the discharge point is a long way from the mouth so that the pollutant has decayed away before reaching the sea. Then we can use as boundary conditions $C = 0$ at $x' = \pm\infty$. The solution is in two parts: upstream of the pollutant source it is

$$C = C_0 \exp[(x' - L')(1 + \sqrt{1 + \alpha})] \qquad (7.40)$$

and downstream it is

$$C = C_0 \exp[(x' - L')(1 - \sqrt{1 + \alpha})], \tag{7.41}$$

where in both cases

$$C_0 = \dot{M}/(Q_f\sqrt{1 + \alpha}) \tag{7.42}$$

is obtained from the requirement that $\int_{-\infty}^{\infty} kCA\, dx = \dot{M}$.

The reader may recall our comments in Section 5.5 about the inappropriateness of an equation similar to Eq. (7.41) because either α is too small to matter or else cross-sectional mixing is too slow to permit use of a one-dimensional analysis. Those comments do not apply here because, whereas in Chapter 5 α was based on the mean river velocity, here it is based on the fresh water discharge velocity; there is no relationship between K and U_f similar to the one between K and $\bar{u}$ used in Chapter 5. In estuaries U_f is often small and α often large. U_f can even be zero in the case of a tidal inlet with no freshwater inflow, for which $\alpha \to \infty$ and the results given above can be rewritten as

$$C = (\dot{M}/A\sqrt{4Kk}) \exp[\pm(x - L)\sqrt{k/K}], \tag{7.43}$$

where the sign is chosen to make the exponent negative.

A more general solution is obtained by imposing the boundary condition $C = 0$ at $x = 0$, which corresponds to complete removal of the pollutant at the estuary mouth. The upstream part of the solution is the same as before Eq. (7.40) because the upstream boundary condition is unchanged. The downstream solution, as given by Stommel (1953) is

$$C = C_0 \frac{\exp[(1 - \sqrt{1 + \alpha})x'] - \exp[(1 + \sqrt{1 + \alpha})x']}{\exp[(1 - \sqrt{1 + \alpha})L'] - \exp[(1 + \sqrt{1 + \alpha})L']}. \tag{7.44}$$

C_0 is also affected by the loss at the mouth, and is now given by

$$C_0 = \dot{M}(1 - \exp(2L'\sqrt{1 + \alpha}))/(Q_f\sqrt{1 + \alpha}) \tag{7.45}$$

Example 7.5: Waste distribution in an estuary. A sewage treatment plant discharges 2 kg/sec of BOD having a decay constant $k = 0.2/\text{day}$ into an estuary whose cross-sectional area is $A = 600$ m^2, fresh water discharge is $Q_f = 10$ m^3/sec, and longitudinal dispersion coefficient is $K = 60$ m^2/sec. Plot the longitudinal distribution of mean concentration for two cases: (a) the discharge point is 30 km from the mouth of the estuary; and (b) the discharge point is 5 km from the mouth of the estuary.

Solution.   The necessary parameters are

$$U_f = Q_f/A = -1/60 \text{ m/sec},$$

$$x' = U_f x/2K = -1.39 \times 10^{-4}\, x \ (x \text{ in meters}),$$

$$L' = U_f L/2K = -0.69 \text{ for case (b) and } -4.14 \text{ for case (a)},$$

$$\alpha = 4Kk/U_f^2 = 1.99.$$

For case (a) L' is large enough that the BOD will almost completely decay before reaching the mouth, and we can use Eqs. (7.40)–(7.42). With x and L in meters the equations become

$$C = C_0 \exp[-3.79 \times 10^{-4}(x - L)] \quad \text{(upstream)}$$

$$C = C_0 \exp[1.01 \times 10^{-4}(x - L)] \quad \text{(downstream)}$$

$$C_0 = 2 \text{ kg}/[(10 \text{ m}^3)\sqrt{2.99}] = 0.116 \text{ kg/m}^3 = 116 \text{ mg/liter}.$$

For case (b) the distribution is affected by the nearness of the mouth and Eqs. (7.44) and (7.45) give

$$C = C_0[\exp(1.01 \times 10^{-4}x) - \exp(-3.79 \times 10^{-4}x)]/(1.66 - 0.15)$$

$$C_0 = 116 \text{ mg/liter}[1 - \exp(2(-0.69)\sqrt{2.99})] = 105 \text{ mg/liter}.$$

The solutions are plotted in Fig. 7.24. ∎

It should be emphasized that the distribution plotted in Fig. 7.24, or any distribution obtained from Eqs. (7.40) to (7.45), does not represent the actual distribution that would be seen at any instant during the tidal cycle. What actually happens is that at slack tide high concentrations build up near the outfall because there is little current to provide dilution. The peak formed at slack tide is advected and dispersed during the following flood or ebb, but may still be apparent as a lesser peak at the end of one tidal excursion. The shape of distribution plotted in Fig. 7.24 is only found by averaging concentrations over the tidal cycle. Analytical solutions for the distribution of BOD and dissolved oxygen throughout the tidal cycle have been given by Li (1974), who solved the one-dimensional dispersion equation using tidally varying quantities,

$$\frac{\partial}{\partial t}(AC) + \frac{\partial}{\partial x}(\bar{u}AC) = \frac{\partial}{\partial x}\left(K_t A \frac{\partial c}{\partial x}\right) + \text{(source or sink terms)}, \quad (7.46)$$

where K_t means the dispersion coefficient appropriate to a tidally varying analysis and $\bar{u}$ is the tidally varying cross-sectional mean velocity. Li's analysis is limited to oscillating flow in a uniform channel sufficiently narrow that cross-sectional mixing is immediate, as are tide-varying one-dimensional numerical analyses. In broad estuaries it is not uncommon that the cross-sectional mixing

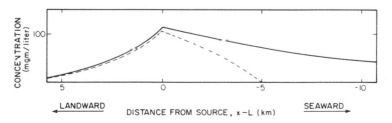

Figure 7.24   Longitudinal concentration distributions for Example 7.5: ——— discharge 30 km from the mouth; ––– discharge 5 km from the mouth.

time is much longer than the tidal cycle; for example, Fischer (1974) found that
the mixing time across the Delaware Estuary near Marcus Hook, where the
channel is approximately 1200 m wide, was approximately 10 days. Therefore,
the time required for initial mixing may be longer than the decay time of the
pollutant or longer than the time required for the pollutant to be carried by the
mean outflow either out of the estuary or into a section of changed characteristics. Use of Eq. (7.46) also raises the question whether K in Eq. (7.38) has the
same magnitude as K_t in Eq. (7.46); some researchers have claimed that K must
be much larger than K_t because K includes the average of variations over the
tidal cycle that are not expressed in K_t, but to the writers this seems an argument
against the adequacy of Eq. (7.46). It appears that the tidally varying one-
dimensional analysis is only useful for tidal flow in narrow, relatively uniform
channels, such as the long braided network of sloughs sometimes found in
deltas or in tidal rivers. In wide and irregular channels it is not reasonable to
try for a tide-varying one-dimensional analytical description for pollutant dispersion, and the best we can appear to do short of a two-dimensional numerical
model or a physical model is to fall back on the tidally averaged Eq. (7.38),
justifying the existence of a bulk dispersion coefficient on the random walk
argument already mentioned.

7.5.4 Calculation of the "Flushing Time"

Instead of computing concentrations, we sometimes wish to compute the
mean time that a particle of tracer remains inside an estuary. Sometimes this
concept is termed the "flushing time," although the proper name for it is the
mean detention time. It is convenient to work in terms of "freshness," rather
than salinity. Freshness is defined as the fraction of any sample of water that is
pure fresh water,

$$f = (S_0 - S)/S_0 \tag{7.47}$$

so that pure fresh water has a freshness of 1 and pure ocean water has a freshness
of 0.

The amount of fresh water in an estuary between a given cross section located
at $x = L$ and the mouth ($x = 0$) can be determined from the total volume of
water times its freshness. Since freshness varies from place to place an integration
is required

$$V = \int_0^L \int_A f \, dA \, dx, \tag{7.48}$$

in which V is the total volume of pure fresh water. The mean detention time
between the cross section at $x = L$ and the mouth is then

$$T_f = V/Q_f, \tag{7.49}$$

which gives the time required for the freshwater discharge to completely replace the fresh water in the estuarine volume seaward of the cross section.

Another way to estimate an approximate replacement time is to use an analysis similar to that for cross-sectional mixing time presented in Section 5.1.3. The time required for a slug of material initially concentrated at one end of a basin to reach an approximately uniform concentration throughout the basin is the same as the time required for mixing across a stream from one side, approximately

$$T = 0.4L^2/K, \qquad\qquad (7.50)$$

where L is the length of the basin and K the dispersion coefficient along it. The replacement time may be imagined to be the time required for complete mixing from the mouth of the estuary up the length, so in an approximate way Eq. (7.50) gives the time required for replacement of a tracer throughout a distance L landward from the mouth in an estuary with dispersion coefficient K.

Example 7.6.   An estuary has a constant cross-sectional area $A = 10,000$ m^2 and a constant longitudinal dispersion coefficient $D = 100$ m^2/sec. The freshwater inflow is 30 m^3/sec. Find the mean detention time according to Eq. (7.49) and the approximate replacement time according to Eq. (7.50), for a volume bounded by the mouth and a cross section 30 km landward from the mouth.

Solution.   Since the cross section and dispersion coefficient are constant, the salinity distribution is exponential and an analytical solution is possible. The salinity distribution is

$$S = S_0 \exp[-(Q_f/A)(x/K)] = S_0 \exp(-3 \times 10^{-5}x)$$

and the corresponding freshness distribution is

$$f = (S_0 - S)/S_0 = 1 - \exp(-3 \times 10^{-5}x).$$

Letting $L = 30,000$ m, the volume of fresh water in the bounded estuary volume is

$$V_f = \int_0^L fA\,dx$$

$$= A \int_0^L [1 - \exp(-3 \times 10^{-5}x)]\,dx$$

$$= 1.02 \times 10^8 \text{ m}^3.$$

The "flushing time" is

$$T_f = \frac{V_f}{Q_f} = \frac{1.02 \times 10^8 \text{ m}^3}{30 \text{ m}^3/\text{sec}} = 3.41 \times 10^6 \text{ sec} = 39.4 \text{ day}.$$

The approximate "replacement time," according to Eq. (7.50), is

$$T = 0.4L^2/K = 3.6 \times 10^6 \text{ sec} = 41.7 \text{ day.} \quad \blacksquare$$

7.5.5   Uses and Limitations of the One-Dimensional Analysis

We use the one-dimensional analysis mostly as a matter of practical necessity; two- and three-dimensional analyses, even using very sophisticated computer codes, are not workable for many practical problems. Use of a one-dimensional model is usually acceptable if the following tests are met:

(1)   The time scale for mixing across the estuary, approximately $0.4\ W^2/\varepsilon_t$, where W is the width and ε_t the transverse mixing coefficient, is significantly less than the time required for the effluent to pass out of the estuary or into a section of greatly changed cross section, or for the substance to decay.

(2)   The estuary is not significantly stratified, so that the effluent can be expected to mix uniformly over the depth (although if the estuary is strongly stratified it is sometimes possible to use separate one-dimensional analyses in each layer).

(3)   Allowance is made in the analysis for the higher concentrations expected near the source, before cross-sectional mixing takes place, and for distributed sources and sinks in the case of a naturally occurring substance such as nutrients, dissolved oxygen, etc.

If these tests cannot be met the engineer may elect to use the one-dimensional analysis anyway, for valid practical reasons, but he should be careful to take account of the possibility for error. The following worked example describes a reasonable practical use of the one-dimensional analysis to study the discharge of an effluent from a point source.

Example 7.7:   Use of the one-dimensional analysis.   An industry plans to discharge 0.5 m^3/sec containing a conservative constituent at a concentration of 1000 ppm into San Pablo Bay, a part of the San Francisco Bay system, at the point shown in Fig. 7.23. Use one-dimensional methodology to make a preliminary estimate of the resulting concentration in the bay. Investigate the accuracy of the one-dimensional method as applied to this specific discharge, and comment on the results. The following may be used as representative mean values of the necessary parameters:

cross-sectional area	$A = 25,000\text{ m}^2$	rms tidal velocity	$U_t = 0.75\text{ m/sec}$
mean depth	$d = 8\text{ m}$	rms shear velocity	$u^* = 0.075\text{ m/sec}$
mean width	$W = 3,125\text{ m}$		

It is anticipated that upstream reservoirs will be operated to maintain a minimum tributary net outflow of 100 m^3/sec. At this outflow, hydraulic model studies (Fischer and Dudley, 1975) indicate that at the proposed discharge point the mean salinities are 24 ppt at the surface and 26 ppt at the bottom.

Solution.   The dilution discharge corresponding to the observed salinity is, by Eq. (7.32)

$$Q_d = 100.5 \text{ m}^3/\text{sec} \times \frac{33}{33 - 25} = 415 \text{ m}^3/\text{sec}.$$

Although not essential to the problem, it is interesting to note the implied value of the tidal exchange ratio. The tidal prism is approximately equal to the rms tidal velocity multiplied by the cross-sectional area and by half the tidal period, i.e., $P \approx U_t A T/2$. This gives

$$Q_0 T/P \approx 2(Q_d - Q_e - Q_f)/U_t A = 2(314.5)/(0.75)(25,000) = 0.034.$$

An approximate expected mean concentration in San Pablo Bay, according to Eq. (7.33), is

$$C = \frac{0.5 \text{ m}^3/\text{sec} \times 1000 \text{ (ppm)}}{415 \text{ (m}^3/\text{sec)}} = 1.2 \text{ ppm}.$$

We now investigate near-source conditions to see within what range the mean concentrations of 1.2 ppm will be exceeded. The first question is whether the vertical stratification will inhibit vertical mixing. The local Richardson number, using bulk properties, is given by

$$Ri = \frac{\Delta \rho}{\rho} \frac{gd}{U_t^2} = \frac{0.0014 \times 9.8 \text{ m/sec}^2 \times 8 \text{ m}}{(0.75 \text{ m/sec})^2} = 0.2.$$

The result is large enough to suggest a reduction, but not elimination, of vertical turbulent mixing. As a rough estimate, assume

$$\varepsilon_v = 0.005 du^* = 0.003 \text{ m}^2/\text{sec}.$$

The time for adequate vertical mixing is approximately

$$T_v = 0.4(d^2/\varepsilon_v) = 8500 \text{ sec} \approx 2.5 \text{ hr}.$$

During this period the effluent will be carried downstream a distance, under average flow conditions, of approximately $U_t T_v = 6.4$ km.

While the effluent plume is spreading vertically, it will also be spreading transversely. A reasonable estimate is that at least close to the source the transverse mixing coefficient will be similar to that for a wide river, $\varepsilon_t = 0.6 du^* = 0.6 \times 8$ m $\times 0.075$ m/sec $= 0.36$ m^2/sec. After $2\frac{1}{2}$ hours the transverse extent of the plume is given approximately by

$$4\sigma_t = 4\sqrt{2 \times 0.36 \times 8500} = 313 \text{ m}$$

and the peak concentration according to Eq. (5.7) will be

$$C = \frac{\dot{M}}{dU_t\sqrt{4\pi\varepsilon_t(x/U_t)}}$$

$$= \frac{0.5 \text{ m}^3/\text{sec} \times 1000 \text{ ppm}}{8 \text{ m} \times 0.75 \text{ m/sec} \sqrt{4\pi \times 0.36 \text{ m}^2/\text{sec} \times 8500 \text{ sec}}}$$

$$= 0.42 \text{ ppm}.$$

This value is less than the mean for the cross section estimated by Eq. (7.33), which results from the continued addition of effluent over many tidal cycles. Therefore we can expect that the plume from the source will become indistinguishable from the background buildup of concentration at some distance less than 6.4 km from the source.

The time required for complete transverse mixing using the coefficient for near-source mixing given above is approximately

$$T_t = 0.4 \frac{(3125 \text{ m})^2}{0.36 \text{ m}^2/\text{sec}} = 125 \text{ day}.$$

Although the mixing time appears to exceed the flushing time, transverse mixing in the tidal flow through the straits between Richmond and San Rafael probably occurs much faster than in 125 days, and the effluent is probably well mixed across the cross section south of this section. In San Pablo Bay the transverse mixing time is probably also much less than 125 days, because of larger scale circulations, but probably much greater than the tidal cycle; hence transverse concentration gradients will persist, meaning that concentrations on the side near the outfall will be higher than the prediction of 1.2 ppm.

Comment: For the condition analyzed, concentration of the effluent in San Pablo Bay will build up to an approximate mean value of 1.2 ppm. However, the concentration will be higher than this value on the side of the bay near the discharge, and lower on the opposite side. Without a more complete analysis we cannot say exactly what the difference will be. In addition, there will be a plume of higher concentration extending from the discharge site in the direction of flow. Within a region extending approximately 5 km in the direction of the tidal flow from the discharge site the plume will contain appreciable vertical and transverse concentration gradients, and the peak concentration will exceed the mean value in the surrounding water.

We conclude that the one-dimensional analysis gives a reasonable preliminary view of what to expect from the proposed discharge. An exact prediction of the concentration distribution within San Pablo Bay is not available, but the main features of the distribution can be inferred from the calculations we have made.

■

Chapter 8

River and Estuary Models

Laboratory measurements and physical scale models have always been an important part of hydraulic engineering because analytical solutions and hand calculations are unable to reproduce all the intricacies of real flows. When computers became available the ability to compute flow fields and transport properties was greatly increased, and for a time some researchers thought that numerical calculations would completely replace the laboratory. At this writing numerical programs are firmly established as a common tool in mixing studies, but so are physical models. A numerical model is an attempt to represent nature by having a computer solve a set of equations that are thought to describe the natural processes; a physical model is an attempt to reduce nature to a small, observable scale. Neither attempt can be completely successful, because nature itself is marvelously complex and defies exact simulation. A model is an abstraction. Certain natural processes are reproduced; others are not. Therefore modeling is an art. The modeler, as an artist, must determine which aspects of the natural process are important, and must be sure that these aspects are simulated correctly in the model. The role of the modeler is much like the role of the landscape painter or photographer, who selects from the natural scene that which can create a pleasing picture and omits the rest. The most important step in the art of modeling is selection of the right model; the primary purpose of this chapter is to provide enough understanding of the working of numerical and physical models so that the reader can make an informed choice. The first part of this chapter describes what types of model are in use and comments on their capabilities; the second and third sections discuss in turn the working of numerical and physical models. We do not give any details of numerical programming methods, however, because these are in a state of rapid change.

We have already discussed a numerical simulation of the initial mixing period in rivers in Section 5.3 and a numerical simulation of reservoirs in Section 6.8. If the reader has no experience in numerical programming a good start would be to study the finite-difference formulation given in Section 5.3 and write a simple numerical program to carry out that scheme.

8.1   CONSIDERATIONS IN CHOOSING A MODEL

The first step must be to develop a clear statement of purpose, because the choice of a model depends crucially on what the model is to do. Next should come a careful "desk" study, in which the investigator assembles and assimulates everything that is known about the water body. If possible the investigator should become personally familiar with the water body by going out on it in the smallest boat that is safe. Then, before venturing near a computer or a model basin, he or she should make all possible computations, being approximate where necessary but seeking a "feel" for what the model will predict. The relative importance of density- and wind-driven currents, pumping, trapping, and shear flow dispersion must be evaluated, and some idea obtained of how the final model will simulate whichever mechanisms are most important. This stage is much like when the artist or photographer studies a subject before beginning work; in the words of the famous landscape photographer Ansel Adams (1965), "Visualization is of utmost importance; many failures occur because of our uncertainty about the final image."

The next step is to choose between a physical and a numerical model, and if a numerical model, which one. Physical models are expensive to build, but often a model of the water body will already exist and if so the running cost of a physical model is usually much less than the cost of building a new numerical model. Physical models have the scaling problems discussed in Section 8.3, but usually if a physical model exists the engineer is well advised to make some use of it. Sometimes a small scale physical model can be used to provide boundary conditions for a numerical model of a local site. If a numerical model is to be used the choice is between a one-dimensional model and one that incorporates two or three spatial dimensions. In estuary studies there is an additional choice between averaging over the tidal cycle and computing the flow within the tidal cycle. One-dimensional models often work, in the sense that they are economical and their results are sufficiently accurate for the purpose at hand. The view of the writers, which we acknowledge is not shared by all practicing engineers or researchers, is that two- and three-dimensional models are still in the research stage. Two-dimensional models can give useful results in some applications, if used with care and understanding. Nevertheless numerical modeling is a field in which bigger is not necessarily better; the simplest model that can solve the problem is the one to use.

In what follows we discuss mostly estuary and coastal models but it should be understood that the same remarks apply to river models. We do not attempt to give a simple set of rules on how to choose a model; the range of possible

TABLE 8.1

Types of Transport Models

Code	Name	Description
1A	One-dimensional tidally averaged	A numerical solution of the one-dimensional tidally averaged dispersion equation [Eq. (7.38)]. May be steady state, meaning that the coefficients are constant in time, or the dispersion coefficient and flow parameters may vary between tidal cycles
1T	One-dimensional tidally varying	A numerical solution of Eq. (7.46), in which the tidal elevation, velocity, and dispersion coefficient vary during the tidal cycle.
1TB	Branching one-dimensional tidally varying	A network of 1T models connected at junctions.
2VA 2HA	Two-dimensional tidally averaged	A numerical solution of a two-dimensional tidally-averaged dispersion equation. V means a model which uses a vertical, x–z plane and H means a model which uses a horizontal x–y plane.
2VT 2HT	Two-dimensional tidally varying	Similar to 2VA and 2HA except that the tidal elevation and flow velocity vary during the tidal cycle.
3A 3T	Three-dimensional	Three-dimensional tidally averaged (A) and tidally varying (T) numerical models.
P	Physical	A small-scale physical replica of the prototype geometry with provisions for generating tidal and river flows.
NP	Hybred numerical physical	A combination of a physical and a numerical model, using one model to generate input information for the other.

problems and water bodies is too great. Rather, we begin by reviewing the dispersion mechanisms discussed in previous chapters, and remark on what types of models replicate which mechanisms. Table 8.1 lists the types of pollutant transport models which are generally available and defines a code which will be convenient for reference. The numerical models require a specification of the flow, which may be provided by a separate similar flow model. The numerical models may also include a section to compute chemical and biological reactions of constituents, but the results of that section usually do not affect the flow and transport computations. Cases where the biological reactions do affect the flow, for example, the growth of algae limiting light penetration and changing the stratification, are beyond the scope of our discussion here.

"*Trapping*" (Section 7.2.2.3) can be modeled by a 2HT model, a physical model, or a 1TB model in which branches represent the "traps." Trapping is probably the dominant dispersion mechanism in many estuaries, and it seems that most cases where 2HT and physical models have been well verified for dispersion have been because of simulation of the trapping mechanism.

Density-driven circulation (Section 7.2.3) can be modeled by 2VA and 2VT numerical models if transverse gravitational circulation is not important, and in theory by 3A and 3T models. A substantial difficulty, however, is that if density-driven currents are important the equations determining the flow and the salinity distribution are coupled. Most models run a program to solve for the flow after a repetition of two or three tidal cycles, and use the result as input to a program to solve for the transport of salinity and pollutants over many tidal cycles. If the programs are coupled they must be run together; this greatly increases the running cost and limits the ability of the model to simulate the large number of tidal cycles usually needed to reach a steady state salinity or pollutant distribution.

Physical models reproduce density-driven circulation, but most physical models are built with distorted scales that may distort the internal flow. No present model seems to be entirely capable of predicting density-driven circulation in a real estuary; the best engineering practice, if density-driven circulation is a dominant mechanism, seems to be to use a physical model well calibrated by prototype data.

Tidal pumping (Section 7.2.2.2) is reproduced by a 2HT or a physical model, but the accuracy of either may be difficult to establish. For instance, 2HT models documented to date do not accurately reproduce jet flows and entrainment of the sort sketched in Fig. 7.6. On the other hand, pumping through a channel as sketched in Fig. 7.7 can be modeled accurately by a 1TB model. The pumping sketched in Fig. 7.8 was computed by a 2HT model: similar results have been obtained for larger bodies of water such as the North Sea using a combination of 2HT and 2HA models (Nihoul and Ronday, 1976).

Shear flow dispersion can be modeled by a 2HT or 2VT model. Physical models include shear flow dispersion, but Fischer and Holley (1971) showed that if the model scales are distorted the magnitude of the dispersion coefficient resulting from this mechanism will probably not be correct.

Wind effects can be included in all the two- and three-dimensional numerical models, and there have been cases where fans were used to generate wind currents in physical models. Wind driven gyres like the one sketched in Fig. 7.3 can be generated by 2HT models, as illustrated in Fig. 8.1.

Rotational effects are easily included in 2HT models. Rotating physical models of lakes and bays have been built, and there is a rotating model of the North

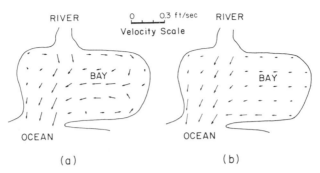

Figure 8.1 A wind-driven gyre computed by a 2HT model. (a) Velocities computed during an ebb tide with a wind at 10 mph from the west. (b) The same tide but with no wind.

Sea at Grenoble, but most estuary models do not rotate. Geophysical fluid dynamicists have constructed a number of 3A models of lakes and coastal areas, where the flow is driven by wind, thermohaline, and rotational effects, but a 3T model of an estuary with important wind, tide, and Coriolis forcing does not seem to be presently practical. It should be noted that in a deep estuary rotation often induces vertical circulations which cannot be computed by a 2HT model.

Catastrophic and seasonal changes have usually been simulated only in 1A and physical models. The physical models of the Delaware Estuary and San Francisco Bay, for example, have been used to simulate year-long periods of draught, and it is common to use a one-dimensional tidally averaged model to simulate a period of a year or more. The expense of running more complicated numerical models has generally limited their use to simulating shorter periods, usually with a constant meteorologic and tidal input.

This brief review has made clear, we hope, that each type of model has its place but that no one type should be used in every place. In the remainder of this chapter we study the use of one-dimensional and physical models, for which there is an extensive body of operating experience. We mention some fundamental aspects of two- and three-dimensional modeling and a few uses. We do not discuss current numerical modeling methods because they are changing so quickly.† The reader who wishes to make immediate use of a two-dimensional numerical model should study the most recent engineering literature; he or she should also study the descriptions of dispersion mechanisms given in the preceding chapters so that whatever model is chosen will simulate the real mechanisms as nearly as possible.

---

† Volume 11 of *Advances in Hydroscience* (V. T. Chow, ed.) contains useful summaries of the state of two- and three-dimensional numerical modeling methods as of 1978 by Liu and Leendertse (1978) and Cheng (1978).

8.2   NUMERICAL MODELS

8.2.1   One-Dimensional Models

Most river problems, and many problems in estuaries, deltas, and marshes, can be analyzed using the one-dimensional transport equation. The one-dimensional equation simply means that all quantities of interest are taken as cross-sectional averages of the true values, rather than variables over the cross section. Solutions will be accurate in any case where the time scale of the process being studied is substantially greater than the time scale for cross sectional mixing. In practical use of one-dimensional models, instantaneous complete cross sectional mixing is usually assumed. The reader may wish to refer back to more detailed discussions of the applicability of one-dimensional concepts in Sections 7.5.3 and 7.5.5.

8.2.1.1   Finite Difference One-Dimensional Models

In Section 7.5.3 we gave the tidally averaged and tidally varying forms of the one-dimensional transport equation as

$$A\frac{\partial C}{\partial t} + Q_\mathrm{f}\frac{\partial C}{\partial x} = \frac{\partial}{\partial x}\left(KA\frac{\partial C}{\partial x}\right) + \text{(source or sink terms)} \qquad (7.38)$$

and

$$\frac{\partial}{\partial t}(AC) + \frac{\partial}{\partial x}(\bar{u}AC) = \frac{\partial}{\partial x}\left(K_t A\frac{\partial C}{\partial x}\right) + \text{(source or sink terms)}. \qquad (7.46)$$

To solve Eq. (7.46) we must be given the cross-sectional area A and the mean velocity $\bar{u}$ as a function of x and t. Sometimes a separate numerical program is used to solve the one-dimensional equation of motion, but sometimes field data can be used or a simple assumption made that will provide an adequate specification. Since this is a book on mixing we do not discuss numerical solutions of the equations of motion; the writers have found that the method of characteristics solution given in the book by Streeter and Wylie (1967) is usually satisfactory, and a use of this method in branching channels is documented with a program listing by Fischer (1970). Finite difference and finite element methods may be equally satisfactory; the choice usually depends on what is most familiar and available to the investigator. To solve Eq. (7.38) we need only specify A, Q_f, and K at each grid point.

A finite-difference transport model sets out to solve either equation by means of a finite-difference representation of each derivative. In all finite difference techniques a grid system is introduced in which values of C are known at a given time t at grid points in the x direction, as shown in Fig. 8.2. The known values are subscripted as $C_{j-1,n}$, $C_{j,n}$, $C_{j+1,n}$, etc., and the unknown values to be computed are $C_{j-1,n+1}$, $C_{j,n+1}$, $C_{j+1,n+1}$, etc. The derivatives in Eq. (7.38) must be written in terms of the known and unknown values of C; the different techniques arise from different ways of expressing the derivatives.

An *explicit* technique is one in which all the derivatives are expressed in terms of known values. For instance, we may use the backward difference operator,

$$\partial C/\partial x \approx (C_{j,n} - C_{j-1,n})/\Delta x \tag{8.1}$$

the forward difference operator

$$\partial C/\partial x \approx (C_{j+1,n} - C_{j,n})/\Delta x \tag{8.2}$$

or the central difference operator

$$\partial C/\partial x \approx (C_{j+1,n} - C_{j-1,n})/2\Delta x. \tag{8.3}$$

All of these operators use only values of C known at time level n. An *implicit* scheme is one which uses some of the unknown values of C, for instance an implicit central difference operator may be written as

$$\frac{\partial C}{\partial x} \approx \frac{1}{2}\left[\frac{C_{j+1,n+1} - C_{j-1,n+1}}{2\Delta x} + \frac{C_{j+1,n} - C_{j-1,n}}{2\Delta x}\right]. \tag{8.4}$$

Explicit schemes are, of course, easier to program because the solution for each unknown can be written entirely in terms of knowns, whereas in an implicit scheme a set of simultaneous equations must be solved to obtain all the values at the new time level at the same time. On the other hand, implicit schemes are generally more stable and a longer time step can be used.

Two important requirements of any numerical scheme are that, given a flow of uniform velocity and dispersion characteristics, the advective velocity of the mean of a tracer cloud be the same as the mean advective velocity of the flow and that the variance of the cloud increase at the rate $2K$ as required by Eq. (2.22). The latter requirement means that the scheme should cause the tracer cloud to disperse at the rate called for by the dispersion coefficient used in the program, not by some other rate induced by numerical processes. Unfortunately, however, most numerical schemes do induce some unwanted numerical spreading. To see how this comes about consider the simple scheme

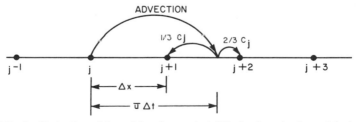

Figure 8.2 An illustration of the origin of numerical diffusion in a simple model, showing the case where $\bar{u}\Delta t/\Delta x = 1\frac{2}{3}$. The mass originating at point j is proportioned $\frac{2}{3}$ to point $j + 2$ and $\frac{1}{3}$ to point $j + 1$.

we used in Section 5.3 to model advection along the stream tubes, which the reader may verify was equivalent to using the backward difference operator. A similar scheme for representing advection is illustrated in Fig. 8.2. The mass represented by the concentration at a grid point is advected forward during a time step a distance $\bar{u} \, \Delta t/\Delta x$ grid points, and then is divided between the two nearest grid points proportionally according to the distance from each. In the figure we show the case where $\bar{u} \, \Delta t/\Delta x = 1\frac{2}{3}$; $\frac{2}{3}$ of the mass originating at point j is assigned to point $j + 2$ and $\frac{1}{3}$ is assigned to point $j + 1$. Division between the grid points is necessary because the numerical scheme has no way of representing a concentration except at a grid point, but notice that a mass originally concentrated at one point is now spread numerically over two points. If we consider a unit mass originally at point j its variance at the beginning of the time step is zero; its variance at the end of the step is

$$\sigma^2 = \int_{-\infty}^{\infty} (x - \bar{x})^2 C \, dx = (\tfrac{2}{3}\Delta x)^2(\tfrac{1}{3}) + (\tfrac{1}{3}\Delta x)^2(\tfrac{2}{3}) = \tfrac{6}{27}(\Delta x)^2. \qquad (8.5)$$

According to Eq. (2.22) this result is equivalent to a diffusive process having a coefficient $2K' = d\sigma^2/dt = [\tfrac{6}{27}(\Delta x)^2 - 0]/\Delta t$, or $K' = \tfrac{3}{27}(\Delta x)^2/\Delta t$. K' is an apparent diffusivity caused by the numerical process, and the result is often referred to as "numerical diffusion." The reader may verify that the maximum value of K' is $0.125(\Delta x)^2/\Delta t$, which occurs if the advective step carries the mass exactly half-way between grid points.

It should also be noted that the scheme illustrated in Fig. 8.2 advects the mean of the mass distribution at exactly the mean velocity of the flow, an important attribute of the backward difference operator that is not exactly shared by the others. In Section 5.3 we used a scheme based on the backward difference operator because we wanted to model differential advection in the stream tubes and needed to have the correct mean transport velocity in each stream tube. We also noted in that section that a substantial amount of numerical diffusion along the stream tubes did not matter, being overwhelmed by longitudinal dispersion caused by the differential rate of advection. A one-dimensional model, on the other hand, is essentially a one-stream tube model in which all of the effect of shear flow is expressed by the dispersion coefficient. Consequently, in using a one-dimensional model it is essential that the numerical diffusion represented by K' be kept much smaller than the dispersion represented by K, the coefficient which the operator thinks represents the real rate of dispersion. An alternate possibility, suggested by Bella and Grenny (1970), is that if the operator can accurately forecast K', the value of K used in Eq. (7.2) can be reduced accordingly. This can work, of course, only if K' generated by the numerical scheme is less than K. It is easy to write a scheme which generates much larger values of K' than the value of K one should expect in the prototype; therefore the investigator should always check the numerical diffusion properties of a proposed scheme. Usually this can be done by setting $K = 0$ in the numerical program and observing the results.

Methods are available to control numerical diffusion. Referring again to Fig. 8.2, in the absence of real diffusion the concentration we would like to place at grid point $j + 2$ at the end of the time step is the concentration $\frac{1}{3}$ of the way from grid point j to point $j + 1$ at the beginning of the step. The problem, then, is to obtain the most accurate estimate possible of the initial concentration at a location between grid points. This can be done by constructing a higher order polynomial using initial values of concentration at several grid points; for example, Hinstrup et al. (1977) use four points to construct a third order polynomial; alternatively, Holly and Preissman (1977) suggest constructing a third order polynomial from the values of concentration and its derivative at two points. The higher order the polynomial the more accurate the scheme, but of course also the more expensive the computation. One scheme that has been used frequently in estuary studies is that of Stone and Brian (1963), as applied by Thatcher and Harleman (1972). Stone and Brian found that the scheme that combined the least error in advective velocity with the least numerical dispersion was to use the implicit central difference operator [Eq. (8.4)] for the advective term and to represent the time derivative in a spread form as

$$\left.\frac{\partial C}{\partial t}\right|_{x=j} = [\tfrac{1}{6}(C_{j-1,n+1} - C_{j-1,n}) + \tfrac{2}{3}(C_{j,n+1} - C_{j,n})$$

$$+ \tfrac{1}{6}(C_{j+1,n+1} - C_{j+1,n})](\Delta t)^{-1}. \tag{8.6}$$

The diffusive term is represented by the Crank–Nicholson approximation,

$$\frac{\partial^2 C}{\partial x^2} \approx \frac{1}{2}\left[\frac{C_{j+1,n+1} - 2C_{j,n+1} + C_{j-1,n+1}}{\Delta x^2} + \frac{C_{j+1,n} - 2C_{j,n} + C_{j-1,n}}{\Delta x^2}\right]. \tag{8.7}$$

For the case studied by Stone and Brian, for which the diffusion coefficient is relatively small, this scheme seems to be the most accurate. There remains a certain amount of numerical dispersion, however, which might be of considerable importance in hydrologic applications.

Stone and Brian's method can be used equally well for a tidally averaged or a tidally varying analysis. The details of a tidally varying analysis are given by Thatcher and Harleman (1972).

Example 8.1: Numerical analysis of estuary dispersion. The following example is sufficiently simple that the reader can follow through the steps in long hand; once this is done computer programming for a larger problem is reasonably simple.

Consider an estuary 4000 ft long in which the mean fresh water velocity U_f is 1.33 ft/sec and the longitudinal dispersion coefficient K is 666 ft^2/sec. The estuary is to be divided into five grid points, one at each end and three in the center, spaced 1000 ft apart, labeled $j = 1, \ldots, 5$. Concentrations at the grid points are denoted $C_{j,n}$, where j is the index for the grid point and n the index for the time step. The boundary conditions are unit concentration at one end and zero concentration at the other end for all time, i.e.,

$$C_{1,n} = 1 \qquad \text{for all} \quad n$$
$$C_{5,n} = 0 \qquad \text{for all} \quad n.$$

The initial condition ($n = 1$) is given to be

$$C_{1,1} = C_{2,1} = C_{3,1} = 1; \qquad C_{4,1} = C_{5,1} = 0,$$

which corresponds approximately to a unit concentration in the first 2500 ft of the estuary and a zero concentration in the remaining 1500 ft. Using a time step of 500 sec compute by long hand the concentration distribution after the first time step (at $n = 2$) using the scheme suggested by Stone and Brian.

Solution. The Stone and Brian scheme for representing Eq. (7.38) about the point $j = 2$ is

$$\frac{1}{\Delta t} \left[\frac{1}{6}(C_{12} - C_{11}) + \frac{2}{3}(C_{22} - C_{21}) + \frac{1}{6}(C_{32} - C_{31}) + \frac{U_f}{4\Delta x} \right.$$

$$\left. \times (C_{32} - C_{12} + C_{31} - C_{11}) \right]$$

$$= \frac{K}{2(\Delta x)^2} (C_{32} - 2C_{22} + C_{12} + C_{31} - 2C_{21} + C_{11}).$$

Analogous equations can be written for the other interior points, $j = 3$ and $j = 4$, providing three equations for the three unknowns C_{22}, C_{32}, and C_{42} in terms of the seven knowns $C_{11}, \ldots, C_{15}, C_{12}$, and C_{52}. Setting for simplicity

$$U_f/4\Delta x = \alpha \qquad K\Delta t/2(\Delta x)^2 = \beta,$$

these equations can be written in matrix form as

$$
\begin{bmatrix}
\frac{2}{3} + 2\beta & \frac{1}{6} + \alpha - \beta & 0 \\
\frac{1}{6} - \alpha - \beta & \frac{2}{3} + 2\beta & \frac{1}{6} + \alpha - \beta \\
0 & \frac{1}{6} - \alpha - \beta & \frac{2}{3} + 2\beta
\end{bmatrix}
\begin{bmatrix}
C_{22} \\
C_{32} \\
C_{42}
\end{bmatrix}
$$

$$
=
\begin{bmatrix}
\frac{1}{6} + \alpha + \beta & \frac{2}{3} - 2\beta & \frac{1}{6} - \alpha + \beta & 0 & 0 & -\frac{1}{6} + \alpha + \beta & 0 \\
0 & \frac{1}{6} + \alpha + \beta & \frac{2}{3} - 2\beta & \frac{1}{6} - \alpha + \beta & 0 & 0 & 0 \\
0 & 0 & \frac{1}{6} + \alpha + \beta & \frac{2}{3} - 2\beta & \frac{1}{6} - \alpha + \beta & 0 & -\frac{1}{6} - \alpha + \beta
\end{bmatrix}
$$

$$
\times
\begin{bmatrix}
C_{11} \\
C_{21} \\
C_{31} \\
C_{41} \\
C_{51} \\
C_{12} \\
C_{52}
\end{bmatrix},
\qquad\qquad (8.8)
$$

where the matrix on the right consists entirely of known quantities and is equivalent to one known vector.

The general solution of Eq. (8.8) requires a matrix inversion, a standard procedure in computer application. For our problem, however, the matrix equation is the equivalent of three simultaneous equations which can be easily solved. Using the given values for the Cs on the right and the computed values $\alpha = -\frac{1}{6}$ (note that U_f is negative) and $\beta = \frac{1}{6}$ gives

$$C_{22} - \tfrac{1}{6}C_{32} = \tfrac{5}{6}$$
$$\tfrac{1}{6}C_{22} + C_{32} - \tfrac{1}{6}C_{42} = \tfrac{1}{2}$$
$$\tfrac{1}{6}C_{32} + C_{42} = \tfrac{1}{6},$$

from which we obtain $C_{22} = \frac{49}{54}$, $C_{32} = \frac{4}{9}$, and $C_{42} = \frac{5}{54}$. The solution is shown in Fig. 8.3. ∎

8.2.1.2 A Lagrangian Transport Model

Another way to minimize numerical diffusion, which is also computationally more efficient than even the simplest finite-difference scheme, is the Lagrangian method developed by Fischer (1972b). The method is similar to the one for reservoirs already described in Chapter 6. The method establishes marked volumes of water, strung out along the channel axis, which are moved along the channel at the mean flow velocity. Numerical diffusion can be almost entirely eliminated, because there is no need to establish grid points and allocate concentrations to specific locations; rather, the "grid" is a set of moving points which represent the centers of the marked volumes. Longitudinal dispersion between marked volumes can be set at whatever rate the programmer thinks appropriate.

Figure 8.4 shows how the method generates and moves volume "elements" for the case of a tidal channel connecting a constant river outflow to a tidal ocean. During each time step a new "element" is created at the river end to contain the water volume and constituent concentration of the river discharge during the time step. At the beginning of a time step the water in the channel is divided into volume "elements" numbered 1, 2, 3, ..., n along the channel axis; each

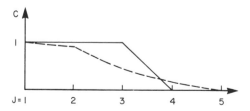

Figure 8.3 ———— The initial condition and –––– the numerical solution after one time step for example 8.1.

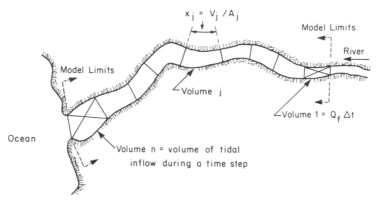

Figure 8.4 Arrangement of volume elements in Fischer's (1972b) Lagrangian transport model.

element containing volume V_j has a concentration assigned to it C_j. When the new river element is added it becomes V_1 with concentration C_1; the index of each element already in the channel is increased by one to allow for the new element. At the ocean end a new element must be added if the tide is flooding; if the tide is ebbing part of one or more of the original elements is lost to the ocean. The volume gained or lost at the ocean is determined by the continuity equation,

$$V_0 = V_r - A_s \Delta h \qquad (8.9)$$

where V_0 is the volume flowing out of the channel into the ocean, $V_r = Q_f \Delta t$ the volume flowing into the channel from the river, A_s the surface area of the channel, and Δh the change in surface elevation during the time step. If V_0 is positive the volume in element $n + 1$ (the original element n) is reduced by V_0, and if element $n + 1$ is eliminated entirely element n is reduced, etc., until enough outflow is provided. If V_0 is negative (tidal inflow) a new volume element, $n + 2$, is created, with volume $V_{n+2} = V_0$ and concentration C_{n+2} corresponding to the concentration of substance in the ocean.

The remarkable feature of this method is that advection is modeled without the need to compute the motion or location of any elements except those flowing into or out of the channel at its ends. After the advective step is complete longitudinal dispersion, chemical reactions, and sources and sinks can be treated as in other models. During these steps the location of a volume element is computed when required, using the simple relationship.

$$x_i = \sum_{j=1}^{i} (V_j/A_j), \qquad (8.10)$$

in which x_i is the distance of the downstream face of the element from the river end of the channel and A_j is the channel cross-sectional area associated with volume element V_j.

A detailed description of how to program the Lagrangian approach is contained in the report by Fischer (1972b); the report also shows how to combine flow in several channels to form a two-dimensional network, as discussed in the next section.

8.2.1.3   One-Dimensional Network Models

Often a water body with a relatively complex plan can be represented by a network of one-dimensional channels. The most obvious example is a delta, in which interconnected channels suggest a model of interconnected one-dimensional segments. Network analyses can also be applied in less obvious circumstances; for instance a set of connected bays and channels might be idealized as one-dimensional segments, even though the flow in the individual bays might have important two-dimensional aspects.

One widely used network model is the Link–Node Model, developed by the firm of Water Resources Engineers, Inc., in the 1960s and published as the "Dynamic Estuary Model" by the Federal Water Quality Administration (the predecessor to the Environmental Protection Agency). The model is equivalent to dividing the water body into a set of continuously stirred tanks, connected by pipes. The "tanks" are zones of surface area layed out on a map of the water body, as illustrated in Fig. 8.5, each represented by a node. The "pipes," called "links" in the model, are real or fictional channels between the nodes. If the

Figure 8.5   Schematization used to model the Suisun Bay, California, using the link–node model.

nodes are layed out in a line along a channel the hydraulic properties of the links are those of the real channel; if the nodes are arrayed to represent a bay fictional hydraulic properties of the links must be invented. In either case the flow in a link is computed by using the manning formula for flow in open channels, the slope being computed in each time step by comparison of water surface elevations at the nodes at each end of the link. Much of the attraction of the link–node method is its ability to combine in one program a quasi-two-dimensional flow in an open portion and a network of one-dimensional channels in a channelized portion of the same water body. The distribution of nodes in a bay can define the equivalent of a two-dimensional grid, although the program is not truly two-dimensional because flow is permitted only in the direction of a link.

A significant disadvantage of the link–node model is that concentration is represented only at the nodes, which are the equivalent of the grid points in a finite difference model. If the nodal spacing is large there can be a substantial amount of numerical diffusion. For instance, if the nodes are 1 km apart, a typical separation used in models of San Francisco Bay, and if a time step of 15 min is used, the numerical dispersion can be estimated to be on the order of 100 m^2/sec. A reasonable order of magnitude of the actual dispersion coefficient in the bay is on the order of 200 m^2/sec, so it is likely that the link–node model with a 15 min time step will overdisperse.

If the flow is entirely confined to a network of one-dimensional channels, or can be schematized as such, a network of the Lagrangian models described in the previous section eliminates most of the problem of numerical diffusion. Most of the complexity of programming is concerned with describing the transfers at channel junctions; otherwise the method is computationally very efficient and flexible. A 40-channel application of the method to model the Suisun Marsh, California, is described by Fischer (1977).

8.2.2 Multidimensional Models

There is a continuing disagreement among experts on the value of multidimensional models. Although numerical techniques and computer hardware improve at a steady pace, the concern is how well the physics of flow and exchange are understood and properly expressed. The mixing coefficients used in numerical models express the net result of all processes whose scale is less than the grid size of the model. These processes include shear flow spatial velocity variations as well as turbulent fluctuations, and the ability of a numerical program to predict flow or pollutant transport is limited by how well the processes that cause local mixing are understood and represented in the mixing coefficients. For some types of model the transport is primarily by turbulent fluctuations and the need is to estimate the turbulent mixing coefficient. Higher order closure schemes for relating the turbulent mixing coefficient to the properties of the flow are leading to improved models of the atmospheric boundary layer, for example. In river and estuary models, however, turbulent transport is often a small fraction of the total transport represented by the mixing coefficient, and efforts to relate the magnitude of the coefficient to the properties of the turbulence are not likely to be useful. For example, in a vertically averaged estuary model the transverse mixing coefficient represents primarily the skewed shear flow of the velocity profile (see Fig. 4.8). A layered model could compute the effect of the shear flow explicitly, but at great extra expense. To take one example, suppose that we wished to make a numerical model of South San Francisco Bay in which the mixing coefficients expressed only turbulent transport. We would probably want approximately ten layers in the vertical, and a horizontal grid spacing on the order of 100 m. The bay is approximately 35 km long and averages 8 km wide, so we would need on the order of 300,000 grid points. Even then we would have to face questions of how accurately the turbulent processes are represented, especially if the water column was stratified. Most practical estuary models have used far fewer than 300,000 grid points (a typical number is on the order of 1000), and average over at least one spatial dimension, over the tidal cycle, or over both. The mixing coefficient represents what has been averaged, and an estimate of the magnitude of the coefficient must be based on an understanding of the physics of the underlying processes. What is in the average depends on the number of dimensions of the model; for example, in Eq. (7.19) we saw how K in the one-dimensional tidally averaged equation represents all the dispersion mechanisms discussed in Section 7.2. In multidimensional models some of the mechanisms are included in the computation of advection, and the mixing coefficients represent correspondingly less mixing.

Three-dimensional models have not had enough use in hydrologic applications to give a basis for evaluation. Two-dimensional models are used frequently. A two-dimensional model averages either over the vertical or across the cross section, depending on whether the two dimensions are in plan or in section. We focus on "in plan" models because for most applications "in section" models that assume lateral homogeneity seem to omit too much of what is important to dispersion to have engineering value. Among "in-plan" models there are two types, those which average over the tidal cycle (type 2HA), and those which attempt to replicate the tidal motion (type 2HT). Type 2HA models utilize pseudodispersion coefficients to account for the effect of the tidal motion. They begin with the equation,

$$\frac{\partial C}{\partial t} + U \frac{\partial C}{\partial x} + V \frac{\partial C}{\partial y} = \frac{1}{d} \left[\frac{\partial}{\partial x} \left(dK_x \frac{\partial C}{\partial x} \right) + \frac{\partial}{\partial y} \left(dK_y \frac{\partial C}{\partial y} \right) \right], \qquad (8.11)$$

in which $\partial/\partial t$ means a change per tidal cycle, U and V are tidal averages of the vertical averaged x- and y-direction velocities, d is the local depth, and K_x and K_y express the results of all the mechanisms that cause mixing within a tidal cycle. Terms like $(\partial/\partial x)(dK_{xy} \partial C/\partial y)$ probably ought to be included, but usually are not because K_x and K_y alone are hard enough to evaluate. Their values are necessarily large, but cannot be predicted by the theory given in earlier chapters because they result from tidal currents not replicated in the model. Moreover their values are likely to change from day to day and week to week in response to storms and tidal conditions, and from season to season in response to changes in stratification of the water column, wind and temperature climate. Thus 2HA models should be used only in conjunction with extensive field data to define the magnitude of the dispersion coefficients.

Type 2HT models are in common use and have the advantage that they represent the important dispersion mechanisms of trapping, pumping, and wind and Coriolis driven circulations. Type 2HT models appear to be practical for smaller bodies of water; the most completely documented example is the study of Jamaica Bay, a roundish bay with a diameter of approximately 3 miles (Leendertse, 1970; Leendertse and Gritton, 1971). On the other hand, 2HT models have practical difficulty representing larger water bodies, unless such a coarse spatial grid is used that the advantages of replicating the tidal cycle may be lost. Perhaps the most important limit on the practicality of 2HT models is that to be useful a transport model must be operated to simulate at least as much real time as is needed to reach an equilibrium distribution of tracer in the water body. A typical equilibrium time in a large estuary is on the order of 100 days. A 2HT model using, say, a one minute time step will require 144,000 time steps for the distribution to approach an equilibrium, which even with anticipated improvements in computers imposes a heavy burden of computer cost per run.

Type 2HT models use the same equation as 2HA models [Eq. (8.11)] except that U and V are tidal velocities and $\partial/\partial t$ means the change per time step. The magnitudes of K_x and K_y are much less, of course, since they represent only the effect of the vertical velocity profile. In fact, terms containing K_x and K_y can sometimes be neglected, because the important dispersive mechanisms are represented by the time-variable advection and numerical diffusion accomplishes the small scale mixing. For instance, Leendertse has found in his studies of Jamaica Bay that the results are essentially unaffected by changes in the dispersion coefficients. This appears to be because most of the dispersion in the Bay comes from trapping in side channels and pumping around a central island.

Type 2HT models have two important limitations. The first is implicit in vertical averaging; the water column must not be sufficiently stratified to inhibit vertical mixing. The second is because in most cases a time-varying flow field must be obtained from a first-stage 2HT flow model. The flow model must produce the flows which lead to trapping, pumping, and the other dispersion mechanisms discussed in Chapter 7. The residual circulations illustrated by Figs. 7.6 and 7.8 are caused by the nonlinear frictional and inertial terms in the equations of motion and by separation at sharp corners. Even such apparently simple separated flows as the eddy in the lee of a jetty or rapid expansion present great difficulties (see Abbott and Rasmussen, 1977). The model presented by

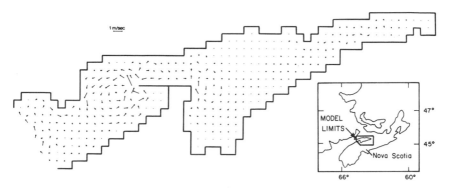

Figure 8.6a The tide-driven residual circulation in the Minas Basin, Nova Scotia, according to Tee's (1976) numerical model. [After Tee (1977).]

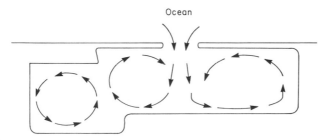

Figure 8.6b Tidal inflow into a model of a small harbor. [After Fischer (1976b).]

Leendertse (1967) included nonlinear friction and inertial terms, and has been widely used in part because of the clarity of Leendertse's description; however, the no-slip boundary condition was not imposed. Tee (1976) has described in detail a model similar to Leendertse's but incorporating the no-slip condition. Tee computed a residual circulation resulting from boundary layer separation, as shown in Fig. 8.6a. Tee (1977) has also reported prototype observations which were in reasonable agreement with the computed flow pattern, and his results are an interesting large-scale demonstration of the type of residual gyre whose source is illustrated by Fig. 7.6. It is not clear, however, whether any present 2HT flow model would be capable of accurately predicting more complex real flows, such as the three-gyre flow illustrated in Fig. 8.6b. Indeed, flow distributions computed by present 2HT models usually seem oversimplified compared to the complex patterns found in nature, and there has been disturbingly little comparison of model outputs to observations in the prototype or even in simple laboratory flows.

We do not give specific codes for any multidimensional model because techniques and methods are changing so rapidly.† At this writing it is not clear

---

† See the footnote on page 283.

whether the finite element method has significant advantages over finite difference methods. Perhaps wholly new approaches will displace both methods. A promising line of research is the use of a 2HT model for one tidal cycle to define the dispersion coefficients needed in a longer run of a 2HA model. The development of three-dimensional models is also sure to be looked on with interest, but until more is understood about the physics of vertical mixing and the formation of density gradients in stratified flow it seems unlikely that three-dimensional models will take their place as dependable tools of engineering practice.

8.3   PHYSICAL MODELS

8.3.1   Introduction

Many important estuaries and some river basins have been modeled by constructing a physical representation at small scale. Building a physical model is expensive, but using one that exists is relatively simple and cheap. For example, to study a proposed waste outfall, a dye (usually Rhodamine WT, which was developed specifically for use in physical models) can be discharged into the model at the point of the proposed outfall in the prototype, and the resulting concentration distribution can be obtained by taking samples at various points of interest. The point of discharge can easily be shifted between alternate sites, and changes of estuarine bathymetry can be studied by rebuilding a part of the model. Not only are the results useful to the engineer; they also have a powerful impact on the lay public. An escorted tour during conduct of an experiment will often mean more to a political decision maker than reams of computer output.

Indeed, physical modeling is so attractive that we are constrained to begin with a severe caution. A physical model is not an exact replica of the prototype. To contain the model in a building of reasonable size the horizontal scale must be very small, typically 1 in the model to 1000 in the prototype. The vertical scale cannot be as small; if it were a typical estuarine depth of say 5–30 m would be represented by a model depth of 5–30 mm, and the flow would be dominated by viscous and surface tension effects. Instead, vertical scales are ordinarily from 5 to 20 times the horizontal scales. The vertical exaggeration converts a typically wide and shallow cross section, as shown in Fig. 8.7a, into the more canyonlike

(a)   (b)

Figure 8.7   (a) A typical estuarine cross section (b) the transformed shape of the same cross section in a 10 to 1 distorted model.

cross section shown in Fig. 8.7b. The conversion serves the essential purpose of making the model flow turbulent, but it also changes the longitudinal slope of the channels and distorts rates of vertical and transverse mixing. The tendency of the model flow to be too fast, because of the increased slope, must be resisted by adding friction to the channels. Figure 8.8 shows a typical physical model of an estuary; the vertical copper strips which are arrayed over the entire channel provide the extra friction needed to counteract the distorted channel and water surface slopes. Of course, the vertical strips also stimulate mixing, but there is no certainty that the mixing rates generated by the strips in the model and the rates generated by tidal currents and bottom friction in the prototype will be the same. During the construction process most models are calibrated against prototype observations of tidal elevation, and often against prototype observations of currents in the main channels. The copper strips are bent up or down until the "fit" of model and prototype data seems acceptable. These calibrations do *not* assure that mixing will be modeled correctly. The cross-sectional distribution of velocity is usually not well enough known in the prototype to be tested in the model, and local rates of transverse and vertical mixing are usually not tested at all. Where a salinity variation exists it is customary to compare model and prototype salinities; many reports of model studies state something like, "the salinity verification was considered to be excellent." We will see later why such statements must be viewed with caution. As a general rule, a physical model should be regarded as unverified with respect to mixing studies until proven otherwise.

Figure 8.8   A section of the San Francisco Bay model operated by the U.S. Army Corps of Engineers. (Photo courtesy Mr. William B. Kirkland, Jr.)

Often a quantitative "proof," i.e., a direct comparison of model versus prototype, will not be possible. The model may still be useful for comparative studies; for instance, comparison of model results for alternate outfall sites may lead to the choice of the best site, even though the absolute concentrations in model and prototype may not be the same. Nevertheless, justification of use of the model results should always be the burden of the investigator; it should not be considered adequate engineering practice to make use of physical model results without careful attention to possible inaccuracies caused by distortion and the reduced scale.

In this section we begin by reviewing the so-called model "laws," i.e., the relationships between model and prototype that are dictated by the hydrodynamic equations. Next we review the extent to which models have been verified for use in mixing studies and the relevant experience. Finally we describe some uses of the physical model of San Francisco Bay to assist in studies of salinity intrusion and the siting and design of a sewage outfall. In this section, however, we limit the discussion to studies of discharges that do not dynamically affect the flow of the receiving water; modeling of thermal discharges, which do often affect the receiving flow, is discussed in Section 10.5.

8.3.2   Model Laws and Scaling Ratios

Henderson (1966) discusses the principles of modeling of open channel flows in general, including movable bed models. Here we limit ourselves to fixed bed estuary models and associated river and coastal sections of the sort commonly used to study dispersion. These models are usually large (100 m across is not unusual), and are sculpted in concrete with copper strips to provide the necessary extra roughness. The designer of such a model has only two independent choices, the length ratio (i.e., the ratio of a length in the model to the same length in the prototype) and the depth ratio. We will use the notation

$$L_r = \text{length ratio} = \frac{\text{horizontal distance in prototype}}{\text{horizontal distance in model}} = \frac{L_p}{L_m}$$

$$d_r = \text{depth ratio} = \frac{\text{depth in prototype}}{\text{depth in model}} = \frac{d_p}{d_m}.$$

The subscript r will be used throughout to indicate the ratio of a prototype quantity to a model quantity; subscripts m and p mean model and prototype quantities, respectively. Once the length and depth ratios have been selected, the ratios of all other quantities are established by physical laws or by attempts to make the model reproduce observed flows in the prototype. An important requirement is that a frictionless small amplitude wave propagates at the correct velocity. The wave celerity is $c = \sqrt{gd}$, so $c_m/c_p = \sqrt{gd_m/gd_p}$. Since gravity cannot be varied the celerity ratio is

$$c_r = d_r^{1/2}. \tag{8.12}$$

The time required for the wave to propagate a horizontal distance L is $t = L/c$. Since $t_m = L_m/c_m$ and $t_p = L_p/c_p$ the time ratio required for similarity of wave propagation is

$$t_r = L_r/c_r = L_r d_r^{-1/2}. \tag{8.13}$$

The velocity ratio is determined by the requirement that a fluid particle that moves a distance $L_p = u_p t_p$ in the prototype, where L is any horizontal distance, must move an equivalent distance $L_m = u_m t_m$ in the model. Thus

$$u_r = L_r/t_r = c_r = d_r^{1/2}. \tag{8.14}$$

Equation (8.14) establishes that the Froude number, $u/(gd)^{1/2}$, is the same in the model and the prototype; most river and estuary models impose this requirement, and are sometimes referred to as "Froude law" models. Froude number equivalence establishes that the ratio of gravitational to inertial forces is the same in the model and in the prototype. The propagation of tidal and flood waves depends on gravitational, inertial, and frictional forces; Froude law scaling assures the proper ratio of gravitational forces, and the copper strips are used, as described later, to obtain the proper ratio of frictional forces.

In density stratified flows it is also necessary to obtain the correct ratio for internal wave velocities. This requires that the internal Froude number, $F_i = u[(\Delta\rho/\rho)gd]^{-1/2}$, introduced in Eq. (6.12) be the same in model and prototype. Equivalence of external and internal Froude numbers at the same time can only be achieved by having the same density ratios in model and prototype. Customarily the ocean salinity is maintained in the model at approximately the value of the prototype ocean and fresh water is used for the river outflows. If model and prototype temperatures are similar this means that the imposed density difference is the same. As with the tidal flows, however, frictional forces are also important to the internal flows, and a verification of the internal structure is necessary as described in the next section.

Other important ratios are derived from the ones already given, and include:

Slope ratio	$S_r - d_r/L_r$,
Width ratio	$W_r = L_r$,
Cross sectional area ratio	$A_r = W_r d_r = L_r d_r$,
Discharge ratio	$Q_r = A_r u_r = L_r d_r^{3/2}$.

Ratios important to mixing analyses include the shear velocity ratio, $u_r{}^* = (gds)_r^{1/2} = d_r/L_r^{1/2}$, and the ratios necessary for correct modeling of vertical and transverse mixing. The correct scaling for a mixing coefficient is derivable from

Eq. (2.25), $\varepsilon = \frac{1}{2}(d\sigma^2/dt)$ which can be subscripted to $\varepsilon_r = \sigma_r^2/t_r$. Thus for turbulent mixing to be modeled correctly it is necessary to have:

$$\varepsilon_{v_r} = d_r^2/t_r = d_r^{5/2}/L_r, \qquad \varepsilon_{t_r} = L_r^2/t_r = L_r d_r^{1/2}. \tag{8.15}$$

Note, however, that $(du^*)_r = d_r^2 L_r^{1/2}$, which is quite different from the required mixing coefficient ratios and suggests the likelihood that turbulent mixing may not be modeled correctly. Some observations of prototype versus model mixing are discussed in the next section.

Length are depth scales of 1000 and 100, respectively, are common for models; for example, the model of San Francisco Bay discussed in Section 3.4 uses these scales. The resulting ratios are as follows:

$$L_r = 1000 \qquad\qquad Q_r = 1,000,000$$
$$d_r = 100 \qquad\qquad u_r^* = 3.13$$
$$u_r = 10 \qquad\qquad (du^*)_r = 313$$
$$t_r = 100$$
$$s_r = \tfrac{1}{10}$$

Of these, the model operator controls only the discharge ratio for the tributary inflows, the height ratio and time ratio for the ocean tides, and the salinity ratio in the ocean. The actual elevations, currents, and salinities occurring throughout the model are determined in part by the frictional characteristic of the model channels and the distribution of the copper strips. Thus an essential phase of model construction is the laborious adjustment of the copper strips to make the model conform as nearly as it can to the stated ratios. This phase is described in the next section.

8.3.3 Model Verification

8.3.3.1 Tidal Elevations and Currents

The first step in model verification is comparison of model tidal depths and currents to the prototype. Sometimes only prototype depths are available, since measuring currents in the prototype is usually difficult. If the model simulates the depths correctly it must simulate the cross-sectional mean currents correctly, because the currents supply the tidal prism. The detailed distribution of currents across the cross section is much more difficult to verify, because adequate prototype data are usually not available. For the model of the Delaware, for example, prototype observations were taken only at three stations across each cross section. For the model of San Francisco Bay only channel centerline observations were used. Currents were measured at three depths, and the vertical distribution of current at the observation stations was modeled with reasonable accuracy. Unfortunately there is no way to know whether the transverse velocity distribution is correct, because there have been no calibration data taken in the wide shallow portions away from the ship channel.

8.3.3.2 Vertical and Longitudinal Salinity Gradients

After the model has been adjusted to yield the correct tides and currents it is hoped that the salinity distribution will also match the prototype. If not, the cure may be worse than the disease. There are three ways of adjusting the salinity distribution without greatly changing the flow calibration. The fresh water discharge from tributary rivers can be changed from its scaled value to push the longitudinal salinity profile up or down estuary; fans can be used to agitate the water surface to stimulate vertical mixing; or air can be bubbled from small orifices in the channel bottom. None of these methods have a theoretical base, but they give the operator ways of making empirical adjustments. An example of the use of fans and bubbles is the U.S. Army Corps of Engineers model of Galveston Bay and the Houston Ship Canal (Bobb et al., 1973). When the model was first operated the flow was observed to be more stratified than in the prototype. Bubbles were used along the centerline of the ship channel, and an array of fans was used to stimulate vertical mixing in the wide expanses of Galveston Bay. With these Herculean measures an adequate match of model and prototype was obtained.

The vertical roughness strips seem to play a crucial role in establishing the vertical salinity profile. In the prototype vertical mixing is caused by wind mixing at the surface, bottom and side shear induced turbulence, and perhaps by breaking of internal waves. In the exaggerated depth of a distorted model these mechanisms are inadequate. It has been found that if the roughness strips are placed horizontally along the bottom, instead of vertically through the water column, a model of a continuously stratified prototype will stratify into two layers with a sharp density step near middepth (Collar and Mackay, 1973; Abraham et al., 1975). The vertical strips cause a wake throughout the water column which assures a distributed field of turbulence and destroys the sharp step; there may also be vertical flows down the face of the strips because of the dynamic pressure gradient, similar to the flow that erodes scour holes in front of bridge piers. On the other hand, there is no theoretical reason to think that the strips will automatically generate the right amount of vertical transport, so in each case the model must be compared to prototype data.

A particularly extensive verification of vertical mixing rates has been obtained as part of a study of the salinity structure of South San Francisco Bay by Imberger et al. (1977). The model was operated to simulate prototype inflows during the period from October, 1972 to October, 1973. Salinity along the channel centerline was observed at high and low slack water during various tidal cycles, in sufficient detail for construction of a two-dimensional plot of the vertical salinity structure. Prototype data obtained by the U.S. Geological survey during the same period were available. The study began at the end of a long period of low discharge, typical of summer conditions in California; the South Bay was nearly totally mixed at a uniform concentration of approximately 31 ppt and the North Bay was vertically well mixed with a longitudinal salinity intrusion of approximately 40 miles. During the period of simulation there was a major rise in the outflow hydrograph to a peak of 188,000 cfs, followed by a recession in February and March and a return to typical summer low flow discharge of approximately 5000 cfs from June to October. The prototype observations showed that during the high flow periods the salinity in the North Bay was repulsed to an intrusion length of approximately 20 miles, the flow in the North Bay became highly stratified, and some of the fresh water from the discharge into the North Bay flowed into the South Bay (as a stratified lens). The South Bay was vertically stratified for a period of approximately a month, after which it became well mixed vertically and the effects of the fresh water overflow were slowly dissipated by tidal flushing. For the most part these features were adequately reproduced by the model. Figure 8.9 shows the model prototype comparison on February 9, 1973 when the Bay was at its most stratified. The vertical salinity structure in the North Bay was well represented. In the South Bay vertical mixing over the San Bruno Shoals seemed to be too fast in the model, and stratification persisted

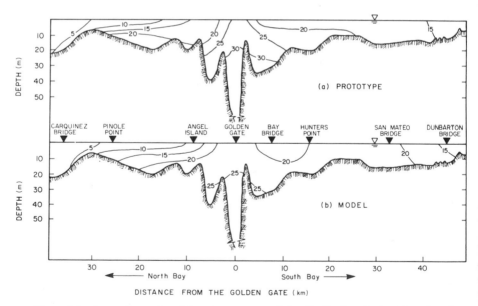

Figure 8.9 Comparison of model and prototype salinity distributions in San Francisco Bay on February 9, 1973. [After Imberger *et al.* (1977).] (a) Prototype observations taken mostly near lower low water except near lower high water in the northern part of the South Bay. (b) Model observations at lower low water. Contours are salinity in parts per thousand.

longer in the prototype than in the model. Otherwise, vertical mixing seemed to be well represented in the model throughout the process of formation and destruction of stratification.

Whether a model will correctly represent longitudinal salinity intrusion is at once a more immediately practical question and one which is less well documented. We would expect a model that is adequately calibrated for the current distribution to represent trapping and pumping effects correctly, to the extent that prototype data are adequate to calibrate the model for transverse velocity profiles. Most models make no attempt to reproduce wind driven circulations, and it is questionable whether density-driven circulation or the shear effect are properly modeled. Thus a model might be adjusted as well as possible for depths and currents, have the right rate of vertical mixing, and still not simulate the correct degree of longitudinal intrusion. Sometimes models appear to simulate intrusion correctly, using the properly scaled value of fresh water inflow, but sometimes it is necessary to alter the fresh water discharge to bring the salinity distribution to where it belongs. Unfortunately, adjustments of this sort are not always made public because of the political sensitivity of the results of many model studies; it appears that an adjustment of the fresh water outflow of a factor of two is not uncommon. In some cases the report of a model study states what was done, but not why, and it is not clear whether an adjustment was made consciously or not. For example, the model of Grey's Harbor, Washington (Brogdan, 1972), was calibrated for salinity intrusion using prototype data collected on October 8–9, 1967 and a fresh water discharge through the various tributaries of 11,400 ft^3/sec. In the prototype the gaged discharge varied from 850 to 11,910 cfs in the 10 day period preceding the salinity survey. The report of the study gives no indication of how the figure of 11,400 ft^3/sec was selected. Similarly, a model of the Hudson estuary (Simmons and Bobb, 1965) was calibrated using a discharge of 20,800 ft^3/sec, although the prototype flow varied rapidly in the month before the salinity survey in the range of 6330–34,800 ft^3/sec. The experience in the Delaware estuary model (Letter and McAnally, 1975) is also of interest. The model was initially calibrated by comparing model and prototype salinity intrusion during the 12 month period March 1931–February 1932 (see Fig. 8.10). During the drought of 1965 the model was used to predict whether salinity would intrude as far as the water supply intakes at Torresdale. Initially the model prediction appeared to be in error. An addition of a constant inflow of 200 ft^3/sec, which was thought to represent ungaged flow of ground water, small streams, and sewage treatment plants, corrected the problem and led to adequate predictions.

We have not been able to discover documentation of any model reaching exactly the same steady state as its prototype, given a prolonged period of steady inflow, but this is in part because in nature prolonged periods of steady inflow are rare and in part because, as the Delaware experience shows, it is devilishly difficult to determine the exact inflow of fresh water into most estuaries. It

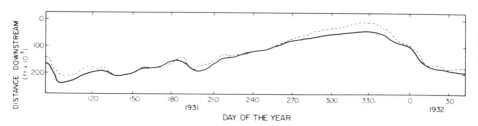

Figure 8.10   Salinity intrusion in the Delaware Estuary, 1931–32 (after U.S. Army, 1956). ——— location of 50 ppm chloride concentration in the prototype; – – – location predicted by the physical model. The vertical scale is distance down channel from Philadelphia in thousands of feet.

appears that models are generally dependable in a comparative sense, when used to indicate the change of salinity given a change in flow conditions. However the absolute relationship between salinity at a point and river outflow cannot be dependably determined by a model.

8.3.3.3   Turbulent Mixing

Even though models are often used to study near-source dispersion problems, it is not customary to attempt verification of local turbulent mixing. In a stratified model verification of the local vertical salinity gradient implies verification of the rate of vertical mixing. It implies nothing about rates of transverse mixing, and we know of no data concerning the ability of a stratified model to simulate transverse mixing.

The situation in an unstratified model is equally unclear. In the model most of the turbulence is generated by the roughness strips, whereas in the prototype it is generated by the bottom shear stress. If we hypothesize that the mixing coefficients are proportional to du^* in the model as well as the prototype, mixing in the model will be too slow vertically and too fast transversely, as can be seen by comparing the ratios $(du^*)_r$, ε_{v_r}, and ε_{t_r} [see Eq. (8.15)]. For the San Francisco Bay model the vertical coefficient would be about three times too small, and the transverse coefficient 36 times too large. On the other hand, Fischer and Hanamura (1975) studied transverse mixing in a rectangular flume with vertical strips and found that ε_t was proportional to the strip width and the mean flow velocity, rather than du^*. They suggested that the transverse mixing coefficient in the San Francisco Bay model might be about right because of a fortuitous choice of strip width. The only direct experimental evidence we know of has been described by Crickmore (1972). Crickmore studied a coastal flow near Heysham, England (see Fig. 8.11), and reported field and model observations of dispersion of a continuous source of tracer. The model plume spread faster transversely and slower vertically than the prototype one, as shown in Fig. 8.12. Crickmore concluded that for coastal studies a model was a valuable way to

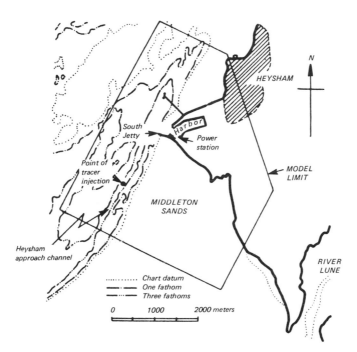

Figure 8.11   Environs of Heysham Harbor, showing the extent of the physical model. [After Crickmore (1972).]

show the mean flow paths prevailing throughout the tide, but that the rate of turbulent mixing had to be estimated by field studies.

8.3.3.4   Tidal "Flushing"

It is generally assumed that a model that correctly simulates the distribution of salinity will also correctly simulate the flushing of any dissolved tracer. The reason is that fresh water is a form of tracer, and the salinity distribution is a measure of the flushing of fresh water (we used this concept in the simple flushing analysis described in Section 7.5.4). This reasoning is only exactly correct if the pollutant to be flushed is introduced into the estuary in exactly the same way as the fresh water; otherwise initial dispersion from a local source may not be modeled correctly because of incorrect modeling of local turbulent mixing or near source advection. Far from the source, however, it is reasonable to expect that pollutant dispersion will be modeled to the same degree of accuracy as salinity intrusion.

It is worth noting that if a given model requires twice the scaled fresh water discharge to simulate salinity intrusion it means that the model "flushes" fresh water twice as fast as it ought to. Presumably any injected tracer will also be

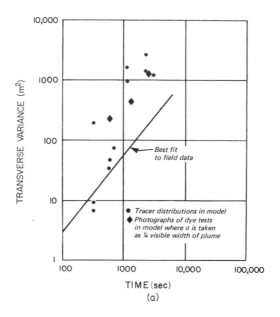

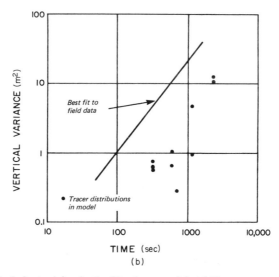

Figure 8.12 Turbulent mixing in the Heysham model. (a) Transverse variance versus travel time from the source. (b) Vertical variance versus travel time from the source. [After Crickmore (1972).]

flushed too fast, and it would be necessary to double the scaled discharge of a waste outfall to obtain concentrations in the model similar to concentrations in the prototype.

8.3.4 The San Francisco Bay Model: A Case Study

The San Francisco Bay model (scales 1000 horizontally and 100 vertically) was constructed in the early 1950s to help evaluate a controversial plan to construct a solid barrier across northern San Francisco Bay. The plan was found impractical, but the model has been used for many other studies. In the 1960s the model was enlarged to include the Sacramento-San Joaquin Delta, as outlined in Fig. 8.13. In recent years the model has been used to study a

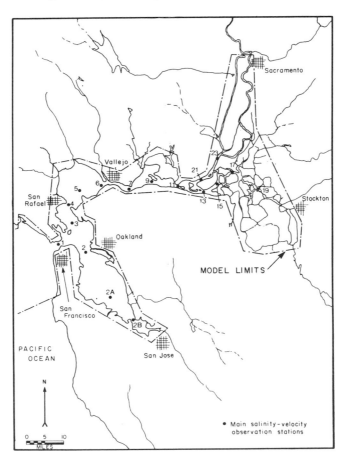

Figure 8.13 The San Francisco Bay and delta system, showing limits of the physical model and main sampling stations.

proposal to deepen the ship channel to Stockton, a proposal to divert Sacramento River water into a canal around the Delta, and a number of proposals to construct municipal and industrial waste outfalls. In this section we review the verification of the model and the results obtained from several of the studies. Our purpose is to give an account of the various ways a typical model can be used, as well as some of the precautions that may be necessary in analyzing the results. We focus on one model to be specific, but most of what is said would apply equally well to other models, and there are many models in laboratories throughout the world in which similar studies have been made.

8.3.4.1   Verification

Initial verification of the model was accomplished by comparing model and prototype tidal elevations, currents, and salinities at stations along the channel centerlines, as identified in Fig. 8.13. The model was operated with outflows of 16,000 and 175,000 ft^3/sec, and compared to prototype observations. A typical comparison is shown in Fig. 8.14. The comparisons at all stations were considered adequate, and the model was deemed verified for the initial studies.

Further verification was required when the model was expanded to include the Delta. Since a steady state is hardly ever reached in the Delta, it was considered necessary to simulate a year of real hydrology. The water year 1967–68 was chosen. Figure 8.15 shows a typical comparison of salinities throughout the year long period, and appears to demonstrate that the model accurately simulates salinity intrusion. Unfortunately, however, there is a difficulty peculiar to this model. The fresh water discharge Q_f downstream of the Delta, is the sum of gaged river inflows less gaged withdrawals to canals and consumptive use by agriculture in the Delta. Consumptive use is not gaged and is difficult to estimate, but in the summer it is a substantial fraction of the river inflow. Therefore, the net outflow which controls salinity intrusion in the prototype is not accurately known; estimates for a specific month have often differed by as much as 2000 ft^3/sec. The salinity verification shown in Fig. 8.15 was obtained after three less

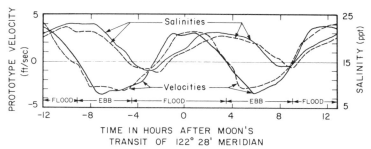

Figure 8.14   Comparison of model versus prototype observations at middepth at station I (between stations 5 and 6) in the San Francisco Bay model at an outflow of 16,000 ft^3/sec. ———— prototype; ––– model. [After U.S. Army (1963).]

successful attempts by using one particular estimate of consumptive use in the Delta. The fact that model-prototype agreement could be obtained does not prove (or disprove) that the model simulates the correct relationship between net delta outflow and salinity intrusion because there is no way of knowing whether the estimate of delta outflow used for the model run was correct.

8.3.4.2   Studies of Salinity Intrusion

One of the most important uses of the model has been to study salinity intrusion. The U.S. Army Corps of Engineers has studied the effect of additional deepening and straightening of the ship channel, and the U.S. Bureau of Reclamation has studied the effect of the proposed peripheral canal. The question raised in the dredging studies was whether deepening the shipping channel by five feet would increase the length of salinity intrusion. The shipping channel is only 600 ft wide, and the only intrusion mechanism likely to be affected by deepening the shipping channel would be density-driven circulation. Fischer and Dudley (1975) report the results of a study designed specifically to identify

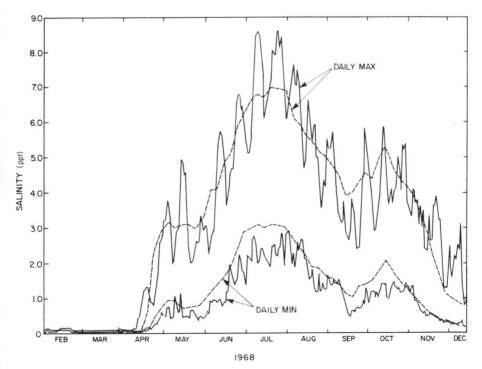

Figure 8.15 Salinities observed in the Sacramento River at Collinsville, California, in 1968; ———— prototype; ––– model during a simulation of what was thought to be the 1968 hydrology. [After U.S. Army (1974).]

the contribution of density-driven circulation during periods of low flow. The model was operated to reach study state with a delta outflow of 4400 cfs (a typical summer low flow) and with the ocean salinity kept at the usual 33 ppt. Then the experiment was repeated with identical conditions except that the ocean salinity was reduced to 11 ppt. Figure 8.16 shows the intrusion of relative salinity; within experimental accuracy there was hardly any change. The experiment shows that during low flow conditions ocean salinity acts almost as a passive tracer of flushing; as far as overall salinity intrusion is concerned there is little dynamic contribution due to density differences. Fischer and Dudley's results suggest that deepening the channel would have little effect on salinity intrusion, gravitational circulation not being of much importance.

The studies of the proposed peripheral canal by the Bureau of Reclamation have been concerned with how water quality can be maintained throughout the Delta while increasing the flow of fresh water to the forebay of the California Aquaduct for transmission southward. Most of the fresh water comes into the Delta from the north via the Sacramento River; the aquaduct forebay is in the South, and without the canal the aquaduct flow must transit the Delta through

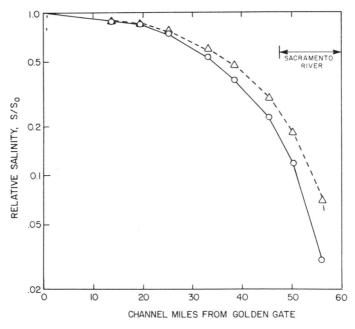

Figure 8.16 Salinity intrusion in San Francisco Bay and the Sacramento-San Joaquin delta. ——— Observed relative salinity in the physical model with $S_0 = 33,000$ ppm (normal ocean salinity); ——— observed relative salinity with $S_0 = 11,000$ ppm (ocean salinity reduced by a factor of three). Test conditions were the same for both cases, except for the ocean salinity. [After Fischer and Dudley (1975).]

the existing channels. The proposal is to route the aquaduct flow around the Delta, while releasing some fresh water into channels all round the periphery. With the proposed canal the freshwater releases for salinity repulsion would converge radially on the outlet and the fresh water for use further south would be isolated from ocean salinity. Figure 8.17 shows a typical comparison of salinity distribution with the canal versus without the canal. The most significant difference is that salinity intrusion into the San Joaquin section of the delta is less with the canal, because the throughflow is routed around the natural channels. The results shown in the figure are an example of why the model is a useful tool even though the correct salinity versus outflow relationship is not known. The results of comparative studies are thought to be dependable, even though absolute results may not be available. The model can be used to compare operational schemes, such as different release patterns from the canal, or changes in channel geometry such as deepening the ship channel, and in most cases we can expect that the results found in the model will be an accurate guide to what to expect in the prototype.

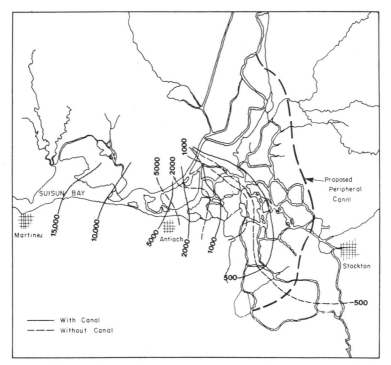

Figure 8.17 Salinity intrusion in the Sacramento–San Joaquin Delta according to results in the physical model. Net delta outflow is 4400 cfs. ———— with the proposed peripheral canal in operation and a proposed pattern of freshwater release at various points around the periphery. ––– without the peripheral canal. [After U.S. Army (1975).]

8.3.4.3   Outfall Siting Studies

A number of proposed discharges of municipal and industrial waste have been studied in the model. As an example, we examine the study of a 90 mgd outfall proposed by the East Bay Dischargers Authority for installation in the South Bay off Oakland Airport.

The model was used to examine which of three alternate sites, shown on Fig. 8.18, should be selected. A predesign study had indicated use of an underwater diffuser (see Chapter 10) with approximately 90 ports over a length of 1000 ft. A model diffuser was set up with nine ports, as called for by the scaling explained in Section 10.5. A continuous discharge in the model of 242 cc/min (92 mgd in the prototype) of fresh water containing Rhodamine WT dye, was injected for 100 tidal cycles at site A, and then for 100 tidal cycles at site C, while the net delta outflow into the model was held at a low flow value of 4400 ft^3/sec. The selection of 100 cycles for approach to steady state was based on previous estimates that approximately 100 days is required for a discharge in the South Bay to come to steady state. The tidal exchange ratio at the discharge site was guessed to be 0.05. It was desired that the concentration in the model near the source would be approximately 100 ppb to provide accurate measurements, so the concentration in the waste discharge was set at 9450 ppb as called for by Eqs. (7.33) and (7.35). Two large vats of fresh water were set up near the model discharge point; the necessary weight of dye was dissolved into the vats, and they were stirred continuously throughout the experiment. Enough dye solution was held in the vats so that no make-up water would be needed during the experiment, to make sure that a constant inflow concentration was used. Concentration in the supply vats was monitored every ten cycles to make sure that there was no loss of dye to the container. The supply tube to the diffuser was a brass tube, as polyethelene tubing is known to take up small amounts of dye. After every tenth cycle the rate of injection was changed briefly and then reset to the desired flow rate; this step is highly recommended, as otherwise a slow drift of the inflow rate may go unnoticed, and the inflow meter may stick at one value while the discharge changes enough to matter.

Samples for measuring dye concentration were taken at high and low slack water at 18 stations in the South and Central Bay. To make sure that samples are taken at exactly slack water the Corps of Engineers has developed a small float which reverses direction at the slack. One sufficiently athletic technician is able to sample approximately five stations, all at exactly slack water, because the time of slack water is different at each station.

Figure 8.18 shows the observed dye distributions at high slack water resulting from the two discharge sites. The more southern site leads to higher concentrations in the South Bay because the tidal exchange ratio is less. The more northern site leads to higher concentration along the shore of the East Bay because the site is on the edge of the submarine valley which leads south, while the southern

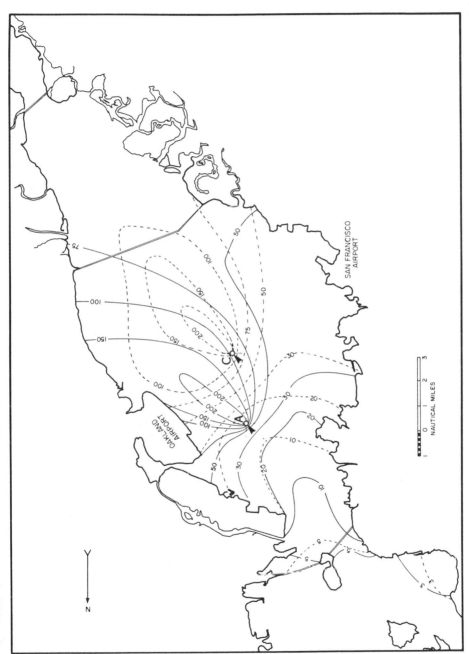

Figure 8.18 Dye distributions observed in the physical model of South San Francisco Bay at high slack water. —— from release at point A. ---- from release at point C. [After Kirkland and Fischer (1976).]

site is on the southern tip. The middle site, which was not tested, is in the middle of the submarine valley and somewhat north of the southern site. It was concluded that the middle site would give the most protection to the East Bay shoreline, while leading to lower concentration in the south part of the Bay: therefore, the middle site was selected for final design of the outfall.

This study is another example of how a model can be used to obtain useful comparisons, even though absolute values may not be exactly correct. Transverse mixing from site C to the east bay shore may not be modeled exactly, and the current patterns in the vicinities of the outfalls have not been verified by field measurements. Nevertheless the comparison of the results from the two sites displays the importance of the submarine canyon and gives a firm basis for selection of the best site. Note that an extremely detailed numerical model would have been required to achieve the same results. In fact, a two-dimensional model of the south bay was tried before the physical model study was decided on, but the numerical model was unable to differentiate between the sites.

8.4   SUMMARY

This chapter has had three parts: some philosophy on what models do and how to choose one; a presentation of what types of numerical models are available, their limitations, and some details of two types of one-dimensional model; and an explanation of how physical models can be used in mixing studies. We have seen that physical models, even though they should not be expected to give an exact duplication of mixing in the prototype and cannot be counted on for absolute answers, are often useful to compare alternative designs or operational strategies. Numerical models can be useful in the same way, but they require more insight into the physical mechanisms that are important in the problem under study because they come in a wider variety and all of them omit or misrepresent one or another physical mechanism. We have stressed that to create a useful numerical model or make proper use of a physical model the modeler must be familiar with the water body and the purpose of the study, and must understand what is causing the mixing that is to be represented in the model. For that reason this chapter should not be read alone; the causes of mixing discussed in Chapters 5–7 must be understood before an investigator considers him or herself competent to undertake a model study.

Chapter 9

Turbulent Jets and Plumes

9.1 INTRODUCTION

Many of society's wastes are naturally ubiquitous in a diluted form. For these wastes, a rapidly diluted discharge to the environment is often the best means of recycling and, as we will see in this chapter, turbulent jets and plumes, because they entrain large volumes of ambient fluid and mix it with the discharge fluid, form an effective mechanism to accomplish this initial dilution.

The actual discharge structure can often be simple, essentially the open end of a submerged pipe, while in other cases much thought and expense must be given to designing a structure to achieve much higher initial dilution in order to minimize the immediate effect of the discharge on the environment. In the next chapter we will discuss one common type of discharge in detail: the submerged multiport diffuser. However, before considering this special type of structure, it is important that the theory of jets and plumes be understood. The purpose of this chapter is therefore to give the design engineer a firm background in the fundamentals of the theory essential to the prediction of how a given discharge system will perform.

We will see that turbulent jet behavior depends on three classes of parameters:

(i)   jet parameters,
(ii) environmental parameters, and
(iii) geometrical factors.

The first group includes the initial jet velocity distribution and turbulence level, the jet mass flux, the jet momentum flux, and the flux of any jet tracer material such as heat, salinity, or contaminant. If the tracer concentration is sufficiently low that the density of the jet efflux is essentially equal to the ambient density level, then tracer concentration, although possibly the most important from an environmental point of view, may not have any effect on the jet dynamics at all.

The second group of variables, the environmental parameters, include such ambient factors as turbulence levels, currents, and density stratifications. These factors usually begin to influence jet behavior at some distance from the actual jet orifice. However, as will be shown later, it is necessary to relate these parameters to the appropriate jet parameters in order to arrive at the actual distances at which the effects begin to be significant.

The geometrical factors that enter into any jet analysis are the jet shape, its orientation and proximity to possible adjacent jets and solid boundaries, the attitude of the jet with respect to boundaries or to the vertical (if the jet has positive or negative buoyancy), and if the jet is submerged, its relationship to any free surfaces.

All of the above factors can enter into a single problem and the unravelling of the influence of each of the above factors is a complex task that even now is not fully complete. The aim of this chapter of the book is to develop an understanding of how each of the above factors does modify the diluting capability of a jet by considering the effect of each factor in turn. We will attempt to take each problem in its simplest possible configuration. Often this will mean that more than one influence on the jet will be present at any one time, but by careful consideration of limiting cases we will attempt to develop a body of knowledge that will permit easy identification of the predominant factors that enter any given problem. In order to develop the ideas we will first study a simple jet emanating from a single point source. The extension of the arguments to plane jets is sufficiently straightforward that only the results will be presented for such jets. Later on we will consider combinations of jets and diffusers comprising rows of simple jets.

Our basic method will be to seek limiting equilibrium asymptotic solutions for simple flows and then combine these solutions into general descriptions of more complex flows. Using this approach we are able to take a complex problem, such as a buoyant jet in a cross flow, and deduce the qualitative influences of buoyancy, momentum, and cross flows on the jet trajectory and rate of dilution. The results of experimental studies then enable us to present detailed methods for calculating dilutions. The technique is extremely simple and is very similar to that used so successfully by Kolmogorov to describe the equilibrium turbulence spectrum, and that used by Monin and Obukhov in their analysis of the atmospheric boundary layer.

9.2 JETS AND PLUMES

A jet is the discharge of fluid from an orifice or slot into a large body of the same or similar fluid. A plume is a flow that looks like a jet, but is caused by a potential energy source that provides the fluid with positive or negative buoyancy relative to its surroundings. For instance, if a nozzle of a garden hose is held under water, we have a jet; an open fire forms a rising plume of smoke and hot gases. The primary difference is that the flow from the hose nozzle is driven by the momentum of the discharged water, whereas the flow above the fire is driven by the air around the fire being warmed and therefore having a reduced density and rising.

Many discharges into the environment are classed as buoyant jets, which are derived from sources of both momentum and buoyancy. The initial flow is often driven mostly by the momentum of the fluid exiting an orifice, but the effluent may be less or more dense than its surroundings and the resulting jet is thus acted upon by buoyancy forces. We will see later in this section that given enough flow distance all buoyant jets eventually act like plumes. In fact, the usual analysis of discharges from wastewater diffusers, discussed in Chapter 10, is based almost entirely on plume formulas. Nevertheless, it is also important that the practicing engineer know something about jets, because in many special cases the initial momentum of the flow is important.

Jets and plumes can be either laminar and turbulent and, as with pipe flow, it is possible to describe a Reynolds number which, if sufficiently large, will guarantee that the flow is fully turbulent. However, in most problems of concern to the engineer, there is generally no question that the flow generated by the discharge will be turbulent so that this is not a worry. Nevertheless, in model studies of particular configurations it is important that the jet Reynolds number be adequate for fully turbulent jet flow.

The jet Reynolds number at which a laminar jet will become turbulent apparently cannot yet be predicted, although several theoretical and experimental studies have been performed. Grant (1974) indicated that the initial velocity profile within the jet is a crucial factor in deciding instability and that it is not possible to establish a general unique Reynolds number for instability. However, in most cases, provided the Reynolds number exceeds 2000 the jet flow will be turbulent, although Labus and Symons (1972) have shown that the turbulent flow does not reach a fully developed state until a Reynolds number of about 4000.

Near the source the flow in a jet or plume is usually controlled entirely by the initial conditions, which include the geometry of the jet, the mean jet exit velocity, the initial density difference between the discharge and the ambient fluid, as well as the turbulence intensity and velocity distribution in the supply pipe. A general and useful description of the factors of prime importance to jet dynamics can be defined as follows:

(i) The *mass flux* of the jet, $\rho\mu$, which is the mass of fluid passing a jet cross section per unit time. The mass flux is given by

$$\rho\mu = \int_A \rho w \, dA, \tag{9.1}$$

in which A is the cross-sectional area of the jet, and must be defined and w the time-averaged jet velocity in the axial direction. μ is called the specific mass flux, or volume flux, of the jet.

(ii) The *momentum flux* of the jet, which is the amount of streamwise momentum passing a jet cross section per unit time. It is given by

$$\rho m = \int_A \rho w^2 \, dA. \tag{9.2}$$

m is called the specific momentum flux and differs from the momentum flux only by the density factor.†

(iii) The *buoyancy flux*, the buoyant or submerged weight of the fluid passing through a cross section per unit time, is defined by

$$\rho\beta = \int_A g\Delta\rho w \, dA, \tag{9.3}$$

where $\Delta\rho$ is the difference in density between the surrounding fluid and the fluid in the jet. β is the specific buoyancy flux, in analogy to the specific momentum flux. The buoyancy flux is related to the flux of the tracer that causes the density variation. It is convenient to define $g \, \Delta\rho/\rho = g'$, the effective gravitational acceleration.

Throughout this chapter we will use the symbols Q, M, and B to refer to the initial values of volume flux, specific momentum flux, and specific buoyancy flux, respectively. For a round jet, Q and M are given by

$$Q = \tfrac{1}{4}\pi D^2 W, \tag{9.4}$$

$$M = \tfrac{1}{4}\pi D^2 W^2, \tag{9.5}$$

in which D is the jet diameter and W the mean outflow velocity assumed uniform across the jet. The initial buoyancy flux is more difficult to define, as it may arise in one of two ways. A plume can be formed by a source of buoyancy, such as a

---

† Note that Eq. (9.1) and Eq. (9.2) exclude the turbulent fluxes of mass and momentum. The justification for this is that, in general, the turbulent velocity fluctuations are less than 0.1 w (see Fig. 9.4 later).

source of heat. The buoyancy imparted to the fluid is determined by the heat added, according to the formula

$$\rho B = \alpha g P / C_p, \tag{9.6}$$

where α is the volume coefficient of thermal expansion, P the heat flux added by the heat source, and C_p the specific heat at constant pressure. Equation (9.6) represents an equivalence between thermal energy flux and a density-deficiency flux since density and temperature are related by an equation of state. For most waste discharge applications, however, the initial buoyancy is generally contained in the discharge. For instance, for a round buoyant jet, the initial specific buoyancy flux is

$$B = g(\Delta \rho_0 / \rho)Q = g_0' Q, \tag{9.7}$$

in which $\Delta \rho_0$ is the difference in density between the receiving fluid and the fluid being discharged, or alternatively, g_0' is the initial apparent gravitational acceleration.

It has been found that Q, M, and B are the primary variables governing the dilution of round turbulent buoyant jets, provided that, as we noted, the Reynolds number (which, for a round jet, can be defined as $M^{1/2}/\nu$) exceeds about 4000. The dimensions of these variables are

$$[Q] = L^3/T, \qquad [M] = L^4/T^2, \qquad [B] = L^4/T^3,$$

in which the square brackets stand for "dimensions of," L stands for length and T for time. The other factors we mentioned, for example, the jet cross-sectional geometry and turbulence level in the supply flow, are of secondary importance and are generally ignored.

For planar jets, formed by flow from a slot rather than an orifice, μ, m, and β are interpreted as specific fluxes per unit length of slot; their initial values are also written as Q, M, and B. The dimensions of these variables are of course reduced in length by one order and, provided it is clear that a planar jet or plume is under consideration, no confusion should arise.

The remainder of this section discusses jets and plumes into stagnant homogeneous receiving waters, for which all of the flow conditions must be determined solely by the initial conditions and the distance from the source. Later sections of this chapter will discuss the effects of receiving-water ambient conditions.

9.2.1 The Simple Jet

The simplest case we can visualize is the discharge from a round nozzle into a body of like fluid. Such jets have been studied extensively so that there is now a reasonably good understanding of how they behave. Figure 9.1 is a Schlieren

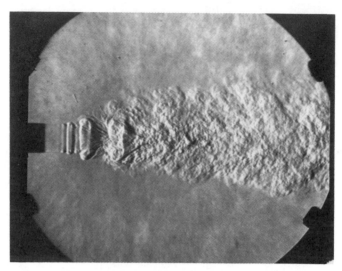

Figure 9.1 Schlieren photograph of a turbulent jet at a Reynolds number of 1.87×10^4. [Courtesy of Steven Crow and Cambridge University Press.]

photograph of such a simple jet. Although this photograph is of an air jet into air, a submerged water jet would not look any different at the same Reynolds number if there were no compressibility effects. It can be seen in the photograph that a shear layer forms between the jet fluid and the ambient fluid and that the shear layer has large cylindrical shaped waves that appear to entrain the ambient fluid in large "gulps" which then break down and mix the two fluids. This is even more evident in Fig. 9.2. The boundary between the ambient and jet fluids is really quite sharp at any instant, and if a measuring probe that can distinguish between ambient fluid and jet fluid is placed in a jet at a fixed point, it shows a rapidly fluctuating environment which may be, at times, purely jet fluid, purely ambient fluid or a mixture of both. However, when a time-averaged measurement is taken it shows an essentially Gaussian distribution of tracer concentration C across the jet, which may be defined by an equation of the form

$$C = C_m \exp[-k(x/z)^2],$$

in which the subscript m refers to the value of C on the jet axis and where z is the distance along the jet axis and x is the transverse (or radial) distance from the jet axis. This is demonstrated in Fig. 9.3 which is a plot of the maximum, minimum, and time-averaged mean concentration distributions in a planar jet. It can be seen that the peak concentration exceeds the mean by almost 70% and that the minimum concentration is still almost zero, even at 36 jet slot widths downstream.

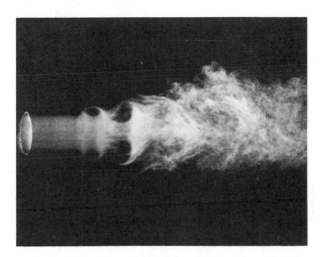

Figure 9.2   Flow visualization of a turbulent jet at a Reynolds number of 10.5×10^4. [Courtesy of Steven Crow and Cambridge University Press.]

This Gaussian distribution is true also for the time-averaged velocity profile across a jet provided that the profile is taken more than about six jet diameters downstream. In the region from the jet orifice to six jet diameters, the shear layer is still "eating" away at the constant velocity core of the jet flow as it comes out of the nozzle; for this reason, this is termed the *zone of flow establishment* (ZFE).

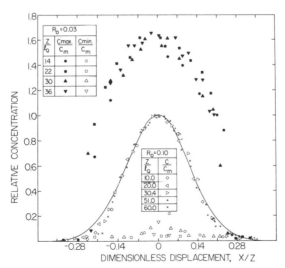

Figure 9.3   Maximum, minimum, and time-averaged mean concentration in a planar turbulent jet, replotted from Kotsovinos (1975). See Eq. (9.10) for definition of l_Q and Eq. (9.47) for R_0.

In actuality though, the turbulent flow within the jet does not reach an equilibrium until about ten jet diameters downstream of the orifice where the turbulence reaches a state of steady decay (see Fig. 9.4).

The flow downstream of the ZFE, in which the jet continues to expand and the mean velocities and tracer concentrations decrease, is often called the *zone of established flow* (ZEF). Mean velocity and concentration profiles in this zone are "self-similar," that is, at any cross section we can express the time-averaged velocity, or tracer distribution, in terms of a maximum value (measured at the jet centerline) and a measure of the width.

For example, the mean velocity distribution across a jet can be represented by a function of the form

$$w = w_m f(x/b_w), \tag{9.8}$$

in which the subscript m specifies the value of w on the jet centerline, x is a coordinate transverse to the jet axis, and b_w is the value of x at which w reduces

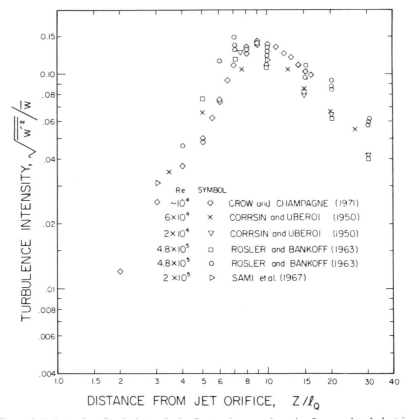

Figure 9.4   Intensity of turbulent velocity fluctuations on the axis of a round turbulent jet.

to some specified fraction of w_m [often chosen to be either 0.5 or 0.37 ($=e^{-1}$)]. The functional form of f is most usually of Gaussian form, so that, for example for tracer concentration, we write

$$C = C_m \exp[-(x/b_T)^2], \qquad (9.9)$$

where b_T, in this case, is the value of x at which C takes the value $0.37C_m$.

By means of such curve fitting to experimental data it is possible to calculate the values of the volume flux and momentum integrals specified by Eqs. (9.1) and (9.2). As noted previously the initial values of the integrals are denoted by Q and M, respectively.

For a simple turbulent round jet we can define a characteristic length scale for the jet in terms of the volume flux Q and momentum flux M. This is given by

$$l_Q = Q/M^{1/2} = \sqrt{A}, \qquad (9.10)$$

where A is the initial cross sectional area of the jet. In particular, for a round jet $l_Q = \sqrt{\pi/4}D$. For a planar jet the equivalent length scale is just the slot width.

If we denote the distance downstream from the jet orifice as z then dimensional analysis implies that all properties of the jet will be a function of z/l_Q, Q, and M. Further, reviewing the experimental turbulence intensity measurements in a jet, plotted in Fig. 9.4, we would also expect to find an asymptotic solution at a distance $z/l_Q > 10$, since this corresponds to the region of the jet where a steady decay of turbulence on the axis occurs.

Applying this reasoning, we can deduce how w_m, C_m, b_w, b_T, μ, and m must depend on the distance from the jet orifice. For example, since w_m has the dimensions length/time we must have that

$$w_m Q/M = f(z/l_Q), \qquad (9.11)$$

where f is some yet-to-be-specified function. The reason for this is that only two independent dimensionless variables can be made from the four variables w_m, Q, M, z. We can even argue much further. First, since we know that as $z \to 0$, $w_m \to M/Q$ we must have that

$$f(z/l_Q) \to 1 \qquad \text{for} \quad z \sim l_Q.$$

Second, if we consider the other extreme where $z/l_Q \gg 1$ we note that this limit is formally equivalent to either

(i) $z \to \infty$ with Q and M fixed, or
(ii) $Q \to 0$ with z and M fixed, or
(iii) $M \to \infty$ with z and Q fixed.

From this equivalence we can see that the further we are from the jet orifice the less important the volume flow is in defining the solution and the more important the momentum flux becomes. Indeed, if we could devise a jet in which Q, the source flow, were zero, but for which the momentum flux M were not, it would, for large z, be indistinguishable from a jet with a given initial flow. This means that for $z \gg l_Q$ all properties of the jet are defined solely in terms of z, the distance from the orifice, and M, the momentum flux (provided, or course, that the Reynolds number $M^{1/2}/v > 4000$). This result indicates that

$$w_m Q/M \to a_1 l_Q/z \qquad \text{for} \quad z \gg l_Q, \tag{9.12}$$

where a_1 is an empirical constant. Figure 9.5 is a graph of $w_m Q/M$ versus z/l_Q for a round jet formed from the experimental results of five investigators. It can be seen that the predicted result is remarkably well confirmed with a value for a_1 of 7.0 ± 0.1, which is also the value suggested by Chen and Rodi (1976) from their evaluation of eight experimental studies.

Dimensional arguments also imply that because l_Q is the only fixed length scale, b_w, and b_T must be specified by functions of the form

$$\frac{b}{l_Q} = f\left(\frac{z}{l_Q}\right),$$

and, as in the previous arguments for $z \gg l_Q$, the form of f must be such as to have Q vanish from the relationship so that $b \sim z$. Table 9.1 is a summary of the

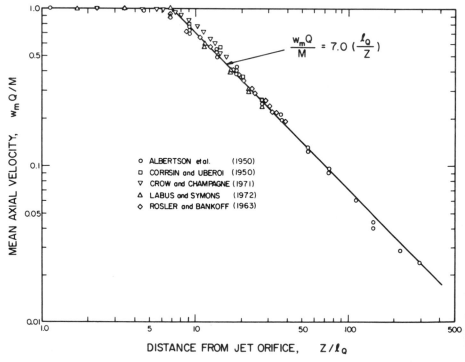

Figure 9.5 Decay of peak time-averaged velocity on the axis of a round turbulent jet.

TABLE 9.1
Width Parameters for Turbulent Round Jets

Investigator	b_w/z	b_T/z
Albertson et al. (1950)	0.114	—
Becker et al. (1967)	—	0.127
Corrsin (1943)	0.100	0.132
Corrsin and Uberoi (1950)	0.114	0.140
	0.130	0.156
Forstall and Gaylord (1955)	0.107	0.115
Hinze and van der Hegge Zijnen (1949)	0.102	0.115
Keagy and Weller (1949)	0.099	0.107
	0.106	0.126
Kizer (1963)	0.099	0.125
Rosenweig et al. (1961)	0.108	0.120
Ruden (1933)	0.103	0.124
Sunavala et al. (1957)	—	0.141
Uberoi and Garby (1967)	0.090	0.101
	0.101	0.114
Wilson and Danckwerts (1964)	0.120	0.156
	0.114	0.138
Mean values	0.107	0.127
	±0.003	±0.004

results of many experimental investigations of round jets which indicate that b_w/z has an average value of 0.107, and $b_T/z = 0.127$ for the mean tracer profile. The ratio of $b_T/b_w = 1.19$. This result has led to unresolved widespread speculation as to why the mean concentration profile should be wider than the mean velocity profile.

Returning to the volume flux in the jet, more dimensional analysis gives

$$\mu = Qf(z/l_Q), \qquad (9.13)$$

where again f is a function whose form is not known. Obviously, as $z/l_Q \to 0$, $f(z/l_Q) \to 1$. Applying the argument that $z \to \infty$ is formally equivalent to $Q \to 0$ gives that for a round jet

$$\mu/Q - c_j(z/l_Q) \qquad \text{for} \quad z \gg l_Q. \qquad (9.14)$$

The value of c_j can be found by using the self-similar velocity profile

$$w = w_m \exp[-(x/b_w)^2], \qquad (9.15)$$

where x is a radial coordinate in three dimensions (or a transverse coordinate in two dimensions). We find that for a round jet

$$\mu = \pi w_m b_w^2, \qquad z \gg l_Q, \qquad (9.16)$$

and therefore, using previously obtained results for w_m and b_w

$$\mu = [7.0\pi(M/Q)(l_Q/z)](0.107z)^2, \tag{9.17}$$

which implies that $c_j = 0.25$ so that

$$\mu/Q = 0.25\,(z/l_Q), \qquad z \gg l_Q. \tag{9.18}$$

A comparison with some experimentally measured values of μ for a round jet is shown in Fig. 9.6.

We are often most interested in the dilution of a tracer material discharged in the jet and in experimental studies it is possible to measure the distribution of the time-averaged tracer concentration C. This has a Gaussian form as given by Eq. (9.9). We can deduce that C_m, the time-averaged concentration measured on the jet axis, is also inversely proportional to z. The argument is as follows: suppose that Y is the rate of supply of tracer mass to the jet. If C_0 is the initial mass concentration of tracer then

$$Y = QC_0. \tag{9.19}$$

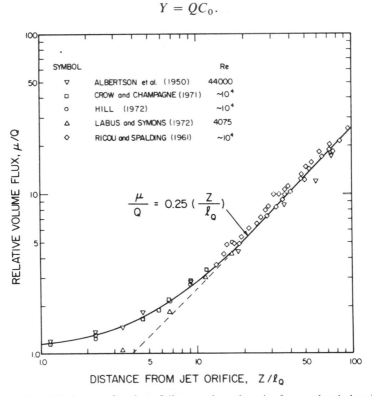

Figure 9.6   Dilution as a function of distance along the axis of a round turbulent jet.

Y therefore has dimensions of mass/time and since C_m has dimensions of mass/(length)^3, C_m/Y has dimensions of time/(length)^3. However, for $z \gg l_Q$, $M^{1/2}$ is the only jet parameter with time involved so that we must have

$$C_m/Y = a_2(M^{1/2}z)^{-1}, \qquad z \gg l_Q, \tag{9.20}$$

or equivalently,

$$C_m/C_0 = a_2(l_Q/z), \qquad z \gg l_Q. \tag{9.21}$$

The value of a_2 has been determined experimentally by many investigators and Chen and Rodi (1976) in their evaluation suggest a value of 5.64 for round jets.

It is also possible to define an average concentration C_{av} for a jet in the following way

$$\mu C_{av} = Q C_0 = Y. \tag{9.22}$$

C_{av} is therefore the equivalent mean concentration that the volume flow would have to carry to advect the same mass flux of tracer. The ratio C_0/C_{av} is called the mean dilution. Note that C_{av} includes the transport of tracer mass by the turbulent fluctuations and we must have, from conservation of mass, that for round jets

$$\mu C_{av} = \int_{jet} 2\pi x w C \, dx + \text{turbulent transport.} \tag{9.23}$$

By assuming the Gaussian forms for w and C the integral in Eq. (9.23) can be evaluated so that

$$\mu C_{av} = \pi w_m C_m \left(\frac{b_w^2 b_T^2}{b_w^2 + b_T^2} \right) + \text{turbulent transport.} \tag{9.24}$$

We can therefore write

$$\frac{\text{turbulent flux of tracer mass}}{\text{total flux of tracer mass}} = 1 - \frac{\pi w_m C_m}{Q C_0} \left(\frac{b_w^2 b_T^2}{b_w^2 + b_T^2} \right)$$

$$= 0.17 \pm 0.12 \tag{9.25}$$

It is difficult to make a better estimate than this because of the relative errors involved in measuring the variables in Eq. (9.25). Nevertheless, it is apparent that the turbulent flux is not zero. From Eqs. (9.21), (9.22), and (9.18) we can see that the flow-weighted average concentration C_{av} is given by

$$C_m/C_{av} = 1.4 \pm 0.1. \tag{9.26}$$

Finally, some readers may note that we have omitted any reference to the virtual origin of the jet. The virtual origin is defined to be the point at which a pure momentum jet would be located to give the flow equivalent to any jet in question. By equating the flow of a pure momentum jet to the flow at the distance from the origin at which a jet is fully developed, according to Fig. 9.4 about $10l_Q$ for a round jet, it is possible to show that the virtual origin is located very close to the actual jet orifice. We will therefore make no further reference to virtual origins.

In summary, it is possible to deduce almost all of the properties of turbulent jets that are of importance to engineers from simple dimensional arguments combined with empirical data. For reference, the results obtained have been collected in Table 9.2 along with the results for planar (two-dimensional) jets. Determining the form of the planar jet formulae is left as an exercise. The reader should also confirm that the formulas given in Table 9.2 are consistent with the result that $m = M$, in other words that the momentum flux is constant in jets.

TABLE 9.2
Summary of Properties of Turbulent Jet

Parameter	Round jet	Plane jet
Initial volume flow rate Q	Dimensions $L^3 T^{-1}$	Dimensions $L^2 T^{-1}$
Initial specific momentum flux M	Dimensions $L^4 T^{-2}$	Dimensions $L^3 T^{-2}$
Characteristic length scale l_Q	$\dfrac{Q}{M^{1/2}}$	$\dfrac{Q^2}{M}$
Maximum time-averaged velocity w_m	$w_m \dfrac{Q}{M} = (7.0 \pm 0.1) l_Q/z$	$w_m \dfrac{Q}{M} = (2.41 \pm 0.04)\left(\dfrac{l_Q}{z}\right)^{1/2}$
Maximum time-averaged tracer concentration C_m	$\dfrac{C_m}{C_0} = (5.6 \pm 0.1)\left(\dfrac{l_Q}{z}\right)$	$\dfrac{C_m}{C_0} = (2.38 \pm 0.04)\left(\dfrac{l_Q}{z}\right)^{1/2}$
Mean dilution μ/Q	$\dfrac{\mu}{Q} = (0.25 \pm 0.01)\left(\dfrac{z}{l_Q}\right)$	$\dfrac{\mu}{Q} = (0.50 \pm 0.02)\left(\dfrac{z}{l_Q}\right)^{1/2}$
Velocity scale of half-width b_w/z	0.107 ± 0.003	0.116 ± 0.002
Concentration scale of half-width b_T/z	0.127 ± 0.004	0.157 ± 0.003
Ratio C_m/C_{av}	1.4 ± 0.1	1.2 ± 0.1

Example 9.1.   A turbulent jet discharges 1 m^3/sec (35.3 ft^3/sec) of liquid at a velocity of 3 m/sec (9.84 ft/sec) into a liquid of the same density. Find the maximum time-averaged velocity, tracer concentration, and the mean dilution at a

distance 60 m from the jet orifice. The initial concentration of tracer is 1 kg/m^3 [10^3 ppm (parts per million)].

$$Q = 1 \text{ m}^3/\text{sec},$$

$$M = 3 \text{ m}^4/\text{sec}^2,$$

$$l_Q = Q/M^{1/2} = 0.58 \text{ m}.$$

$$\text{At } 60 \text{ m, } z/l_Q = 104.$$

From Fig. 9.5 or Table 9.2

$$w_m = \frac{7}{104}\frac{M}{Q} \text{ m/sec} = 0.20 \text{ m/sec}.$$

From Table 9.2

$$C_m/C_0 = 5.6/104$$

$$C_m = 54 \text{ ppm}.$$

Mean dilution from Fig. 9.6 or Table 9.2

$$\mu/Q = 26. \quad \blacksquare$$

9.2.2 The Simple Plume

The pure plume is easier to analyze than the pure jet because in the pure plume there is no initial volume or momentum flux (visualize the smoke plume above a fire, for instance). This means that all flow variables for a plume must be functions only of B, the buoyancy flux, z the distance from the origin, and v the viscosity of the fluid. For example, the time-averaged vertical velocity on the axis of the plume is given by

$$w_m = f(B, z, v).$$

Since there are only four variables, there are two dimensionless groups, so that for a round plume from a point source

$$w_m(z/B)^{1/3} = f(B^{1/3}z^{2/3}/v). \tag{9.27}$$

The term on the right-hand side is a form of Reynolds number and provided it is sufficiently large, that is $z \gg v^{3/2}/B^{1/2}$, the flow is fully turbulent and the effect of viscosity becomes essentially absent. In this case the term on the left becomes constant and we have that

$$w_m = b_1(B/z)^{1/3}, \tag{9.28}$$

where experiments by Rouse et al. (1952) give a value of 4.7 for b_1.

The fluid within the plume is less (or possibly more) dense than its surroundings, so the force of gravity acts to change the momentum in the flow. This means that the flux of momentum, as defined by Eq. (9.2), increases along the axis of the plume, in contrast to the jet for which the momentum flux is approximately constant. The momentum flux at any cross section can only be a function of B and z. Using dimensional analysis for the form of the dependence, and the results of the previously quoted experiments for the value of the constant, we find

$$m = b_2 B^{2/3} z^{4/3}, \qquad (9.29)$$

in which b_2 is found experimentally to be approximately 0.35 for a round plume.

In a similar fashion the volume flux in a round plume can be found to be given by

$$\mu = b_3 B^{1/3} z^{5/3}, \qquad (9.30)$$

and b_3 has an experimentally determined value of 0.15.

Eq. 9.30 can be put into a form similar to the form for jets [Eq. (9.14)] by dividing by the square root of Eq. (9.29) to yield

$$\mu = c_p m^{1/2} z, \qquad (9.31)$$

where $c_p = b_3/b_2^{1/2}$ is the plume coefficient analogous to c_j, and has the experimentally determined value, according to the data available, of 0.254. Thus the volume flux of a plume is given by the same equation as that for a jet, except that the *local* momentum flux must be used in place of the initial momentum flux. This means that for distances far from the source in a plume the flux μ (and the dilution) increase as the $\frac{5}{3}$ power of z, because in a plume the momentum flux is constantly increasing. By comparison in a jet the flux μ grows only as the first power of z since momentum is conserved.

The volume flux of a plume can be expressed solely in terms of m and B by eliminating z from Eqs. (9.29) and (9.30) to give, for a round plume,

$$R_p = \mu B^{1/2}/m^{5/4} \qquad (9.32)$$

in which $R_p = b_3 b_2^{-5/4} = 0.557$. R_p is called the plume Richardson number.

It should also be noted that Eq. (9.31) actually specifies a local mean width of the plume, for if one fits a Gaussian profile to the mean velocity, as in Eq. (9.15), then by evaluating the momentum and volume fluxes Eq. (9.31) becomes

$$\sqrt{2\pi} b_w = c_p z. \qquad (9.33)$$

c_p and c_j are therefore the respective growth coefficients of plumes and jets. Note that the small difference in the values of b_w/z for jets and plumes implies a small difference in the values of c_p and c_j but given the data in Fig. 9.6 a value of 0.25 is recommended.

The rate of decrease of the time-averaged maximum tracer concentration, C_m, in a buoyancy-driven discharge can be deduced in the same way as for a jet. Suppose that Y is the mass flux or equivalent mass flux of tracer (we will explain later what we mean by equivalent mass). The quotient C_m/Y must have dimensions time/length^3 and can only be specified by B the buoyancy flux, and z the distance from the source. We therefore have, from this dimensional argument, that

$$C_m/Y = b_4/B^{1/3}z^{5/3}, \tag{9.34}$$

where b_4 is an empirical constant which is found to have a value of 9.1 (see Chen and Rodi, 1976).

In some cases, the tracer is also responsible for the variation in density causing the buoyancy. For example, if the buoyancy source is due to a heat flux P, the buoyancy flux is given in Eq. (9.6) and the temperature C_m on the axis of the plume will be obtained from

$$Y = P/\rho C_p, \tag{9.35}$$

where C_p is the specific heat of the fluid. The tracer concentration in this case is the temperature anomaly equal to heat/(unit mass $\times$ C_p).

We can also define a flow-weighted average concentration as in Eq. (9.22) and again relate C_m to C_{av} by using Eqs. (9.30) and (9.34) to obtain $C_m/C_{av} = b_3 b_4 = 1.4$.

It will be noted that just as for a pure momentum jet there is no characteristic length scale. But if the source of buoyancy is derived from a volume flow, so that B is defined by Eq. (9.7), then there is now a length scale defined by $Q^{3/5}/B^{1/5}$. This length scale is the distance from the flow source at which buoyancy influences the flow. However, in almost all flow situations in which a buoyant discharge occurs, there is initial momentum, so we must consider how this works into the problem.

Similar results to those presented for round plumes can be deduced for planar plumes and the results are given in Table 9.3. The determination of the formulas given are left as exercises for the reader. The numerical constants given are from the review by Chen and Rodi (1976).

The results given in Table 9.3 for the plane plume are based largely on the detailed experimental study of Kotsovinos (1975), who found that the turbulent transport of tracer in the axial direction is not negligible, but may be a significant fraction of the transport by the mean flow [cf. Eq. (9.23)]. The coefficients in the table imply that 35% of the flux is turbulent transport while 65% is due to the mean flow. Because of this we arrive at the curious result that $C_m/C_{av} = 0.81$, i.e., the flow-weighted average value of concentration is actually larger than the maximum profile value.

The round plume results are based on the earlier work of Rouse et al. (1952) without direct measurements or consideration of turbulent fluxes. A new

TABLE 9.3
Summary of Plume Properties

Parameter	Round plume	Plane plume
Initial buoyancy flux B	Dimensions $L^4 T^{-3}$	Dimensions $L^3 T^{-3}$
Maximum time-averaged velocity w_m	$w_m = (4.7 \pm 0.2)B^{1/3}z^{-1/3}$	$w_m = 1.66B^{1/3}$
Maximum time-averaged tracer concentration C_m	$C_m = (9.1 \pm 0.5)YB^{-1/3}z^{-5/3}$	$C_m = 2.38YB^{-1/3}z^{-1}$
Volume flux μ	$\mu = (0.15 \pm 0.015)B^{1/3}z^{5/3}$	$\mu = 0.34B^{1/3}z$
Velocity scale of half-width b_w/z	0.100 ± 0.005	0.116 ± 0.002
Concentration scale of half-width b_T/z	0.120 ± 0.005	0.157 ± 0.003
Ratio C_m/C_{av}	1.4 ± 0.2	0.81 ± 0.1

detailed investigation of the round plume case could possibly lead to revised coefficients, perhaps making C_m/C_{av} closer to or even less than 1.0 also.

Since this matter has not been fully resolved, the implications of this new finding have not been carried forward into the integral analyses (Section 9.4), or Chapter 10, where some coefficients may ultimately be subject to adjustment as our understanding improves.

Example 9.2.   A freshwater discharge of 1 m³/sec (35.3 ft³/sec) is to be located at a depth of 70 m (230 ft) in the coastal ocean. The discharge has a temperature of 17.8°C (64°F) and the sea is assumed to be well mixed with a temperature of 11.1°C (52°F) with a salinity of 32.5 ‰. What will be the maximum time-averaged concentration of tracer and mean dilution 10 m (32.8 ft) below the surface if the initial concentration of tracer is 1 kg/m³ (10^3 ppm).

$$\text{Seawater density at } 11.1°C \text{ and } 32.5‰ \quad = 1024.8 \text{ kg/m}^3,$$

$$\text{Fresh water density at } 17.8°C \qquad = 998.6 \text{ kg/m}^3,$$

$$\text{Density deficiency } (\Delta\rho_0) \qquad = 26.2 \text{ kg/m}^3,$$

$$g_0' = g\frac{\Delta\rho_0}{\rho} = 9.8 \times \frac{26.2}{998.6} \qquad = 0.257 \text{ m/sec}^2.$$

The buoyancy flux

$$B = g_0'Q = 0.257 \text{ m}^4/\text{sec}^3.$$

Mass flux of tracer $Y = QC_0 = 1$ kg/sec.

From Table 9.3

$$C_m = 9.1 Y B^{-1/3} z^{-5/3} \text{ kg/m}^3$$
$$= 9.1 \times 1 \times (0.257)^{-1/3} \times 60^{-5/3} \text{ kg/m}^3$$
$$= 0.0156 \text{ kg/m}^3$$
$$= 16 \text{ ppm.}$$

Volume flux in plume, from Table 9.3

$$\mu = 0.15 B^{1/3} z^{5/3}$$
$$= 87.7 \text{ m}^3/\text{sec.}$$

Mean dilution $\mu/Q = 87.7$.   ∎

9.2.3   Buoyant Jets

A buoyant jet is a jet whose density initially differs by an amount $\Delta\rho_0$ from the density of the receiving water. $\Delta\rho_0$ may be either positive or negative so that it becomes important to consider the orientation of the jet with respect to the vertical. In this section we focus on a jet discharged vertically upward which is slightly less dense than its surroundings so that it continues to travel upward. Other jet orientations will be discussed later.

A buoyant jet has jetlike characteristics depending on its initial volume and momentum fluxes, and plumelike characteristics depending on its initial buoyancy flux. Far enough from the source the plumelike characteristics always win out, that is, a buoyant jet will always turn into a plume if given enough free distance. To see why, first recall that if the receiving water is stagnant and homogeneous the only parameters that can determine the flow in the jet or plume are the initial fluxes of volume, momentum and buoyancy, Q, M, and B, respectively, and the distance from the source point z. Dimensional analysis therefore indicates that, for a round jet, two independent dimensionless parameters are

$$M^{1/2}z/Q \quad \text{and} \quad B^{1/2}z/M^{3/4}.$$

The first of these we recognize as z/l_Q, the second we define as z/l_M. There are, of course, other choices, such as Bz^5/Q^3, but any such choice can be made from a suitable combination of the two given. The first of the parameters given was found to be important in the analysis of jets, and the second we see includes the effect of buoyancy. We can therefore write that any flow variable (suitably non-dimensionalized) must be a function of these two variables. For example,

$$w_m = \frac{M}{Q} f\left(\frac{z}{l_Q}, \frac{z}{l_M}\right),$$

which is, however, not a very convenient function to evaluate, since z is involved in both independent parameters. In order to see how to find a convenient formulation, we again consider limiting solutions.

Suppose we consider a flow that has both M and B, but no initial volume flux Q. The only characteristic length for a round jet is given by

$$l_M = M^{3/4}/B^{1/2}, \tag{9.36}$$

and the solution for w_m for such a round jet must be of the form

$$w_m \frac{M^{1/4}}{B^{1/2}} = f\left(\frac{zB^{1/2}}{M^{3/4}}\right). \tag{9.37}$$

But, we know that for $B \to 0$, w_m must be independent of B so that the form of f must be such as to have B vanish. However, $B \to 0$ is formally identical with $z \to 0$ or $M \to \infty$, so that

$$w_m \frac{M^{1/4}}{B^{1/2}} \to c_1\left(\frac{M^{3/4}}{zB^{1/2}}\right), \qquad \text{for} \quad z \ll \frac{M^{3/4}}{B^{1/2}}. \tag{9.38}$$

Similarly,

$$w_m \frac{M^{1/4}}{B^{1/2}} \to c_2\left(\frac{M^{3/4}}{zB^{1/2}}\right)^{1/3}, \qquad \text{for} \quad z \gg \frac{M^{3/4}}{B^{1/2}}, \tag{9.39}$$

where c_1 and c_2 are empirical constants.

It is apparent from these results that the controlling parameter for whether a buoyant jet is jetlike or plumelike is the ratio of z and l_M. For $z \gg l_M$ the flow is like a plume, and for $z \ll l_M$ the flow is like a jet.

Now we must also consider the scale l_Q. Recall that if $z \gg l_Q$ the flow is a fully developed jet and if $z \sim O(l_Q)$ then the flow is still controlled by the jet exit geometry. Thus if l_M and l_Q are of the same order then the flow will be very similar to a plume from the outset. The ratio of l_Q/l_M is called the jet Richardson number and for a round jet we define R_0 by

$$R_0 = \frac{l_Q}{l_M} = \frac{QB^{1/2}}{M^{5/4}} = \left(\frac{\pi}{4}\right)^{1/4}\left(\frac{g_0'D}{W^2}\right)^{1/2} = \left(\frac{\pi}{4}\right)^{1/4}\frac{1}{F_d}. \tag{9.40}$$

F_d is commonly called the jet densimetric Froude number. We prefer to use the Richardson number because it leads to simpler expressions, has a value between 0 and 1, and because we can give it a physical explanation in terms of the ratios of characteristic length scales.

Now that we have the asymptotic solutions for a buoyant jet and the criteria for their application we find it convenient to define dimensionless values of the volume flux and distance from the jet orifice. Making use of the previously defined plume coefficients of c_p and R_p we write (for round jets)

$$\bar{\mu} = \frac{\mu B^{1/2}}{R_p M^{5/4}} = \frac{\mu}{Q}\left(\frac{R_0}{R_p}\right),$$

(9.41)

and

$$\zeta = \frac{c_p}{R_p}\frac{z}{l_M} = c_p\left(\frac{z}{l_Q}\right)\left(\frac{R_0}{R_p}\right).$$

(9.42)

Then Eq. (9.18) for the jet volume flux becomes quite simply

$$\bar{\mu} = \zeta, \qquad \zeta \ll 1.$$

(9.43)

Similarly, Eq. (9.30) for the volume flux in a plume becomes

$$\bar{\mu} = \frac{0.15 R_p^{2/3}}{c_p^{5/3}}\zeta^{5/3} = \zeta^{5/3}, \qquad \zeta \gg 1.$$

(9.44)

These two remarkably simple results can be plotted as straight lines on a logarithmic graph as in Fig. 9.7. The only question remaining is where is the jet orifice located?

The answer is given by replotting Fig. 9.6, in the region of flow establishment, with the variables scaled by the value of R_0/R_p. Examples are shown in Fig. 9.7 for $R_0/R_p = 0.5, 0.1$, and 0.01. It is apparent that any flow for which $R_0/R_p \approx \frac{1}{2}$ can be considered as a fully developed plume at the end of the zone of flow establishment. The experimental data is from Ricou and Spalding (1961).

Example 9.3. Consider Example 9.2 but now suppose the discharge velocity is 3 m/sec (9.84 ft/sec) vertically upward. In this case the flow is a buoyant jet with

$$Q = 1 \text{ m}^3/\text{sec}$$
$$M = 3 \text{ m}^4/\text{sec}^2$$
$$B = 0.257 \text{ m}^4/\text{sec}^2$$

$$l_Q = \frac{Q}{M^{1/2}} = 0.577 \text{ m}$$

$$l_M = \frac{M^{3/4}}{B^{1/2}} = 4.5 \text{ m} \qquad \text{(\textit{The flow becomes plumelike very quickly})}$$

$$R_0 = \frac{l_Q}{l_M} = 0.128.$$

From Eq. (9.42)

$$\zeta_{60} = c_p \frac{z}{l_Q}\left(\frac{R_0}{R_p}\right) = 0.25\left(\frac{60}{0.577}\right)\left(\frac{0.128}{0.557}\right)$$

$$= 6.0.$$

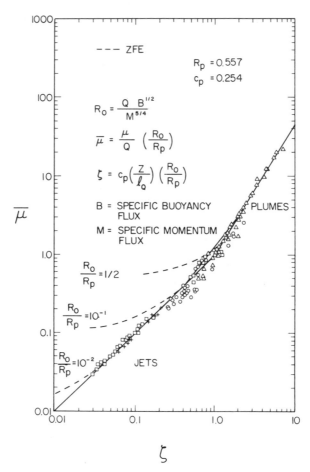

Figure 9.7   Asymptotic solutions for dilution in a vertical round turbulent buoyant jet compared to experimental data of Ricou and Spalding (1961).

From Fig. 9.7

$$\bar{\mu}_{60} = 20$$

$$\frac{\mu}{Q} = 20\left(\frac{R_p}{R_0}\right) = 20\left(\frac{0.557}{0.128}\right) = 87$$

which is almost identical with the plume result (Example 9.2), and comparing this result with that from Example 9.1 we see that a jet with buoyancy gets significantly more dilution than one without.   ■

Similar results can be developed for planar turbulent buoyant jets by defining dimensionless normalized variables appropriate for two-dimensional flow

$$\bar{\mu} = \frac{\mu B^{1/3}}{R_p^{1/2} M} = \frac{\mu}{Q}\left(\frac{R_0}{R_p}\right)^{1/2}, \tag{9.45}$$

$$\zeta = \frac{c_p}{R_p}\frac{z}{l_M} = \frac{c_p z B^{2/3}}{R_p M} = c_p\left(\frac{z}{l_Q}\right)\left(\frac{R_0}{R_p}\right). \tag{9.46}$$

R_p is now the plume Richardson number defined as the asymptotic value of the local Richardson number for a plane buoyant jet,

$$R = \frac{\mu^2 \beta^{2/3}}{m^2}, \tag{9.47}$$

and so $R_0 = Q^2 B^{2/3}/M$. Similarly, c_p is the asymptotic value of the width parameter defined as

$$c = \frac{\mu^2}{mz}. \tag{9.48}$$

Kotsovinos (1975) measured both R and c experimentally over a range of values of z/l_M covering both jets and plumes. His results are shown in Figs. 9.8 and 9.9.

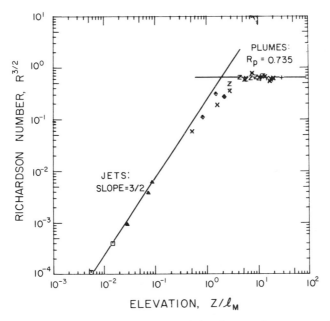

Figure 9.8 Richardson number in a planar turbulent buoyant jet as measured by Kotsovinos (1975).

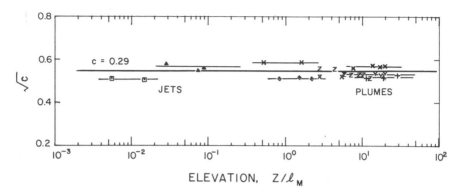

Figure 9.9 Plane jet width parameter c for turbulent jets and plumes showing a constant value of 0.29.

The values of R_p and c_p found were 0.735 and 0.29. Kotsovinos (1975) also measured the dilution, on the axis of a plane vertical jet, and found remarkably good agreement with the predictions

$$\bar{\mu} = \zeta^{1/2}, \qquad \zeta \ll 1, \tag{9.49}$$

$$\bar{\mu} = \zeta, \qquad \zeta \gg 1, \tag{9.50}$$

as shown in Fig. 9.10.

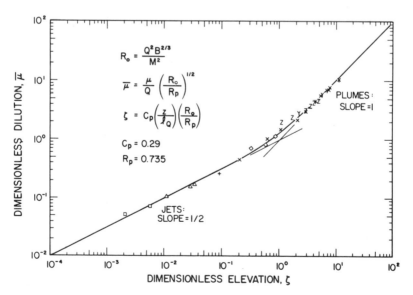

Figure 9.10 Mean dilution in turbulent buoyant plane jets and plumes compared to experiments by Kotsovinos (1975), see Eqs. (9.45) and (9.46) for definitions of $\bar{\mu}$ and ζ, respectively.

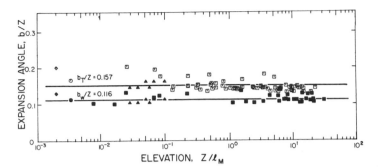

Figure 9.11 Expansion angle of concentration and velocity profiles in plane turbulent jets and plumes. Open symbols are for concentration measurements, closed symbols for velocity. [From Kotsovinos and List (1977).]

One further point deserves mention in regard to planar turbulent buoyant jets. There has been some discussion as to whether the values of b_w and b_T vary between jets and plumes and whether b_T is greater than or less than b_w. Figure 9.11 is a plot of experimental values of b_T (open symbols) and b_w (solid symbols) as a function of z/l_M. While there is some scatter in the data, it is apparent that $b_T/z > b_w/z$. The values chosen as a best fit are $b_T/z = 0.157$, $b_w/z = 0.116$ and these form the basis for the entries in Tables 9.2 and 9.3.

9.2.4 Angle of Jet Inclination

So far we have considered only buoyant jets that are aimed vertically. However, in many design configurations it is advantageous to use horizontal jets, or jets at other inclinations. For example, if a liquid jet is discharged into a shallow body of water, then to maximize the dilution the path length of the jet can be extended by aiming the jet horizontally or at a shallow angle. Similarly, for jets in which the discharge is denser than the ambient fluid, it is generally desirable to avoid having the jet fall back on itself, as will occur with a negatively buoyant jet aimed vertically upward. That this will in fact occur can be seen in Fig. 9.12 which is a photograph of a jet with a strong negative buoyancy flux.

As we have seen, the behavior of a jet is defined in large measure by the jet Richardson number

$$R_0 = QB^{1/2}/M^{5/4}, \tag{9.51}$$

where now the Richardson number must also implicitly include a specification of the jet discharge direction although a negatively buoyant jet pointing vertically upward will be identical to a positively buoyant jet pointing vertically downward, and vice versa. Keeping this in mind, the terminal height of rise z_t

Figure 9.12 Negatively buoyant turbulent jet angled up 60° from horizontal. Richardson number $R_0 = 0.14$.

of a negatively buoyant jet directed vertically upward must satisfy a relationship of the form

$$z_t M^{1/2}/Q = f(R_0), \qquad (9.52)$$

and where the initial volume flux is not important $f(R_0) \to R_0^{-1}$ so that

$$z_t/l_M = \frac{z_t B^{1/2}}{M^{3/4}} = \text{const.} \qquad (9.53)$$

Experiments for vertical jets (Abraham, 1967; Turner, 1966) show that the constant has a value between 1.5 and 2.1. For nonvertical jets insufficient data exist to specify the dependence on the jet angle although a few tests indicate that a reasonable estimate is given by

$$z_t \sim (M \sin \theta)^{3/4}/B^{1/2}, \qquad (9.54)$$

where θ is the angle from horizontal since it is only the vertical component of the momentum flux that is destroyed by the buoyancy flux.

If the ambient fluid is not density stratified, then a dense jet will, after direction reversal, continue in motion until a boundary is met. On some occasions a boundary may be met before a terminal height is reached and this hastens the reversal process. Figure 9.13 illustrates this occurrence. A shallow vertical cold water discharge intersected the free surface and fell back to the bottom where it proceeded to spread as a density current.

The dilution in both positive and negatively buoyant jets discharged at an angle to the horizontal is a function of the initial jet angle, the initial Richardson number and the distance from the jet orifice. At the present time it is not possible to deduce general relationships for the form of the dilution function. However, it

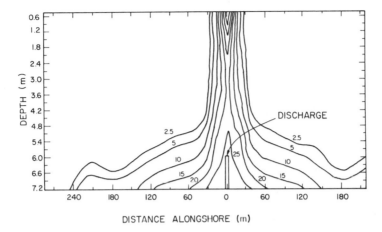

Figure 9.13 Isopleths of temperature resulting from a laboratory model of a shallow submerged vertical cold water discharge. Exit temperature decrement 100%. Richardson number $R_0 = 0.47$, $l_Q = 20$ ft (6.1 m), $l_M = 43$ ft (13.1 m).

is possible to show by consideration of the equations of motion, that a horizontal buoyant jet actually rises exponentially with distance from the source, and that the scaling length is l_M. Thus for distances less than l_M from the jet orifice, the simple jet solutions may be used to predict an initial dilution. Later on in Section 9.4 we will develop a technique for calculating both the trajectory and dilution more precisely; for example, Figure 9.27 gives computed dilutions for horizontal buoyant jets.

9.3 ENVIRONMENTAL PARAMETERS

In Section 9.2 we discussed the effect of changing jet parameters on jet dilution and mechanics and found that the predominant influences were jet momentum, buoyancy and angle of discharge. In this section we study the effects of environmental factors which include density stratification, ambient currents, and ambient turbulence. We will consider the effect of each of these factors acting alone on pure jets, pure plumes and buoyant jets and will show that, as before, there are ranges of the salient parameters over which each of the factors may predominate over all others. In this way we can get an understanding of their relative importance in any particular problem. We will see that in some problems more than one factor may be of significance simultaneously so that interactions can occur, such as the case with strong buoyancy and strong cross flows. Similarly, strong ambient density stratification may completely rule out the necessity to consider ambient turbulence.

9.3.1 Ambient Density Stratification

Ambient density stratifications can and do occur frequently both in the atmosphere and oceans. In the atmosphere the stratification arises from a variation in temperature and humidity of the air; in the ocean it is commonly associated with variations in salinity and temperature of different water masses. Tables exist for computing the density ρ given the temperature and salinity of the seawater (see Appendix A).

In order to study the effect of ambient density stratification on jets and plumes we will write the vertical density distribution as

$$\rho = \rho_0(1 - \varepsilon(z)), \tag{9.55}$$

where $\varepsilon(z)$ is the density anomaly so that

$$\frac{1}{\rho_0}\frac{d\rho}{dz} = \frac{-d\varepsilon}{dz} = -\varepsilon'(z) \tag{9.56}$$

and ρ_0 is the density at $z = 0$. For a statically stable environment with z increasing in the upward direction $\varepsilon'(z)$ must be positive. Values of $\varepsilon'(z)$ commonly encountered in the ocean and in lakes are of the order of 10^{-4}–$10^{-5}\,\mathrm{m}^{-1}$. It can be seen that $(d\varepsilon/dz)^{-1}$ is therefore a characteristic length associated with the intensity of density stratification.

Now consider a point source of momentum with specific momentum flux M directed vertically upward. The effect of this momentum flux will be to carry entrained dense fluid to where the ambient fluid is less dense. We should, therefore, expect such a jet to have a terminal height of rise. Furthermore, this terminal height of rise will depend only on M, ε' (assumed constant for this example) and g the gravitational acceleration. However, because the density stratification causes a gravitationally induced force which is derived from a mass distribution specified by $\varepsilon(z)$, ε', and g must be combined as $g\varepsilon'$.

The terminal height of rise h_M of a round simple momentum jet with specific momentum flux M is therefore given by

$$h_M \sim (M/g\varepsilon')^{1/4}, \tag{9.57}$$

and the constant of proportionality is about 3.8, as determined from the data of Fan (1967) and Fox (1970). Likewise, the terminal height of rise for a round simple plume with specific buoyancy flux B is specified by

$$h_B \sim B^{1/4}/(g\varepsilon')^{3/8}, \tag{9.58}$$

with a coefficient of proportionality, as determined from the data of Crawford and Leonard (1962), Morton et al. (1956), and Briggs (1965), of about 3.8.

It can be seen that the terms on the right-hand sides of Eqs. (9.57) and (9.58), respectively, define characteristic length scales for jets and plumes in density-stratified environments with a constant density gradient. The ratio of these two length scales (raised to the eighth power) we write as

$$N = M^2 g\varepsilon'/B^2, \tag{9.59}$$

and this becomes the defining parameter for buoyant jets in a linearly density-stratified environment, along with the initial jet densimetric Froude, or Richardson number, as defined by Eq. (9.40).

From these simple results we can write down asymptotic functional relationships for the terminal height of rise ζ_T of a buoyant jet:

$$\zeta_T \to \begin{cases} d_p N^{-3/8}, & N \ll 1, \\ d_j N^{-1/4}, & N \gg 1, \end{cases} \tag{9.60}$$

where d_p and d_j are coefficients of proportionality and are both equal to 1.7 ± 0.2, according to the coefficients used in Eqs. (9.57) and (9.58).

That such a terminal height of rise does exist can be seen in Fig. 9.14, which is a photograph of a round buoyant jet in a linearly stratified environment. Note that there is some overshoot.

The photograph in Fig. 9.14 shows quite clearly that the flow undergoes a radical change of direction near the terminal height of rise. It is still possible, however, to speak of a volume flux in the neighborhood of the terminal height of

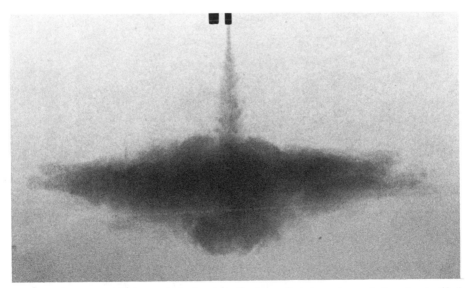

Figure 9.14 A vertical negatively buoyant jet descending in a stagnant, linearly stratified environment $R_0 = 0.052$, $N = 3.2$. [From Fan (1967).]

rise since there is a well-defined flow through a control volume around the terminal point. Simple dimensional analysis again enables us to predict that the mean dilution at the terminal level $\bar{\mu}_T$ must be given by the asymptotic results

$$\bar{\mu}_T \to \begin{cases} e_p N^{-5/8}, & N \ll 1, \\ e_j N^{-1/4}, & N \gg 1. \end{cases} \tag{9.61}$$

The coefficients e_p and e_j can be computed from the work presented in Section 9.4. The values obtained are $e_p = 1.5 \pm 0.2$ and $e_j = 1.2 \pm 0.2$.

The results given in Eqs. (9.60) and (9.61), when combined, indicate that the terminal dilutions and height of rise will actually lie on lines parallel to the asymptotic lines in Fig. 9.7 (see later, Fig. 9.26).

In most field situations, where the theory is to be applied for design purposes, the density stratification is not linear but an empirically measured function of elevation. Later on we will present techniques to calculate terminal dilutions and heights of rise for such stratifications and we will see that the above asymptotic results are well confirmed and, furthermore, that a simple extension of Fig. 9.7 exists to include linearly stratified environments (see Section 9.4).

Asymptotic solutions for two-dimensional, i.e., plane, jets and plumes in density-stratified environments can be determined in a completely analogous fashion to round jets and plumes. The parameter N remains as in Eq. (9.59).

The coefficients of proportionality in the asymptotic relationships presented in Table 9.4 do not have a large number of experimental results supporting them and should be used with some caution; the sources for the data available are Brooks (1973) and Bardey (1977).

TABLE 9.4

Dimensionless Terminal Height of Rise and Dilution for Plane Jets and Plumes in a Linearly Stratified Environment

Variable	$N \ll 1$	$N \gg 1$
h_M	—	$4.0(M/g\varepsilon')^{1/3}$
h_B	$2.8B^{1/3}/(g\varepsilon')^{1/2}$	—
ζ_T	$1.1N^{-1/2}$	$1.6N^{-1/3}$
$\bar{\mu}_T$	$1.0N^{-1/2}$	$k^a N^{-1/6}$

^{a} Not known.

Example 9.4. In the previous example (Example 9.3) we assumed a homogeneous ocean over a depth of 70 m (230 ft). This is seldom realistic since even in wintertime there is generally some temperature change over this depth. Suppose that a uniform temperature gradient exists over the lower 60 m (197 ft) of ocean at the discharge site so that the temperature increases from 11.1 C

(52°F) at a depth of 70 m (230 ft) to 17.8°C (64°F) at a depth of 10 m (33 ft) and is constant above this depth. The freshwater discharge remains at 1 m³/sec with a temperature of 17.8°F and there is essentially no momentum of discharge. Will the plume reach the surface if the ocean salinity is 32.5‰?

From Appendix A:

$$\text{Seawater density at } 17.8°C \text{ and } 32.5‰ = 1023.4 \text{ kg/m}^3$$

$$\text{Seawater density at } 11.1°C \text{ and } 32.5‰ = 1024.8 \text{ kg/m}^3$$

$$\text{Density gradient } \varepsilon' = \frac{1024.8 - 1023.4}{1024.8 \times 60}$$

$$= 2.28 \times 10^{-5} \text{ m}^{-1}.$$

$$g\varepsilon' = 22.3 \times 10^{-5} \text{ sec}^{-2}.$$

Buoyancy flux

$$B = g_0' Q$$

$$= 9.8 \times \frac{(1024.8 - 998.6)}{998.6} \times 1$$

$$= 0.257 \text{ m}^4/\text{sec}^3.$$

From Eq. (9.58)

$$h_B \approx 3.8 \frac{B^{1/4}}{(g\varepsilon')^{3/8}} \text{ m}$$

$$= 63 \text{ m}.$$

The terminal height of rise does not exceed the depth but the plume could possibly surface from an overshoot. To see what happens if the discharge velocity were 3 m/sec, see Example 9.7. ∎

The buoyancy of a discharge in the ocean may be associated with both an increase in temperature and a reduced salinity. In this circumstance, when the ambient density stratification may be primarily temperature induced, the buoyant jet may reach a level of neutral buoyancy and yet have both temperature and salinity different from the ambient fluid at that level. This can give rise to the double diffusive convection phenomenon wherein the difference in diffusivities of heat and salinity results in a greatly enhanced vertical mixing in the laterally spreading plume. For a discussion of this topic see the paper by Fischer (1971) or the excellent monograph of Turner (1973).

9.3.2 Ambient Crossflows

Just as with ambient density stratification, practical problems involving jets very frequently also have cross flows. In most cases, this cross flow will have shear, although not often does the shear become as extreme as that shown in Fig. 9.15! In this photograph the taller stack is 500 ft (150 m) high, the shorter stacks are 250 ft (75 m) high and it is a cold February morning in Massachusetts with a strong density-stratified shear flow.

While the description of turbulent buoyant jet behavior in shear flows is still a topic for active research there are some important results that exist for jets in uniform cross flows, and these will be discussed.

There are a large number of experimental laboratory and field studies described in the research literature. These studies range from describing in great detail the flow in the immediate neighborhood of the jet orifice (see, for example, Chaissang *et al.*, 1974; Moussa *et al.*, 1977), to the region downstream where the ambient turbulence becomes the dominant feature. In this section we will concentrate on classifying the available results according to the characteristic length scales of the problem. In order to see what these length scales are, we consider, as before, some elementary limiting problems. For example, if there is a vertical momentum flux source M in a uniform cross flow U, then the only characteristic length scale is $z_M = M^{1/2}/U$. We therefore expect all the properties of the flow to be described in terms of z/z_M. This means that $U \to 0$ is formally equivalent to $z \to 0$ or $M \to \infty$, so that for $z/z_M \ll 1$ we should expect solutions in which M dominates U. Alternatively for $z/z_M \gg 1$, solutions should be such that the effect of U dominates M. For example, for the vertical velocity on the axis of the jet we would have

$$w_m/U = f(z/z_M), \tag{9.62}$$

Figure 9.15 Turbulent buoyant jets in a density-stratified shear flow. [Photo by Ralph Turcotte, *Beverly* (Massachusetts) *Times*; see text for details.]

with possibly different asymptotic forms of the function f depending on the relative magnitudes of z and z_M.

Finding these asymptotic formulas can be accomplished using simple dimensional arguments, as described by Scorer (1959). However, we believe it is instructive to consider how the solutions can be developed from the equations of motion by the use of similarity solutions. For the reader unfamiliar with similarity arguments this section will introduce the basic idea of how local scaling reduces velocity and density distributions to forms which apply over an extended section of a jet. The Gaussian concentration profile given in Eq. (9.9) is one example of a self-similar distribution.

The techniques applied in this section are used in a broad range of problems in fluid mechanics and hydraulic engineering and for this reason alone it is believed worthwhile spending some time getting involved in somewhat more detailed analysis than has previously been the case.

The basic idea applied to a momentum jet in a crossflow is to assume that there are two asymptotic states of the jet corresponding to $z \ll z_M$ and $z \gg z_M$. In the first state, the jet is assumed to be almost unaffected by the cross flow and in the second, the jet is dominated by the cross flow. The jet can be imagined to be as shown schematically in Fig. 9.16, with the first region specified by flow through the plane $A(z)$, the second by flow through the plane $A(x)$. The motion is described by the following two time-averaged momentum equations[†]

$$\frac{\partial}{\partial x}\left(\bar{u}^2 + \overline{u'^2} + \frac{\bar{p}}{\rho_0}\right) + \frac{\partial}{\partial y}(\overline{u'v'}) + \frac{\partial}{\partial z}(\bar{u}\bar{w} + \overline{u'w'}) = 0, \qquad (9.63)$$

$$\frac{\partial}{\partial x}(\bar{u}\bar{w} + \overline{u'w'}) + \frac{\partial}{\partial y}(\overline{w'v'}) + \frac{\partial}{\partial z}\left(\bar{w}^2 + \overline{w'^2} + \frac{\bar{p}}{\rho_0}\right) = \left(\frac{\rho_a - \bar{\rho}}{\rho_0}\right)g, \qquad (9.64)$$

where overbars denote time-averaged values of velocities, primes the deviations from these time averages and $\bar{p}$ includes the hydrostatic pressure distribution. ρ_a is the ambient fluid density and ρ_0 a reference density. It is assumed that, so far as the inertia of the flow is concerned, $\rho_0 \approx \rho_a \approx \bar{\rho}$, and viscous stresses are negligible.

The conservation of volume and mass require that

$$\frac{\partial \bar{u}}{\partial x} + \frac{\partial \bar{w}}{\partial z} = 0, \qquad (9.65)$$

and that (provided ρ_a is independent of z)

$$\frac{\partial}{\partial x}(\bar{u}(\bar{\rho} - \rho_a) + \overline{u'\rho'}) + \frac{\partial \overline{v'\rho'}}{\partial y} + \frac{\partial}{\partial z}(\bar{w}(\bar{\rho} - \rho_a) + \overline{w'\rho'}) = 0, \qquad (9.66)$$

[†] In this section we introduce overbars to denote time-average variables.

where ρ' is the deviation of the fluid density from its time-averaged value and molecular diffusive transport is ignored.

Now consider each of the two possibilities: the first where the jet is in a predominantly vertical motion, the second where the trajectory is significantly bent over. In these two cases we can either average these equations across a horizontal plane, such as cross section $A(z)$ in Fig. 9.16, or across a vertical plane normal to the crossflow velocity, such as cross section $A(x)$ in Fig. 9.16.

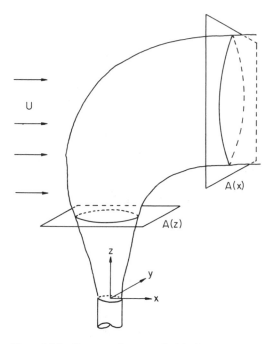

Figure 9.16   Geometry for a vertical jet in a crossflow.

For example, integrating Eq. (9.64) across the jet cross section $A(z)$ we have that

$$\int_{A(z)} \frac{\partial}{\partial x} (\bar{u}\bar{w} + \overline{u'w'}) \, dx \, dy + \int_{A(z)} \frac{\partial}{\partial y} (\overline{w'v'}) \, dx \, dy$$

$$+ \int_{A(z)} \frac{\partial}{\partial z} \left(\bar{w}^2 + \overline{w'^2} + \frac{\bar{p}}{\rho_0} \right) dx \, dy = \int_{A(z)} \left(\frac{\rho_a - \bar{\rho}}{\rho_0} \right) g \, dx \, dy. \quad (9.67)$$

We can define the boundary of the jet cross section in a variety of ways. For example, we could select it as the perimeter of the jet beyond which jet-induced turbulent stresses vanish. Alternatively, it could be chosen as the boundary beyond which mean vertical velocities vanish. These two boundaries do not necessarily coincide, but so far as this analysis is concerned it makes no difference. This being so, then it is relatively easy to see that Eq. (9.67) reduces to

$$\int_{A(z)} \frac{\partial}{\partial z} \left(\bar{w}^2 + \overline{w'^2} + \frac{\bar{p}}{\rho_0} \right) dx \, dy = \int_{A(z)} \left(\frac{\rho_a - \bar{\rho}}{\rho_0} \right) q \, dx \, dy, \qquad (9.68)$$

Similarly, Eq. (9.66) becomes

$$\int_{A(z)} \frac{\partial}{\partial z} (\bar{w}(\bar{\rho} - \rho_a) + \overline{w'\rho'}) \, dx \, dy = 0, \qquad (9.69)$$

when integrated over the same cross section.

Equations (9.68) and (9.69) will form the basis of our analysis of buoyant jets in the vertical flow regime. Equation (9.68) in effect states that the rate of change of vertical flow force in a vertical direction is equal to the buoyancy force. Miller and Comings (1957) have shown that in two-dimensional jets the contributions to the flow force from $\overline{w'^2}$ and $\bar{p}/\rho_0$ are small and are opposite in sign. Although this result is not necessarily true for other flow configurations, such as we consider here, we will assume it to be the case and rewrite Eq. (9.68) as

$$\int_{A(z)} \frac{\partial}{\partial z} \bar{w}^2 \, dx \, dy = \int_{A(z)} \left(\frac{\rho_a - \bar{\rho}}{\rho_0} \right) g \, dx \, dy. \qquad (9.70)$$

Similarly, in Eq. (9.69) we will ignore the turbulent transport term $\overline{w'\rho'}$, although Kotsovinos (1975) has shown this to be a poor assumption in buoyancy-driven flows. Nevertheless, the zero order description of the flow we are seeking here should not be greatly compromised, so we write

$$\int_{A(z)} \frac{\partial}{\partial z} (\bar{w}(\rho_a - \bar{\rho}) \, dx \, dy = 0. \qquad (9.71)$$

Equations (9.70) and (9.71) are the equations we will use for the description of predominantly vertical flows. Before proceeding with their use, however, consider the integration of Eqs. (9.64) and (9.66) across a vertical plane, $A(x)$ in Fig. 9.16. Making the same kind of simplifications as in the integration across the horizontal plane we have the results

$$\int_{A(x)} \frac{\partial}{\partial x} (\bar{u}\bar{w}) \, dy \, dz = \int_{A(x)} \left(\frac{\rho_a - \bar{\rho}}{\rho_0} \right) g \, dy \, dz, \qquad (9.72)$$

$$\int_{A(x)} \frac{\partial}{\partial x} (\bar{u}(\rho_a - \bar{\rho}) \, dy \, dz = 0. \qquad (9.73)$$

Equations (9.72) and (9.73) will be used to define the horizontal flow regimes. Equation (9.72) states that the rate of change of the *horizontal* flux of *vertical* momentum is equal to the buoyancy force acting in a vertical plane. Equation (9.73) merely states that the horizontal flux of buoyancy is conserved.

Let us first take a jet without buoyancy. We will consider it to be only a source of volume flux Q and momentum flux M, and at first consider only the vertical flow regime so that Eqs. (9.70) and (9.71) are appropriate. Since the jet has no buoyancy the right-hand side of Eq. (9.70) will be zero and in Eq. (9.71) we take $\rho_a - \bar{\rho}$ to be the excess concentration of some tracer material in the jet.

We imagine the flow to be fully developed, that is $z \gg l_Q$, and assume self-similarity of the velocity and tracer profiles. Accordingly we write

$$\bar{w}(x, y, z) = w_m(\bar{z})\phi\left(\frac{x}{\bar{z}}, \frac{y}{\bar{z}}\right), \tag{9.74}$$

$$\frac{\rho_a - \bar{\rho}}{\rho_0} = \theta(\bar{z})\psi\left(\frac{x}{\bar{z}}, \frac{y}{\bar{z}}\right), \tag{9.75}$$

where $\bar{z}$ is the z coordinate of the jet axis and is a function of x; ϕ and ψ are undefined functions describing the lateral distribution of velocity and tracer. Substituting Eqs. (9.74) and (9.75) into Eqs. (9.70) and (9.71) and remembering that the right-hand side of Eq. (9.70) will be zero for nonbuoyant motion gives

$$\frac{d}{dz}\int_{A(z)} \bar{z}^2 w_m{}^2(\bar{z})\phi^2 \, d\left(\frac{x}{\bar{z}}\right) d\left(\frac{y}{\bar{z}}\right) = 0, \tag{9.76}$$

and

$$\frac{d}{dz}\int_{A(z)} \bar{z}^2 w_m(\bar{z})\theta(\bar{z})\phi\psi \, d\left(\frac{x}{\bar{z}}\right) d\left(\frac{y}{\bar{z}}\right) = 0, \tag{9.77}$$

where the differentiation can be moved outside the integral because ϕ and ψ are assumed to vanish at the perimeter of the cross section of integration. These two equations imply that

$$\bar{z}^2 w_m{}^2(\bar{z}) \sim M, \tag{9.78}$$

and

$$\bar{z}^2 w_m(\bar{z})\theta(\bar{z}) \sim B/g, \tag{9.79}$$

because the integrals are independent of $\bar{z}$ by virtue of the fact that the integrands vanish outside $A(z)$ and the actual value of the integrals is irrelevant provided that we are only concerned with proportionalities. The parameter on the right-hand side of Eq. (9.79) must be B/g since the left-hand side is proportional to the flux of excess tracer mass in the jet.† From Eqs. (9.78) and (9.79) we see that

$$w_m(\bar{z})/U \sim M^{1/2}/(\bar{z}U), \tag{9.80}$$

$$\theta(\bar{z}) \sim B/(gM^{1/2}\bar{z}), \tag{9.81}$$

or rewriting

$$w_m(\bar{z})/U \sim (z_M/\bar{z}), \tag{9.82}$$

$$Mg\theta(\bar{z})/UB = D_1(z_M/\bar{z}), \tag{9.83}$$

where $z_M = M^{1/2}/U$ and D_1 is a constant of proportionality. These solutions are valid where $w_m(\bar{z}) \gg U$ or equivalently, $\bar{z} \ll z_M$. It is apparent that z_M is the vertical height at which the vertical velocity in the jet has decayed to the order of the crossflow velocity.

For a jet in a crossflow it seems reasonable that the slope of the jet trajectory is specified by

$$w_m(\bar{z})/U = d\bar{z}/dx. \tag{9.84}$$

Then Eq. (9.82) implies that the jet trajectory is given by

$$\bar{z}/z_M = C_1(x/z_M)^{1/2}, \qquad \bar{z} \ll z_M, \tag{9.85}$$

for some "constant" C_1 which may be a function of the ratio z_M/l_Q. Thus, for a momentum-dominated jet with a "weak" crossflow we have obtained solutions for the maximum vertical velocity, tracer concentration and jet trajectory, plus a criterion for their application.

We will now consider the case when the jet is in a bent-over region and Eqs. (9.72) and (9.73) are appropriate. We should expect self-similarity again and in this case we can write

$$\bar{w}(x, y, z) = w_m(\bar{z})\phi\left(\frac{z - \bar{z}}{\bar{z}}, \frac{y}{\bar{z}}\right), \tag{9.86}$$

$$\frac{\rho_a - \bar{\rho}}{\rho_0} = \theta(\bar{z})\psi\left(\frac{z - \bar{z}}{\bar{z}}, \frac{y}{\bar{z}}\right), \tag{9.87}$$

$$\bar{u} \simeq U, \tag{9.88}$$

because the similar profile will be centered on $\bar{z}$. Then, provided ϕ and ψ vanish outside the perimeter of the jet, we have

$$\frac{d}{dx}\int_{A(x)} \bar{z}^2 U w_m(\bar{z})\phi \, d\left(\frac{z - \bar{z}}{\bar{z}}\right) d\left(\frac{y}{\bar{z}}\right) = 0 \tag{9.89}$$

$$\frac{d}{dx}\int_{A(x)} \bar{z}^2 U \theta(\bar{z})\psi \, d\left(\frac{z - \bar{z}}{\bar{z}}\right) d\left(\frac{y}{\bar{z}}\right) = 0, \tag{9.90}$$

† Note that it is convenient to retain the symbol B for the flux of tracer even though we assume the buoyancy flux has no dynamical effect here, i.e., the right-hand side of Eq. (9.70) is zero.

which imply that

$$w_m(\bar{z})/U \sim (z_M/\bar{z})^2, \qquad \bar{z} \gg z_M, \qquad (9.91)$$

$$Mg\theta(\bar{z})/UB = D_2(z_M/\bar{z})^2, \qquad \bar{z} \gg z_M, \qquad (9.92)$$

and D_2 is an empirical constant. These solutions will only apply in the bent-over region where $w_m(z) \ll U$ implying that $\bar{z} \gg z_M$. Again the trajectory can be deduced from Eqs. (9.84) and (9.91) to be

$$\bar{z}/z_M = C_2(x/z_M)^{1/3}, \qquad \bar{z} \gg z_M, \qquad (9.93)$$

for some constant C_2. It is again apparent that z_M is the vertical height at which the jet will begin to appear appreciably bent.

We now consider the case of a pure plume in a crossflow, that is we assume the flow is produced solely by a source of buoyancy flux B and that a "vertical" and a "horizontal" flow region occur as before. This time, because of the buoyancy, it is not possible to ignore the right-hand side of Eqs. (9.70) and (9.71), so that assuming self-similarity in the "vertical" region leads to the results

$$(d/dz)[\bar{z}^2 w_m^2(\bar{z})] \sim g\bar{z}^2\theta(\bar{z}), \qquad (9.94)$$

and

$$(d/dz)[\bar{z}^2 w_m(\bar{z})\theta(\bar{z})] = 0. \qquad (9.95)$$

From these two results and using Eq. (9.84) it may be easily shown that if $dz \simeq d\bar{z}$, which seems reasonable, then

$$w_m(\bar{z})/U \sim (z_B/\bar{z})^{1/3}, \qquad \bar{z} \ll z_B, \qquad (9.96)$$

$$\left(\frac{z_B}{z_M}\right)^2 \frac{gM\theta(\bar{z})}{UB} = D_3\left(\frac{z_B}{\bar{z}}\right)^{5/3}, \qquad \bar{z} \ll z_B, \qquad (9.97)$$

$$\bar{z}/z_B = C_3(x/z_B)^{3/4}, \qquad \bar{z} \ll z_B, \qquad (9.98)$$

where C_3 and D_3 are proportionality constants to be determined empirically and $z_B = B/U^3$ is the characteristic length scale for this problem. These solutions are therefore only valid in the region $\bar{z} \ll z_B$. z_B is the vertical distance along the jet trajectory where the vertical velocity of the plume decays to the order of the cross-flow velocity.

The final case to consider is the plume in a bent-over flow. In this region Eqs. (9.72) and (9.73), and similarity forms such as Eqs. (9.86) and (9.87), imply that

$$(d/dx)[\bar{z}^2 U w_m(\bar{z})] \sim g\bar{z}^2\theta(\bar{z}), \qquad (9.99)$$

$$(d/dx)[\bar{z}^2 U \theta(\bar{z})] = 0. \qquad (9.100)$$

Again using Eq. (9.84) it may then be shown, with some algebra, that

$$w_m(\bar{z})/U \sim (z_B/\bar{z})^{1/2}, \qquad \bar{z} \gg z_B, \qquad (9.101)$$

$$\left(\frac{z_B}{z_M}\right)^2 \frac{gM\theta(\bar{z})}{UB} = D_4 \left(\frac{z_B}{\bar{z}}\right)^2, \qquad \bar{z} \gg z_B, \qquad (9.102)$$

and that

$$\bar{z}/z_B = C_4(x/z_B)^{2/3}, \qquad \bar{z} \gg z_B. \qquad (9.103)$$

with C_4 and D_4 constants of proportionality.

With these results for pure jets and pure plumes we have established four possible trajectories and associated dilution rates. The question we now address is what path will a turbulent buoyant jet follow given the initial volume flux Q, the momentum flux M, and the buoyancy flux B. The solution can best be seen by considering the two cases corresponding to z_M greater than and less than z_B assuming, of course, that both are greater than l_Q.

Figure 9.17 is a schematic logarithmic graph of the normalized vertical velocity $w_m(\bar{z})/U$ plotted against $\bar{z}$ for both pure plumes and pure jets with $z_M < z_B$. Provided that z_M is not too small then the flow will begin as a jet, or momentum-dominated flow, and the trajectory will have the form $\bar{z} \sim x^{1/2}$. At a vertical height of about l_M, the flow will become plumelike and follow the trajectory $\bar{z} \sim x^{3/4}$. Subsequently, at an elevation of about z_B, the flow will

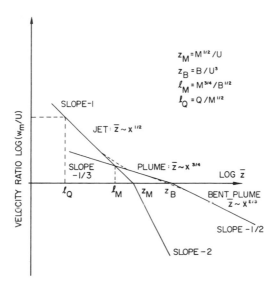

Figure 9.17 Vertical to horizontal mean velocity ratio in a turbulent buoyant jet in a uniform cross flow ($z_M < z_B$). Trajectory relations also shown.

transform to that of a bent-over plume with a trajectory $\bar{z} \sim x^{2/3}$. In this problem the two appropriate length scales are obviously l_M and z_B, but only provided $z_M < z_B$. In actuality the path appropriate to any jet will include the effect of the ratio of l_Q/l_M, or the initial Richardson number of the jet.

The case when $z_M > z_B$ is shown in Fig. 9.18. This corresponds to the case when the buoyancy flux is weak and the flow begins as a jet, the jet becomes bent over and follows a trajectory given by $\bar{z} \sim x^{1/3}$ and then ultimately transforms to a bent-over plume, as in the previous example. It can be seen that the characteristic lengths of importance in this case are actually z_M and z_C, since these are the order of the elevations at which significant changes in the jet trajectory will occur, provided that $z_B < z_M$.

The actual jet trajectories predicted by the analyses are represented diagrammatically in Figs. 9.19 and 9.20 for the two cases $z_M < z_B$ and $z_M > z_B$, respectively. From Figs. 9.19 and 9.20 it is apparent that unless the correct normalization is used to plot experimental or field data, and in addition the relative orders of magnitude of the various length scales are recognized, then distinguishing the different regimes of flow would be extremely difficult.

9.3.2.1 Normalized Descriptions of Jets in Crossflows

For design purposes the above results are best written in normalized form since it then becomes a straightforward process to determine the sensitivity of a design to changes in parameters. It must be remembered, however, that the results presented here are *asymptotic* results. This is not to say that they are not useful, quite to the contrary. Asymptotic results are, in general, very useful for

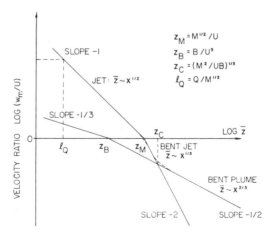

Figure 9.18 Vertical to horizontal mean velocity ratio in a turbulent buoyant jet in a uniform cross flow ($z_M > z_B$). Trajectory relations also shown.

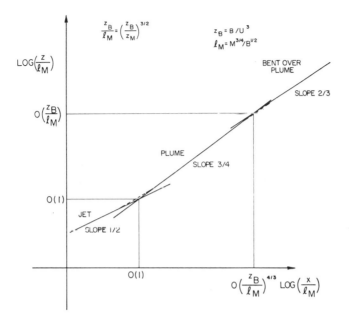

Figure 9.19 Jet trajectory when $z_M < z_B$.

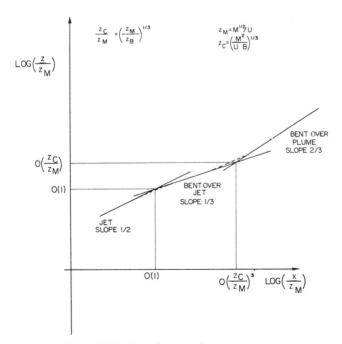

Figure 9.20 Jet trajectory when $z_M > z_B$.

design purposes because they allow a rapid categorization of the type of problem under consideration.

A convenient set of dimensionless normalized variables to describe jets in cross flows is presented in Table 9.5. This set of variables enables us to write all the jet solutions, Eqs. (9.83), (9.85), (9.92), and (9.93), in a particularly simple form. Similarly, the plume solutions, Eqs. (9.97), (9.98), (9.102), and (9.103), can can also be reduced.

The functional forms for the trajectories and dilutions given in Tables 9.6 and 9.7 can also be represented graphically for different values of the parameters ξ_c and $\hat{\xi}_c$, as shown in Figs. 9.21 and 9.22 for the trajectories and dilutions, respectively. These figures indicate the role of the parameters ξ_c and $\hat{\xi}_c$ in fixing the form of the jet trajectory.

It must be remembered, of course, that a number of empirical constants remain to be specified. Fortunately, there is a reasonable amount of laboratory

TABLE 9.5

Dimensionless Variables for Asymptotic Solutions for a Turbulent Buoyant Jet in a Crossflow[a]

Dimensionless variable	Case: $z_M > z_B$	Case: $z_M < z_B$
Horizontal distance	$\xi = \dfrac{x}{z_M}\left(\dfrac{C_1}{C_2}\right)^6$	$\hat{\xi} = \left(\dfrac{x}{z_M}\right)\left(\dfrac{z_B}{z_M}\right)\left(\dfrac{C_3}{C_1}\right)^4$
Vertical elevation	$\zeta = \dfrac{\bar{z}}{z_M}\dfrac{1}{C_1}\left(\dfrac{C_1}{C_2}\right)^3$	$\hat{\zeta} = \left(\dfrac{\bar{z}}{z_M}\right)\left(\dfrac{z_B}{z_M}\right)^{1/2}\left(\dfrac{C_3}{C_1}\right)^2\dfrac{1}{C_1}$
Dilution	$S = \left(\dfrac{\mu U}{M}\right)\bigg/\left(\dfrac{1}{D_1}\dfrac{C_2^3}{C_1^2}\right)$	$\hat{S} = \left(\dfrac{\mu U}{M}\right)\bigg/\left(\dfrac{1}{D_1}\left(\dfrac{z_M}{z_B}\right)^{1/2}\dfrac{C_1^3}{C_3^2}\right)$

[a] Constants C_i given in Table 9.8.

TABLE 9.6

Asymptotic Solutions for Trajectories and Mean Dilutions for a Vertical Turbulent Buoyant Jet in a Uniform Crossflow[a]

Dimensionless variable	$\hat{\xi} \ll 1$	$1 \ll \hat{\xi} \ll \hat{\xi}_c$	$\hat{\xi}_c \ll \hat{\xi}$
$\hat{\zeta}$	$\hat{\xi}^{1/2}$	$\hat{\xi}^{3/4}$	$\hat{\kappa}(z_B/z_M)^{1/6}\hat{\xi}^{2/3}$
$\hat{S}$	$\hat{\xi}^{1/2}$	$\hat{\xi}^{5/4}$	$(z_M/z_B)^{1/6}\hat{\xi}^{4/3}/\hat{\kappa}$

[a] When $z_M < z_B$, with

$$\hat{\xi}_c = \hat{\kappa}^{12}(z_B/z_M)^2, \quad \hat{\kappa} = (C_4/C_3)(C_3/C_1)^{1/3}.$$

TABLE 9.7

Asymptotic Solutions for Trajectories and Mean Dilutions for a Vertical Turbulent Buoyant Jet in a Uniform Crossflow[a]

Dimensionless variable	$\xi \ll 1$	$1 \ll \xi \ll \xi_c$	$\xi_c \ll \xi$
ζ	$\xi^{1/2}$	$\xi^{1/3}$	$\kappa(z_B/z_M)^{1/3}\xi^{2/3}$
S	$\xi^{1/2}$	$\xi^{2/3}$	$\kappa^2(z_B/z_M)^{2/3}\xi^{4/3}$

[a] When $z_M > z_B$, with

$$\xi_c = (1/\kappa^3)(z_M/z_B), \quad \kappa = (C_4/C_1)(C_2/C_1).$$

and field experimental data available from which the unknown constants $C_1 - C_4$ and $D_1 - D_4$ can be determined. The readily available results are presented in Table 9.8. It should also be noted that it can be argued that not all the constants $C_1 - C_4$ are independent, since the two solutions specified by $\xi_c = 1$ and $\xi_c = 1$ must coincide. This provides a check on the consistency of the experimental data given in Table 9.8. Furthermore, because the mean dilution must coincide at the transition points on the trajectory only one dilution constant D_1, say, needs to be specified.

TABLE 9.8

Constants Used in Asymptotic Trajectory and Dilution Laws for a Buoyant Jet in a Uniform Crossflow

Investigator(s)	Constant C_1
Hoult et al. (1969)	1.8–2.5
Wright (1977)	1.8–2.3
	Constant C_2
Briggs[a] (1975)	1.8–2.1
Wright (1977)	1.6–2.1
Chu and Goldberg (1974)	1.44
	Constant C_3
Wright (1977)	1.4–1.8
	Constant C_4
Briggs[a] (1975)	1.1 (0.82–1.3)
Wright (1977)	$(0.85–1.4)(z_M/z_B)^2$
Chu and Goldberg (1974)	1.14
	Constants $D_1 - D_4$
Wright (1977)	~2.4

[a] Summary of 14 investigations.

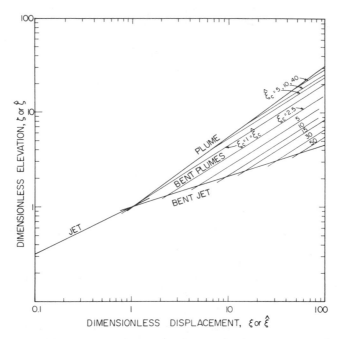

Figure 9.21 Possible trajectories for round turbulent buoyant jets in a uniform crossflow (see Tables 9.4–9.7 for symbols).

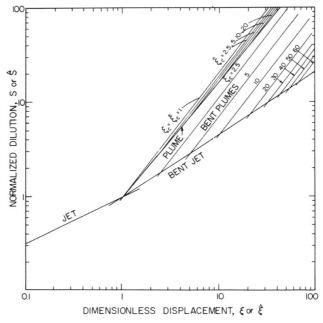

Figure 9.22 Mean dilution in round turbulent buoyant jets in a uniform crossflow (see Tables 9.4–9.7 for symbols).

Finally, it is important to remember that the relationships given should only be regarded as providing *order of magnitude* estimates for trajectories and dilutions. In many circumstances there actually will be factors that will modify the predictions of the asymptotic theory, for example, shearing of the velocity profile, or localized density stratifications and geometrical influences. The values of the constants given for the trajectories may also vary depending upon whether the trajectory is defined optically or with a concentration measurement. Indeed, as Turner (1960) and Scorer (1959) have vividly shown, a strongly buoyant plume in a cross flow may actually bifurcate into a pair of vortices leading to two concentration maxima as shown in Fig. 9.23.

The bifurcation shown in Fig. 9.23 may become quite strong with the two concentration maxima diverging at an angle of 8–10°. Turner (1960) has shown that these maxima are associated with line vortices which appear to be the trailing arms of a horseshoe vortex generated by the interaction of the cross flow and the discharge.

The problem of negatively buoyant jets in uniform cross flows is also an important one in many applications. The negatively buoyant jet may be aimed vertically upward, although this is generally not desirable from a design point of view, since in the absence of a cross flow the effluent falls directly back onto the discharge orifice, as shown in Fig. 9.13. Alternatively, the discharge may be

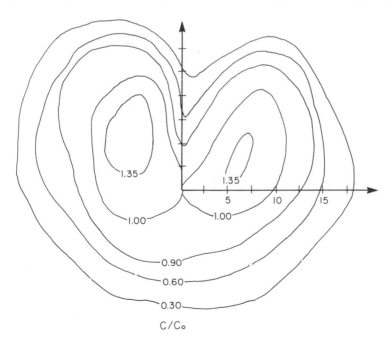

Figure 9.23 Concentration isopleths showing bifurcation in a turbulent buoyant jet in a cross-flow $z_M/z_B = 7$, $x/l_M = 1.43$, $z/l_M = 1.0$. (Axes are in arbitrary units.) [From Wright (1977).]

aimed up from the horizontal at an angle as in Fig. 9.12, and the cross flow may
be parallel or transverse to the vertical plane containing the jet axis. The parameters describing such jets are, for a fixed jet angle, l_Q, l_M, z_M, and z_B just as for
positively buoyant jets. There is now, for any fixed jet angle, a terminal height
of rise to the jet z_t say, whose specification, provided $l_M \gg l_Q$, will be of the form

$$z_t = l_M f(z_M/z_B).$$

Similarly, the dilution along the jet will be a function of l_M/l_Q and z_M/z_B. However,
further experiments and analyses must be carried out before generally applicable
results can be presented.

   As an indication of the difficulties associated with the consideration of turbulent buoyant jets in cross flows consider the following example.

Example 9.5.   A vertical buoyant discharge with $Q = 1$ m^3/sec and a discharge velocity of 3 m/sec is located on the sea floor. The ocean is homogeneous
with a temperature 11.1°C (52°F) and salinity 32.5‰. The discharge is freshwater and has a temperature of 17.8°C (64°F). There is a uniform cross flow of
0.25 m/sec (0.49 knot). What form will the jet trajectory take?

   From previous Examples 9.3 and 9.4

$$Q = 1 \text{ m}^3/\text{sec}, \qquad M = 3 \text{ m}^4/\text{sec}^2,$$
$$B = 0.257 \text{ m}^4/\text{sec}^3, \qquad U = 0.25 \text{ m/sec}.$$

The characteristic length scales are

$$l_Q = Q/M^{1/2} = 0.58 \text{ m}, \qquad l_M = M^{3/4}/B^{1/2} = 4.5 \text{ m},$$
$$z_M = M^{1/2}/U = 6.9 \text{ m}, \qquad z_B = B/U^3 = 16.4 \text{ m}.$$

   We see that $z_B > z_M > l_M$ so that we would expect a jet trajectory as shown in
Fig. 9.19 with

$$\bar{z}/z_M = C_1(x/z_M)^{1/2}, \qquad l_Q \ll \bar{z} \ll l_M,$$
$$\bar{z}/z_B = C_3(x/z_B)^{3/4}, \qquad l_M \ll \bar{z} \ll z_B,$$
$$z/z_B = C_4(x/z_B)^{2/3}, \qquad z_B \ll \bar{z},$$

as given by Eqs. (9.85), (9.98), and (9.103).

   The two points of intersection of the two asymptotic trajectories are given by

$$x_1 = z_M(z_M/z_B)(C_1/C_3)^4$$
$$z_1 = z_M C_1(z_M/z_B)^{1/2}(C_1/C_3)^2$$

and

$$x_2 = z_B(C_4/C_3)^{12},$$
$$z_2 = z_B C_3(C_4/C_3)^9.$$

[Check that these points (x, z) correspond to $(1, 1)$ and $(\hat{\xi}_c, \hat{\xi}_c^{3/4})$ as defined in Tables 9.5 and 9.6.]

Selecting values of C_1, C_3, and C_4, respectively, of 2.0, 1.6, and 1.1 from Table 9.8 we see that

$$z_1 = 6.9(2.0)\left(\frac{2.0}{1.6}\right)^2\left(\frac{6.9}{16.4}\right)^{1/2} \text{m}$$
$$= 14 \text{ m},$$

$$z_2 = 16.4(1.6)\left(\frac{1.1}{1.6}\right)^9$$
$$= 1 \text{ m}.$$

What has gone wrong here? According to the theory z_2 should be greater than z_1!

The difficulty arises because the trajectories of slope $\frac{3}{4}$ and $\frac{2}{3}$ are very close when plotted logarithmically (see Fig. 9.19) so that a small error in the constant C_3 or C_4 makes a very large shift in the elevation of the intersection point z_2. In this case we first assume a plume solution

$$z/z_B = C_4(x/z_B)^{3/4}$$

and find that for $z = 60$ m

$$x_{60} = \left[\frac{1}{1.1}\left(\frac{60}{16.4}\right)\right]^{4/3} 16.4 \text{ m}$$
$$= 81 \text{ m},$$

or alternatively, a bent plume solution

$$z/z_B = C_3(x/z_B)^{2/3}$$

$$x_{60} = \left[\frac{1}{1.6}\frac{60}{16.4}\right]^{3/2} 16.4 \text{ m}$$
$$= 57 \text{ m}.$$

We therefore estimate the plume will be approximately 60 m from the bottom at between 60 and 80 m from the discharge point.

The approximate dilution at this point is given by the following:

$$\hat{\zeta}_{60} = \frac{60}{z_M}\left(\frac{z_B}{z_M}\right)^{1/2}\left(\frac{C_3}{C_1}\right)^2\frac{1}{C_1}$$

$$= \frac{60}{6.9}\left(\frac{16.4}{6.9}\right)^{1/2}\left(\frac{1.6}{2.0}\right)^2\frac{1}{2.0}$$

$$= 4.3.$$

From Fig. 9.21 $\hat{\xi}_{60} \approx 9$ (little dependence on $\hat{\xi}_c$), and from Fig. 9.22 $\hat{S}_{60} \approx 19$ (little dependence on $\hat{\xi}_c$). From Table 9.5.

$$\hat{S}_{60} = \frac{\mu U}{M} \bigg/ \left(\frac{1}{D_1}\right)\left(\frac{z_M}{z_B}\right)^{1/2} \frac{C_1{}^3}{C_3{}^2}$$

$$\left(\frac{\mu}{Q}\right)_{60} = 19 \frac{M}{QU}\left(\frac{1}{D_1}\right)\left(\frac{z_M}{z_B}\right)^{1/2}\frac{C_1{}^3}{C_3{}^2}$$

$$= 192 \quad \blacksquare$$

Comparing this with the dilution from a simple jet (Example 9.1) and a simple plume (Example 9.3), we can see the effectiveness of a cross flow in increasing the dilution.

9.3.3   Jets with Ambient Crossflows and Stratification

When density stratification exists along with a crossflow then the parameters involved are Q, M, B, $g\varepsilon'$, U, which suggests that three dimensionless parameters will govern the solutions. Since we have already introduced six length scales: l_Q, l_M, h_M, h_B, z_M, and z_B, the three parameters are simply the ratio of these length scales taken in pairs, for example, l_Q/l_M, h_M/h_B, and z_M/z_B. The solution of the problem, represented by the jet trajectory and the dilution along the trajectory, will therefore be represented by this three parameter family. However, as we have seen, the parameter l_Q/l_M (the Richardson number) essentially fixes the origin (Fig. 9.7) so that the two parameters of primary importance are the stratification parameter N, specified by Eq. (9.59), and the crossflow parameter $M^{1/2}U^2/B$ specified by z_M/z_B.

So far as the terminal height of rise of a buoyant jet is concerned this can be specified from the previous solutions for jets in crossflows by noting that there is one length scale that involves only the crossflow and the stratification, this is

$$\lambda = U/(g\varepsilon')^{1/2}. \tag{9.104}$$

This length scale has a simple interpretation: $(g\varepsilon')^{-1/2}$ is the resonant period of oscillation of any particle located at a position of neutral density, which means that λ is simply the horizontal wavelength of the vertical oscillations of the moving plume that are supported by the density stratification. Figure 9.24 is a photograph of a laboratory simulation of a turbulent buoyant jet in a density-stratified crossflow generated by moving a dense jet through a quiescent tank of density-stratified water; the wavelike nature of the plume oscillations can be clearly seen (see also, Fig. 9.15). Furthermore, since λ is the only *horizontal* length scale in the problem, the terminal height of rise for any asymptotic solution will be *roughly* specified by replacing x by λ in the appropriate tra-

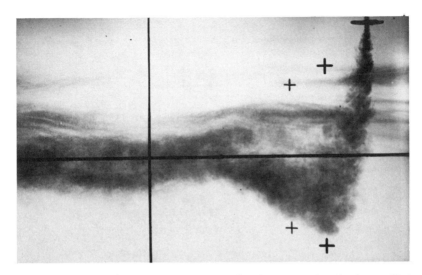

Figure 9.24 Turbulent negatively buoyant jet descending into a moving density-stratified environment $l_Q \ll z_M \ll z_B$. [From Wright (1977).]

jectory equation. The terminal heights of rise will therefore be as given from Tables 9.5–9.7 by replacing ξ by ξ_T and $\hat{\xi}$ by $\hat{\xi}_T$, where

$$\xi_T \sim \lambda/z_M = U^2/(g\varepsilon'M)^{1/2}, \tag{9.105}$$

and

$$\hat{\xi}_T \sim \left(\frac{\lambda}{z_M}\right)\left(\frac{z_B}{z_M}\right) = N^{-1/2}. \tag{9.106}$$

Using these definitions it can be shown that the asymptotic solutions given in Table 9.9 are valid for the ranges shown.

The values of the coefficients in the solutions given in Table 9.9 have been determined in some cases. For example, if both $U^2/(g\varepsilon'M)^{1/2} \ll 1$, and $N^{-1/2} \ll 1$, it is apparent that neither the crossflow nor the buoyancy has much influence on the jet and the terminal height of rise is specified by Eq. (9.57). The value of E_1 is therefore about 3.8 as shown by Briggs (1975) using Fan's (1967) data. For $1 \ll N^{-1/2} \ll (z_B/z_M)^2$, the flow becomes plumelike and reaches a terminal height of rise before being significantly bent over. The constant of proportionality E_3 has a value of about 3.8 [see Eq. (9.58) and Wright (1977)].

The other two cases given in Table 9.9 correspond to a momentum jet in a strong crossflow and a plume in a strong crossflow. A value of the constant E_2 does not appear to have been given in the literature. Briggs (1975), on the basis of ten studies in the atmosphere, suggests a value of 3.8 for E_4. Again, the important point in Table 9.9 is the range of the various dimensionless parameters

TABLE 9.9

*Asymptotic Heights of Rise for a Vertical Turbulent Buoyant Jet
Discharging into a Density-Stratified Crossflow[a]*

Case	$\dfrac{U^2}{(g\varepsilon'M)^{1/2}} \ll 1$	$1 \ll \dfrac{U^2}{(g\varepsilon'M)^{1/2}} \ll \dfrac{z_M}{z_B}$	$\dfrac{z_M}{z_B} \ll \dfrac{U^2}{(g\varepsilon'M)^{1/2}}$
$z_M > z_B$	$\dfrac{z_T}{h_M} = E_1$	$\dfrac{z_T}{z_M^{2/3}\lambda^{1/3}} = E_2$	$\dfrac{z_T}{z_B^{1/3}\lambda^{2/3}} = E_4$

Case	$N^{-1/2} \ll 1$	$1 \ll N^{-1/2} \ll \left(\dfrac{z_B}{z_M}\right)^2$	$\left(\dfrac{z_B}{z_M}\right)^2 \ll N^{-1/2}$
$z_M < z_B$	$\dfrac{z_T}{h_M} = E_1$	$\dfrac{z_T}{h_B} = E_3$	$\dfrac{z_T}{z_B^{1/3}\lambda^{2/3}} = E_4$

[a] See text for values of E_i.

for which these solutions are appropriate. For any given problem Table 9.9 can be quickly used to establish the order of magnitude of the terminal height of rise of any discharge in a linearly stratified uniform crossflow.

The interactions of crossflows and stratification for chimney plumes in the atmosphere have been well described in qualitative terms by Scorer (1959). It is to be expected that many of the features observed with chimney plumes will be reproduced with buoyant discharges in the coastal ocean. The computation of the dilutions to be expected for the general types of stratification and crossflows that occur in the ocean must be performed using the methods to be presented in Section 9.4.

Example 9.6.   A freshwater discharge of temperature 17.8 C (64°F) is released at a rate of 1 m³/sec (35.3 cfs) with a velocity of 3 m/sec (9.84 ft/sec) at a depth of 70 m (230 ft). There is a uniform ambient temperature stratification as in Example 9.4 and a uniform crossflow of 0.25 m/sec (0.49 knot). Will the discharge plume reach the surface?

From Examples 9.4 and 9.5

$$Q = 1 \text{ m}^3/\text{sec}, \qquad M = 3 \text{ m}^4/\text{sec}^2, \qquad B = 0.257 \text{ m}^4/\text{sec}^3,$$

$$l_Q = 0.58 \text{ m}, \qquad l_M = 4.5 \text{ m}, \qquad z_M = 6.9 \text{ m}, \qquad z_B = 16.4 \text{ m}.$$

Since $z_M < z_B$ use Table 9.9 with N being the critical parameter

$$N = M^2 g\varepsilon'/B^2 = 0.03$$

$$N^{-1/2} = 5.7 > 1$$

$$(z_B/z_M)^2 = 5.7.$$

Since $N^{-1/2} \approx (z_B/z_M)^2$, the asymptotic solutions will provide only a rough guide

$$z_T = E_3 h_B = E_3 B^{1/4}/(g\varepsilon')^{3/8} = 16.7E_3,$$

$$z_T = E_4 z_B^{1/3} \lambda^{2/3} = E_4(B/Ug\varepsilon')^{1/3} = 16.6E_4.$$

Since the coefficients E_3 and E_4 both have values of 3.8, the terminal height of rise will be 63 m. ∎

9.3.4 Shear Flows and Ambient Turbulence

Although some work has been done on the influence of ambient shear flows and ambient turbulence on turbulent buoyant jet behavior (see Slawson and Csanady, 1971), almost all applications have been to air pollution in atmospheric flows. The great difficulty in considering turbulent buoyant discharges in the ocean is that, in general, the distributions of ambient mean velocity and turbulence are so poorly known.

While there has been a large amount of effort devoted to measuring the turbulence and shear properties of the atmosphere, very few results exist to define the turbulent state of the ocean. Until better data are available for the distribution of shear and turbulence in the ocean, on such a scale that their interactions with turbulent buoyant jets can be appreciated and understood, their effects are probably best ignored. Since it is unlikely that the effect of shearing and turbulence will be to *decrease* dilutions, this is probably not a bad approach if only the immediate dilution is of interest. Further away from the discharge source, where all effects of the initial discharge parameters are in effect forgotten, the methods of Chapters 2 and 3 may be applied (with a great deal of circumspection) directly. The major difficulty is that the influence of density stratification on ambient turbulence levels and mixing is not well understood and is an active research topic. Presentation of generally applicable results must await the development of this knowledge.

9.4   BUOYANT JET PROBLEMS AND THE ENTRAINMENT
HYPOTHESIS

Probably the first major practical advance in the calculation of dilutions and trajectories of buoyant jets was made in the now classical paper by Morton *et al.* (1956). We have deliberately delayed presenting this approach until after the reader has reviewed the previous sections, which described, in general terms, the influences of jet and ambient parameters. There is a very good reason for this. As will be seen in the presentation that is to follow, jet parameters and environmental parameters enter the problem as initial conditions, boundary conditions, and coefficients in systems of differential equations. Few mathematicians, or engineers, can easily and readily predict the form of the solutions of systems of differential equations without solving the equations directly. However, now that we have already presented what are, in effect, the asymptotic solutions of these equations, it becomes possible to perform rough checks on whether the solutions obtained in any context by the entrainment theory approach are reasonable for the data provided. It cannot be emphasized too strongly how important it is to perform these rough checks on the outcome of any computer-derived solution using the methods about to be presented.

The basis of the entrainment hypothesis method is to relate the rate of inflow of diluting water to the local properties of the jet, specifically its local mean velocity; this, in general terms, was Taylor's original contribution. In other words, Taylor hypothesized that the velocity of inflow of diluting water into any jet would be proportional to the maximum mean velocity in the jet at the level of inflow. For a round jet, for example, the entrainment hypothesis states that

$$d\mu/dz = 2\pi b_w \alpha w_m, \qquad (9.107)$$

where b_w is the radius of the jet (defined in Eq. 9.8) and αw_m is the entrainment velocity at that radius. Equation (9.107) states that the rate of change of volume flux in the jet with distance along the jet is equal to the rate of inflow by entrainment. (For a planar jet the relationship is $d\mu/dz = 2\alpha w_m$.) The constant of proportionality α is the entrainment coefficient. We will see how α is related to the coefficients used to describe jets and plumes (given in Tables 9.2 and 9.3) subsequently.

9.4.1   Equations of Motion

It is instructive to see how Eq. (9.107) is related to the usual time-averaged† volume conservation equation

$$\frac{1}{r}\frac{\partial(r\bar{u})}{\partial r} + \frac{\partial \bar{w}}{\partial z} = 0. \qquad (9.108)$$

Integrating Eq. (9.108) over the jet to some radius $b(z)$ (not equal to b_w) gives

$$\frac{d}{dz}\int_0^{b(z)} 2\pi r\bar{w}\,dr = -\lim_{r\to b(z)}[2\pi r\bar{u}] + \frac{db(z)}{dz}2\pi b(z)\bar{w}(b(z),z). \qquad (9.109)$$

If now we assume that as $b(z)$ becomes large, $\bar{w}(b(z),z) \to 0$, then the second term on the right-hand side becomes negligible and we are left with

$$d\mu/dz = -\lim_{r\to b(z)}[2\pi r\bar{u}], \qquad (9.110)$$

† We use an overbar to denote time-averages in this section.

since the integral is just the volume flux μ defined in Eq. (9.1). The term on the right-hand side is the rate of entrainment. But, while Taylor's hypothesis enables us to relate this term to the jet or plume properties, this equation alone is not sufficient to solve buoyant jet problems and we must also consider the jet momentum flux and the flux of any tracer materials.

The vertical momentum equation in cylindrical polar coordinates can be integrated across the jet and the result is

$$\frac{d}{dz} \int_0^{b(z)} \left[\bar{w}^2 + \overline{w'^2} + \frac{\bar{p} - p(\infty)}{\rho_0} - \frac{\tau_{zz}}{\rho_0} \right] 2\pi r \, dr = - \lim_{r \to b(z)} \left[2\pi r (\bar{u}\bar{w} + \overline{u'w'} - \frac{\tau_{rz}}{\rho_0} \right]$$

$$+ \frac{db(z)}{dz} 2\pi b(z) \left[(\bar{w}^2 + \overline{w'^2} + \frac{\bar{p} - p(\infty)}{\rho_0} - \frac{\tau_{zz}}{\rho_0} \right]_{b(z)}$$

$$- \int_0^{b(z)} 2\pi r g \left(\frac{\bar{\rho} - \rho_a}{\rho_0} \right) dr, \qquad (9.111)$$

where τ_{zz} and τ_{rz} are the viscous stresses, $\bar{p} - p(\infty)$ is the dynamic pressure distribution, w' the difference in fluid velocity from the mean velocity $\bar{w}$, ρ_0 a reference density, and ρ_a the ambient density distribution.

In order to use this equation we must recognize each of its terms so that we can make reasonably intelligent approximations. The term on the left-hand side of the equation is the rate of change of flow force of the jet. The integral represents the force that a jet would exert on an invisible flat plate placed perpendicular to the jet axis. The first term on the right-hand side is the flux of momentum into the jet by the radial flow through its boundary, i.e., the momentum carried into the jet by the entrainment flow. The second term is the flux of axial momentum through the sloping sides of the jet as defined by $db(z)/dz$. The third term is the accelerating force per unit mass of fluid in the jet arising from the distribution of density across the jet.

This momentum equation is generally simplified by assuming that the first two terms on the right-hand side are negligible† and that, in addition, the local specific momentum flux

$$m(z) = \int_0^{b(z)} 2\pi r \bar{w}^2 \, dr \gg \left| \int_0^{b(z)} 2\pi r \left(\overline{w'^2} + \frac{\bar{p} - p(\infty)}{\rho_0} - \frac{\tau_{zz}}{\rho_0} \right) dr \right|. \quad (9.112)$$

The second of these assumptions states that the net advective flux of axial momentum is much greater than the transfer either by the turbulent stresses, or viscous stresses, or pressure gradient force. We are, therefore, left with a simple vertical momentum equation [see Eq. (9.70)]

$$\frac{dm(z)}{dz} = - \int_0^{b(z)} 2\pi r g \left(\frac{\bar{\rho} - \rho_a}{\rho_0} \right) dr, \qquad (9.113)$$

† Note that if there is an external cross flow there will be momentum entrained into the jet.

which states that the rate of change of vertical momentum flux must equal the vertical buoyancy force acting per unit height of jet. If there are no buoyancy forces and $\bar{\rho} = \rho_a$ everywhere, then the equation states that the momentum flux is conserved. The differential equation has a very simple solution in this case, $m(z) = M$.

Equation (9.113) has introduced a further variable into the set of equations, the difference in density between the fluid in the jet and the ambient fluid. We need another equation to specify how this variable must change. This is given by a conservation of tracer equation.

Suppose that $\bar{C}$ is the absolute concentration of tracer in a buoyant jet, then integrating the time-averaged tracer conservation equation across the jet gives†

$$\frac{d}{dz} \int_0^{b(z)} 2\pi r \bar{w}\bar{C} \, dr = -\lim_{r \to b(z)} (2\pi r \bar{u}\bar{C})$$

$$+ \frac{db(z)}{dz} 2\pi b(z) [\bar{w}\bar{C}]_{b(z)}. \quad (9.114)$$

Now if $C_a(z)$ is the ambient concentration of tracer, which can now be a function of elevation, then we can write from the volume flux equation Eq. (9.109)

$$\frac{d}{dz} \int_0^{b(z)} 2\pi r \bar{w} C_a \, dr = -\lim_{r \to b(z)} (2\pi r \bar{u} C_a)$$

$$+ \frac{db(z)}{dz} 2\pi b(z) [\bar{w} C_a]_{b(z)}$$

$$+ \frac{dC_a}{dz} \int_0^{b(z)} 2\pi r \bar{w} \, dr. \quad (9.115)$$

Subtracting Eq. (9.115) from (9.114), we see that

$$\frac{d}{dz} \int_0^{b(z)} 2\pi r \bar{w}(\bar{C} - C_a) \, dr = \frac{-dC_a}{dz} \int_0^{b(z)} 2\pi r \bar{w} \, dr \quad (9.116)$$

since by definition $\bar{C} \to C_a$ as $r \to b(z)$.

It is usual to relate tracer concentrations to density variations by a formula of the form

$$(\bar{\rho} - \rho_0)/\rho_0 = \gamma(\bar{C} - C_0)/C_0, \quad (9.117)$$

where ρ_0 is some convenient reference density, C_0 some convenient tracer concentration, and γ a coefficient of proportionality, usually assumed constant, although Kotsovinos (1975) has shown that if the tracer is temperature, then

---

† Note that we have ignored the turbulent transport of tracer material, that is, the contribution from $\overline{w'C'}$. This is not correct but for engineering purposes the results obtained appear to be sufficiently accurate and are generally used.

for moderate temperature differences, corrections should be made for the variation in γ with temperature.

The reader should also be aware that in some circumstances two different tracer concentrations may influence the density distribution. It is left as an exercise to show how the analysis should be modified; we will only consider a single tracer here.

If the ambient fluid is density stratified, then we write

$$\bar{\rho} = \rho_a - \rho_0 \theta(r, z, t), \tag{9.118}$$

where

$$\rho_a = \rho_0(1 - \varepsilon(z)) \tag{9.119}$$

is the ambient density stratification, and $\theta(r, z)$ the time-averaged density anomaly caused by the jet, and we can rewrite the equation of tracer conservation as

$$\frac{d}{dz} \int_0^{b(z)} 2\pi r \bar{w} \theta \, dr = \frac{-d\varepsilon(z)}{dz} \mu(z). \tag{9.120}$$

We define the specific buoyancy flux of the jet as

$$\beta = \int_0^{b(z)} 2\pi r g \bar{w} \theta \, dr, \tag{9.121}$$

and Eq. (9.120) becomes

$$d\beta/dz = -g(d\varepsilon/dz)\mu. \tag{9.122}$$

We note that for a stable density gradient, for which $\varepsilon'(z)$ is positive, the buoyancy flux must decrease as the volume flux in the jet increases. Note also that for a buoyant jet rising under the influence of a jet density deficiency, $\theta(r, z)$ is positive, thereby giving a positive buoyancy flux.

In summary, the equations of motion for a vertical turbulent buoyant jet in a density-stratified environment are therefore

$$d\mu/dz = - \lim_{r \to b(z)} (2\pi r u), \tag{9.123}$$

$$dm/dz = \int_0^{b(z)} 2\pi r g \theta \, dr, \tag{9.124}$$

$$d\beta/dz = -g(d\varepsilon/dz)\mu, \tag{9.125}$$

with

$$\mu = \int_0^{b(z)} 2\pi r \bar{w} \, dr, \tag{9.126}$$

$$m = \int_0^{b(z)} 2\pi r \bar{w}^2 \, dr, \tag{9.127}$$

$$\beta = \int_0^{b(z)} 2\pi r g \bar{w} \theta \, dr. \tag{9.128}$$

9.4.2 Application to Density-Stratified Environments

The entrainment hypothesis, plus some judicious application of experimental results, enables us to find a solution to the set of Eqs. (9.123)–(9.128).

First, we have seen that Taylor's entrainment hypothesis implies that

$$- \lim_{r \to b(z)} (2\pi r \bar{u}) = 2\pi \alpha b_w w_m,$$ (9.129)

where α is the entrainment coefficient. Second, we have seen that

$$\bar{w} = w_m \exp\left[-\left(\frac{r}{b_w}\right)^2 \right],$$ (9.130)

and

$$\theta = \theta_m \exp\left[-\left(\frac{r}{b_T}\right)^2 \right],$$ (9.131)

are good approximations to the distribution of time-averaged velocity and mass concentration in jets and plumes. Furthermore, the ratio b_T/b_w is constant for both jets and plumes with

$$b_T/b_w = \lambda = 1.2.$$ (9.132)

Substituting Eqs. (9.129)–(9.132) into Eqs. (9.123)–(9.128) and assuming $b(z) \to \infty$ gives a set of three ordinary differential equations for the three variables w_m, θ_m, b_w,

$$(d/dz)(\pi b_w^2 w_m) = 2\pi \alpha b_w w_m,$$ (9.133)

$$\frac{d}{dz}\left(\frac{\pi}{2} b_w^2 w_m^2 \right) = \pi g \lambda^2 b_w^2 \theta_m,$$ (9.134)

$$\frac{d}{dz}\left(\frac{\pi g \lambda^2 b_w^2 w_m \theta_m}{1 + \lambda^2} \right) = -g \frac{d\varepsilon}{dz} \pi b_w^2 w_m.$$ (9.135)

It must be recognized that initial conditions are required to solve this set of equations and initial values of b_w, w_m, θ_m must be specified. These are obtained by solving the set of algebraic equations

$$[\pi b_w^2 w_m]_0 = Q,$$ (9.136)

$$[\tfrac{1}{2}\pi b_w^2 w_m^2]_0 = M,$$ (9.137)

$$\left[\pi g \frac{\lambda^2}{1 + \lambda^2} w_m b_w^2 \theta_m \right]_0 = B,$$ (9.138)

where the zero subscript implies initial values.

In keeping with our comments at the beginning of this section we should check that the asymptotic solutions obtained by dimensional analysis and experiments, and given in Tables 9.2 and 9.3, are indeed solutions of these equations.

First, substituting for the values of b_w and w_m for both jets and plumes we find

$$\alpha_{jets} = 0.0535 \pm 0.0025, \tag{9.139}$$

$$\alpha_{plumes} = 0.0833 \pm 0.0042, \tag{9.140}$$

which immediately indicates that entrainment rates are different in jets and plumes. This comes as no surprise since we already know from Section 9.2 that the dilution rate of a plume is higher than that for a jet even if the local momentum fluxes are equal.

Priestley and Ball (1955) found that α is proportional to the square of the local Richardson number of the jet, so that a more appropriate entrainment function is given by

$$\alpha = \alpha_{jet} - (\alpha_{jet} - \alpha_{plume})(R/R_p)^2, \tag{9.141}$$

where R_p is the plume Richardson number shown, in Section 9.2, to be constant with a value of 0.557. Recall that

$$R = \mu\beta^{1/2}/m^{5/4}, \tag{9.142}$$

so that rewriting in terms of b_w, w_m, and θ_m we have

$$R = \left[\frac{4\sqrt{2\pi}\,\lambda^2}{(1 + \lambda^2)} \left(\frac{gb_w\theta_m}{w_m^{\,2}} \right) \right]^{1/2}. \tag{9.143}$$

Even this approximation to the entrainment coefficient has difficulties when the environment is density stratified. The problem is as follows. When a turbulent buoyant jet is rising in a density-stratified environment the fluid entrained at a given level is denser than that at a higher level so the effect of the vertical rise is to reduce the buoyancy flux β. At some point, the buoyancy flux actually becomes negative, so that if Eq. (9.141) is used, the entrainment actually ceases when R^2 attains a value equal to $\alpha_j R_p^2/(\alpha_j - \alpha_p)$, i.e., about -0.56. A suggested alternative entrainment function for density-stratified flows is

$$\alpha = \alpha_j \exp\left[\ln\left(\frac{\alpha_p}{\alpha_j}\right)\left(\frac{R}{R_p}\right)^2 \right], \tag{9.144}$$

which also gives the correct values of α for jets and plumes and does not become negative.

Calculations of the dilution in turbulent buoyant jets in a linear density stratification and using the entrainment functions specified by Eq. (9.141),

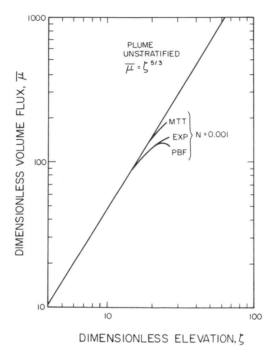

Figure 9.25 Predicted differences in terminal height of rise and dilution for different entrainment coefficients [MTT: Eq. (9.140); EXP: Eq. (9.144); PBF: Eq. (9.141).]

Eq. (9.144), and a constant entrainment coefficient are shown in Fig. 9.25. The difference is not of great engineering significance. Calculations of such jets using different values of the stratification parameter N are shown in Fig. 9.26. It is to be noted that the asymptotic prediction of the terminal heights of rise and dilution given in Section 9.3.1 are confirmed.

9.4.3 Other Applications

The foregoing development was carried out for a vertical jet in a density-stratified environment. However, there have been many attempts to extend the analysis to include curved buoyant jets, resulting from either the presence of a crossflow, or an initial jet discharge other than vertical. Probably the most recent and comprehensive of these is the study by Schatzmann (1976) which includes both arbitrary density stratification and shear flows. Others have also applied the technique to free surface jets and negatively buoyant jets that include flow reversals. There are many computer codes available for general application.

The primary difficulty with all of the computer-derived solutions is the values of the coefficients actually used in the entrainment relations. For example, Schatzmann (1976) has four coefficients to account for such factors as the influence of buoyancy, flow curvature, external crossflows, and ambient turbulence. His entrainment function (our α) includes specific terms to describe the effects of these factors. Each of the coefficients in the entrainment relation is based on an asymptotic solution in the absence of the other factors. While it

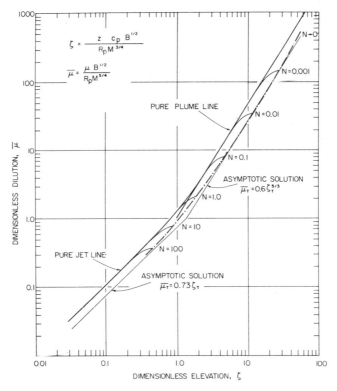

Figure 9.26 Dilution in turbulent buoyant jets with a linearly stratified environment calculated with α specified by Eq. (9.144). $N = M^2 g \varepsilon' / B^2$. (cf., Fig. 9.7)

can be argued that this does not include the possibility of combined effects modifying the coefficients, the agreement obtained with experiments is quite reasonable and certainly adequate for engineering design purposes.

We will not present any details of these numerical models in this text and will instead refer the reader to the many publications, including those by Fan and Brooks (1969), Fan (1967), Abraham (1963), Hirst (1971, 1972), Hoult and Weil (1972), Morton *et al.* (1956), Koh and Brooks (1975), Fox (1970), Schatzmann

(1976), and Schatzmann and Flick (1977). The reader should remember the admonition contained in the introduction of this section—check the results of the computer code against a "rough" asymptotic analysis.

To conclude this section we present some useful graphs of dilution and jet geometry obtained from the application of entrainment theory to round horizontal turbulent buoyant jets. The entrainment coefficient used was that given by Eq. (9.141) or Eq. (9.144) (the numerical results are graphically indistinguishable). Figure 9.27 gives the jet trajectory and diameter in terms of normalized dimensionless coordinates. It is assumed that the zone of flow establishment is of length $7\,l_Q$ and that the mean dilution at that point is 2. Figure 9.28 presents the normalized dilution along the jet axis as a function of the normalized horizontal coordinate. The design engineer particularly interested in the zone of flow establishment with strongly buoyant jets is referred to Hirst (1972).

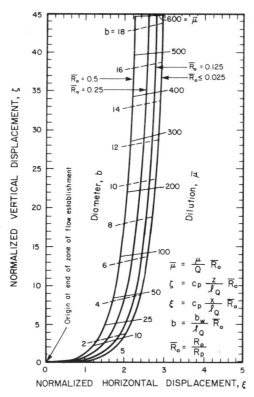

Figure 9.27 Normalized jet trajectory for a horizontal turbulent buoyant jet with jet diameter given on the trajectory.

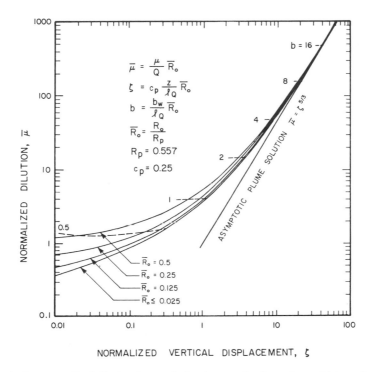

Figure 9.28 Normalized dilution for a turbulent buoyant horizontal round jet as a function of vertical displacement.

Example 9.7. A vertical freshwater discharge of temperature 17.8°C (64°F) is released at a rate of 1 m³/sec (35.3 ft³/sec) with a velocity of 3 m/sec (9.84 ft/sec) at a depth of 70 m (230 ft) in a linearly stratified coastal ocean. The stratification is as specified in Example 9.4. What is the terminal height of rise and dilution in the absence of any cross flow?

We have

$$Q = 1 \text{ m}^3/\text{sec}, \qquad M = 3 \text{ m}^4/\text{sec}^2,$$

$$B = 0.257 \text{ m}^4/\text{sec}^3, \qquad g\varepsilon' = 22.3 \times 10^{-5} \text{ sec}^{-2}.$$

$$l_Q - Q/M^{1/2} = 0.58 \text{ m}, \qquad l_M = M^{3/4}/B^{1/2} = 4.5 \text{ m},$$

$$h_B = B^{1/4}/(g\varepsilon')^{3/8} = 16.7 \text{ m}, \qquad h_M = (M/g\varepsilon')^{1/4} = 10.8 \text{ m},$$

$$R_0 = l_Q/l_M = 0.13, \qquad N = (h_M/h_B)^8 = 0.031.$$

From Fig. 9.26 (estimating for $N = 0.03$)

$$\bar{\mu}_T = 18, \qquad \zeta_T = 7.5.$$

From Eqs. (9.41) and (9.42)

$$\left(\frac{\mu}{Q}\right)_T = \left(\frac{R_p}{R_0}\right)\bar{\mu} = \left(\frac{0.56}{0.13}\right)18$$

$$= 77$$

$$z_T = \left(\frac{R_p}{R_0}\right)\frac{l_Q}{c_p}\zeta_T = \left(\frac{0.557}{0.13}\right)\left(\frac{0.58}{0.25}\right)7.5$$

$$= 75 \text{ m.}$$

Plume will reach the surface. Calculate dilution at 60 m

$$\zeta_{60} = c_p\left(\frac{60}{l_Q}\right)\left(\frac{R_0}{R_p}\right) = 6.0.$$

Estimating from Fig. 9.26 with $N = 0.03$

$$\bar{\mu}_{60} \approx 16$$

$$\mu/Q = (R_p/R_0)16 = 68. \quad\blacksquare$$

Example 9.8.   Compare the dilution 10 m below the sea surface obtained when
$1 \text{ m}^3/\text{sec}$ at a temperature of 17.8°C (64°F) is discharged at a depth of 70 m into
a uniform sea environment at 11.1°C (52°F) at a discharge velocity of $4 \text{ m}^3/\text{sec}$
(i) vertically and (ii) horizontally.

From Example 9.3

$$Q = 1 \text{ m}^3/\text{sec} \qquad B = 0.257 \text{ m}^4/\text{sec}^3$$

$$M = 4 \text{ m}^4/\text{sec}^2$$

$$l_Q = Q/M^{1/2} = 0.5 \text{ m}, \qquad l_M = M^{3/4}/B^{1/2} = 5.58 \text{ m}$$

$$R_0 = l_Q/l_M = 0.0896, \qquad R_0/R_p = 0.0896/0.557 = 0.161.$$

Horizontal.   Referring to Figs. 9.27 and 9.28

$$\zeta_{60} = 0.25\left(\frac{60}{0.5}\right)0.16$$

$$= 4.8$$

$$\xi_{60} = 2.2$$

$$\bar{\mu}_{60} = 22$$

$$\left(\frac{\mu}{Q}\right)_{60} = \left(\frac{22}{0.16}\right) = 138$$

Vertical. Referring to Fig. 9.7

$$\zeta_{60} = 0.25 \frac{60}{0.5} 0.16$$

$$= 4.8$$

$$\bar{\mu}_{60} \approx 13.6$$

$$\left(\frac{\mu}{Q}\right) = \frac{13.6}{0.16} = 85.$$

There is a large difference between horizontal and vertical discharge. ■

9.5 BOUNDARY EFFECTS ON TURBULENT BUOYANT JETS

Our discussion of the mechanics and dilution of jets and plumes has so far excluded the effects of the fluid boundaries. When a rising jet reaches the water surface, or when a sinking jet contacts the bottom, the flow must turn and flow horizontally as a buoyant layer. Sometimes the jet is at the water surface to begin with, as in the case of an open channel discharge of heated water into a river, and the resulting flow is usually called a surface jet. A large scale example of a surface jet is shown in Fig. 9.29, a photograph of two rivers discharging into the ocean. The river waters are made visible by a high level of turbidity, and it can be seen that they float on the heavier ocean water because of their buoyancy. An example of a rising jet contacting the water surface and subsequently flowing as a surface layer is shown in Fig. 9.30. We will return to the details of the flow shown in this figure later on.

The types of flows shown in these illustrations are difficult to describe since they involve three-dimensional buoyant motions. All of the additive effects we have seen previously occur—discharge momentum, entrainment, crossflows, and density stratification; now, in addition, there is not only a fluid interface at the free surface, but also frequently an internal density interface. It is the presence of these interfaces that adds to the complexity. For example, if the spreading layer becomes thin, then interfacial friction between the spreading layer and the liquid beneath may become an important feature of the flow. If the buoyancy of the discharge results from an elevated temperature, then the possibility of buoyancy loss from surface heat transfer may occur. Surface spreading layers are also subject to wind stresses that add a further possible dynamic feature. A complete description of the entire class of problems is still beyond us, and it must be recognized that many problems are still very much subjects of research.

It should also be made clear that not all boundary effect problems involve surface spreading. Of recent engineering interest are the cold water plumes

Figure 9.29   Two small turbid rivers flow into the coastal ocean. Note the sharp boundary between turbid river and clear ocean water.

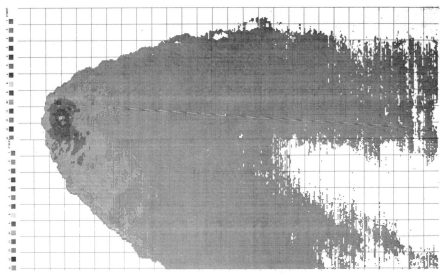

Figure 9.30   Thermal discharge ($Q = 21$ m^3/sec) from a coastal power plant made visible with a scanning infrared radiometer. Vertical discharge is 5 m below the surface and the peak temperature difference is 11.1°C. There is a coastal current of 0.27 knot (0.14 m/sec) and grid spacing is 30.5 m (100 ft). Note bifurcation of the plume.

formed by liquefied natural gas vaporization terminals. The liquefied natural gas (LNG) must be revaporized to be of use and a common procedure is to use seawater for the heat required. The cooled seawater, when returned to the ocean, is denser than ambient water and sinks forming a bottom spreading layer of cool water (Fig. 9.13). The same features of friction, density stratification, entrainment, and crossflows are of importance in describing the motion of these bottom spreading layers.

Another point worth raising is the inherent difference between steady state and starting motions. For example, the front formed by the release of a finite volume of buoyant or dense fluid may be quite different from the flow established by a continuously operating discharge. We therefore should consider both transient and steady motions. In an environment like the ocean, where conditions are continuously changing, the description may be quite different from that in a river where variations in current speed, and possibly direction, are on a much slower time scale.

It will be recalled that in the description of submerged jets we were able to place confidence in general mathematical models because a significant number of asymptotic limiting solutions had been thoroughly investigated experimentally and could be used to confirm the mathematical models. The great difficulty of general models of surface jets is that the corresponding limiting solutions to describe the interactions of important influences such as friction, density stratification, momentum, buoyancy, entrainment, loss of buoyancy, wind stresses, crossflows, and ambient turbulence, by and large do not exist. Because of this absence of general theories it is often necessary to make use of carefully planned hydraulic models coupled with order of magnitude analyses to arrive at sensible results. We will now consider some of the difficulties.

9.5.1   Momentum Effects

Consider a source of horizontal momentum located in the free surface of a liquid that is very deep. We can argue from symmetry that the flow picture will look like half of a regular three-dimensional jet flow resulting from a momentum source located at the boundary. We would expect to see a zone of flow establishment about $7l_Q$ long, where $l_Q = Q/M^{1/2}$, just as for a regular round jet. If the ambient fluid has a finite depth, then the entrainment flow to the lower half of the jet may be strongly influenced by the presence of the bottom. If the depth is shallow, then it simply may not be possible for the entrainment demand of the jet to be met, and the jet becomes attached to the bottom and all of the entrainment flow enters from the sides. In that case the jet flow becomes a complex three-dimensional motion with bottom friction important because of the steep velocity gradients next to the boundary. An example is shown in Fig. 9.31, a laboratory photograph of a surface jet attached to the bottom; the jet shown in

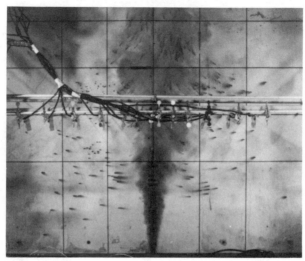

Figure 9.31 A horizontal buoyant jet discharging into a laboratory flume. In the range shown the jet is attached to the bottom and momentum effects are dominant. [Photograph by B. Safaie.]

the photograph is also buoyant, but within the range shown momentum is dominant.

Even the simple momentum jet is difficult to analyze because the limited depth has two major influences: it restricts entrainment from below and imposes a frictional effect on the jet. At present we do not know how to describe either of these effects exactly. In the absence of bottom effects we can predict that the velocity on the axis of the jet will decay like x^{-1}, where x is the distance along the axis. If M is the specific momentum flux, then for a distance along the jet such that $x \gg Q/M^{1/2}$, we know that the maximum time-averaged velocity

$$u_{\mathrm{m}} \sim M^{1/2}x^{-1}, \tag{9.145}$$

and that the volume flux in the jet will have the form

$$\mu \sim M^{1/2}x. \tag{9.146}$$

When the jet reaches the bottom and lower entrainment ceases then we would expect $\mu \sim x^{1/2}$ as for a plane jet.

The problem with this analysis is that such simple jets seldom occur in practice. In most circumstances there are bottom effects, crossflows, and buoyancy. In the following sections we will attempt to give some qualitative description for the effects of each of these possible influences.

9.5.2   Buoyancy Effects

In many surface jets, buoyancy is present; usually the discharge has a different temperature or salinity (or both) from the ambient liquid. The surface jet can therefore be regarded as a source of specific buoyancy flux $B = g_0'Q$, where g_0' is the initial reduced gravity and Q the initial volume rate of flow. We again have characteristic length scales $l_Q = Q/M^{1/2}$ and $l_M = M^{3/4}/B^{1/2}$. The first scale, as we have seen, gives a measure of the distance from the jet origin within which the effect of the initial geometry is important. The length scale l_M has the same interpretation as previously, namely the distance at which the effects of buoyancy become evident, but there is an important difference from when the jet was directed vertically. For a jet directed vertically upward, the buoyancy acts in the same direction as the transport flow and the buoyancy serves to increase the momentum flux of the jet (provided, of course, that the buoyancy is positive). For a jet directed horizontally, however, the buoyancy forces still act vertically but are translated into a horizontal radial pressure gradient which acts uniformly in all horizontal directions and tends to spread out the buoyant liquid at the surface. The effect of this horizontal pressure gradient is most easily seen in the case of a spreading surface layer, but it also occurs in internal spreading layers when the ambient fluid is density stratified. Figure 9.32 is an excellent demonstration of the relative effects of initial momentum, buoyancy, and density stratification. The horizontal buoyant jet in the figure has initial horizontal momentum which is conserved. The buoyancy provides the vertical momentum and the jet keeps rising until a level of neutral buoyancy is attained. However, as the pool forms at the level of neutral buoyancy, a radial pressure gradient is built up because the pool has a finite thickness. The initial horizontal momentum causes the pool to drift to the right but the radial pressure gradient is strong enough to provide some flow to the left as well.

Figure 9.32   Horizontal turbulent buoyant jet in a density-stratified environment. $R_0 = 0.036$, $N = 0.56$. [From Fan (1967).]

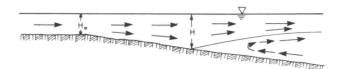

Figure 9.33 Approximate flow directions in a vertical cross section along the axis of a horizontal buoyant jet.

A buoyant surface jet discharged at the edge of a water body is likely to remain attached to the bottom for some distance out from the side, and then to spring clear from the bottom because of buoyancy as sketched in Fig. 9.33. Fig. 9.34 is a photograph of a surface jet for which l_M is relatively short and the lift off point is not far out from the point of discharge. The spots in the picture are crystals of potassium permanganate which were dropped onto the bottom and allowed to dissolve; the directions of the streaks from the spots indicate the direction of the bottom current and identify the location of the lift-off point. Beyond the lift-off point the jet spreads almost radially due to buoyancy; nevertheless it is also possible to identify a core of darker color continuing outward which is caused by residual momentum. The photograph is from a study by Safaie (1978) who found that

$$H = 1.5H_0(l_M/l_Q)^{1/2}, \tag{9.147}$$

where H and H_0 are defined in Fig. 9.33.

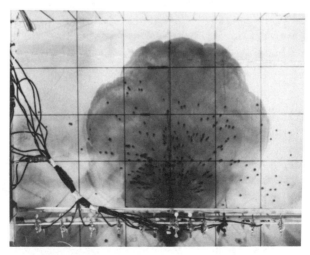

Figure 9.34 A horizontal buoyant jet with low momentum discharging into a laboratory flume (compare the jet with high momentum shown in Fig. 9.31). [Photo by B. Safaie.]

9.5.3   Crossflows

Buoyant surface jets into crossflows involve four parameters, Q, M, B, and the crossflow velocity U. In addition to the length scales l_M and l_Q we also have the length scale $M^{1/2}/U$ which is roughly related to the distance from the origin at which the jet velocity decays to U. A number of efforts have been made to compute the trajectory and dilution in these jets by means of numerical programs, generally based on the integral analysis as, for example, the studies of Prych (1972), Shirazi and Davis (1974), Adams *et al.* (1975), and McGuirk and Rodi (1976). Programs of this type have been found to give accurate temperature predictions in some cases, but problems often arise because of the unknown influence of the finite depth and crossflows on the entrainment function, the difficulty in predicting whether the jet will attach to the bottom, and three dimensional effects not adequately represented in the integral analysis. Two such effects are illustrated by Figs. 9.35 and 9.36.

Figure 9.35 is a photograph of a laboratory model of the discharge shown in Fig. 9.30. Apparently an internal hydraulic jump forms upstream of the discharge and there appear to be two trailing vortices forming a "horseshoe vortex." In the prototype a vertical discharge is located 5 m below the surface, $l_Q = 6.13$ m, $Q = 20.7$ m^3/sec, $B = 0.37$ m^4/sec^3, and $U = 0.14$ m/sec. A splitting of the plume can be quite clearly seen in both the laboratory model and the prototype. This obviously has a great deal in common with the bifurcated vertical plumes discussed previously (Fig. 9.23) and described by Scorer (1959) and Turner (1960).

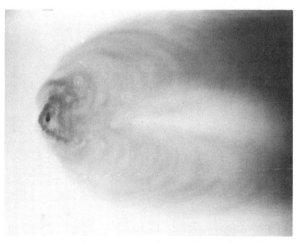

Figure 9.35   Laboratory model of flow shown in Fig. 9.30 made visible with colored dye. Note curved vortex seen upstream of the discharge point and plume bifurcation. Note also that dye intensity is not related to surface temperatures seen in Fig. 9.30.

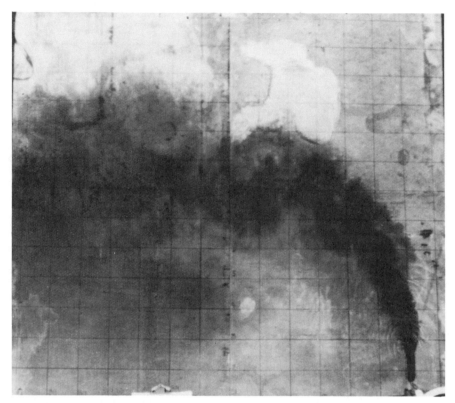

Figure 9.36 A horizontal buoyant jet discharging into a crossflow. Note the large trailing eddy. [Photo by V. Harms.]

Figure 9.36 shows a turbulent buoyant jet discharged perpendicular to the crossflow. In the case shown in the photograph the bottom of the main channel slopes down from the side where the discharge is being released, and the jet is initially attached to the bottom. The exact location of the lift-off point is not known, but it is probably close to the exit. The most striking feature is the large trailing eddy in which some of the discharged material is returned and re-entrained into the jet. At present it does not seem possible to predict the details of this reentrainment, and we cannot even give a criterion for when reentrainment will occur. The photograph makes clear, however, that strong interactions between a buoyant jet and a crossflow should be expected, and that simplified analyses which neglect the possibility of reentrainment should not be expected to be adequate in many prototype cases.

In most problems of the type illustrated in Figs. 9.35 and 9.36 the possible complications that can occur from the interaction of the various parameters make it obligatory that a hydraulic model study be performed. For example, there is no known theory to predict the mixing that occurs in a horseshoe vortex, nor is there theory that predicts the existence of plume bifurcation. Until such time as the theories are developed that will explain such phenomena quantitatively we must continue to rely on hydraulic model studies for both the form and dilution of such flow fields.

9.5.4 Multiple Point Discharges

A problem where the influence of flow boundaries and flow geometry is important is that involving long multiport diffusers in relatively shallow water. For example, a multiport diffuser structure may be 800–1000 m long and be located in water from 10 to 80 m deep. Since the location of the diffuser is fixed, but ocean currents are variable in direction, the orientation of the flow with respect to the diffuser can be important. Furthermore, in any multiport diffuser there is a region in which the flow from individual ports merges to form an agglomerated flow. All of these features involve combinations of two- and three-dimensional jet and plume flows and interactions between these flows and possible cross flows and density stratifications. The analysis of an actual prototype design will be discussed more extensively in Chapter 10. Here we discuss some of the common features of multiport diffusers and the problem of their orientation to the flow field.

A typical problem may involve a multiport diffuser located on the sea floor and used to dilute an effluent from a waste water treatment plant. The discharge from each port of the diffuser forms a turbulent buoyant jet which can be analyzed by the methods described in Sections 9.1–9.3. However, each jet increases in size with distance from the exit port so that each of the jets will interact, merging one with another, as shown in Fig. 9.37. Seen from some distance, the net effect of the diffuser ports will be the production of an essentially two-dimensional flow field such as would be produced from a line buoyant jet.

As the buoyant jet rises from the diffuser it will be subjected to the effects of crossflows, which may, depending upon the orientation of the ocean current, be parallel or perpendicular, or even be at some angle, to the diffuser. Later on, as the rising plume reaches a level of neutral buoyancy, or the free surface, the flows will become nearly horizontal as shown in Fig. 9.32, although buoyancy may continue to cause spreading in a direction transverse to the mean current.

The orientation of a diffuser in relation to the current, its relative depth of submergence and level of buoyancy output, all combine to make an interesting problem for analysis. The dependent variable of importance is generally the minimum surface dilution S_m. Roberts (1977, 1979) has recently investigated this problem and has found that the relevant variables are the crossflow velocity U, the depth of submergence H, the diffuser length L, the buoyancy flux per unit length of the diffuser $b = B/L$, and the volume flux per unit length $q = Q/L$.

Figure 9.37   A model section of a multiport diffuser. Each discharge port is angled 20° up from horizontal; ports alternate ±25° from diffuser axis. Model scale 49.3 : 1, $R_0 = 0.03$.

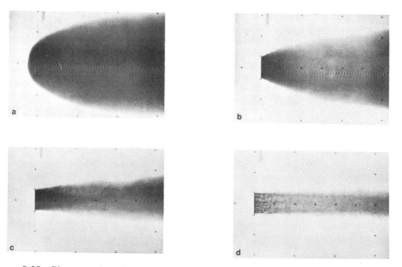

Figure 9.38   Photographs of surface plumes at different values of the Froude numbers F (a) $F \approx 0.1$, (b) $F \approx 1.3$, (c) $F \approx 12$, and (d) $F \approx 120$. [From Roberts (1977, 1979).]

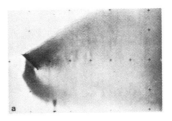

Figure 9.39 Photographs of surface plume in a 45° current at different values of the Froude number F (a) $\approx F \approx 1$, (b) $F \approx 10$, and (c) $F \approx 100$. [From Roberts (1977, 1979).]

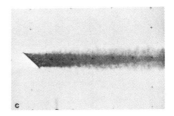

Roberts has found that the Froude number $F = U^3/b$ is the dominant parameter in determining the shape of the flow field and the dilution. For example, Figs. 9.38 and 9.39 show flow patterns corresponding to different Froude numbers for diffusers perpendicular and at 45° to the flow, respectively. The influence of the buoyancy flux is quite dramatic. Figure 9.40 summarizes Roberts' experimental results for minimum surface dilution. For low Froude numbers, Roberts showed

$$S_m q/uH = 0.27F^{-1/3} \tag{9.148}$$

where H is the depth of the diffuser. This is equivalent to

$$\mu = 0.27B^{1/3}H, \tag{9.149}$$

which is similar to the volume flux result given in Table 9.3 for a planar plume at depth H. Note that the constant has been reduced to 0.27 from 0.34 to account for the nondiluting surface layer of previously discharged fluid.

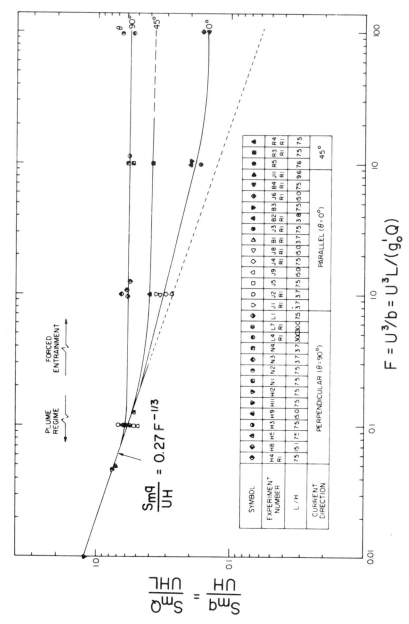

Figure 9.40 Experimental measurements of minimum surface dilution for a finite line source of buoyancy flux in a current. [From Roberts (1977, 1979).]

388

9.6 SUMMARY

By the time the reader has reached this point in the chapter it should be apparent that solving practical dilution problems involving jets is generally not a trivial matter. Some of the factors we have briefly discussed are momentum, buoyancy, density stratification, flow boundaries, loss of buoyancy, crossflows, bottom friction, internal friction, internal hydraulic jumps, and ambient turbulence. It should be further apparent that the concept of finding general models to include all of these possible factors is quite beyond the current state of knowledge. However, we have seen that by categorizing the influences as jet parameters, environmental parameters and boundary effects, and using defined characteristic lengths for the first two groups, we can make order of magnitude analyses possible. By carefully ordering the length scales so specified, the jet dilution in any problem can be roughly evaluated. When geometric influences become dominant, then this no longer becomes possible and, in all but the simplest cases, we must look to a hydraulic model to provide the results we may need.

Chapter 10

Design of Ocean Wastewater Discharge Systems

Release of a large quantity of wastewater from a coastal city, or of heated water from a coastal power plant, raises questions of siting and initial dilution beyond what we have yet discussed. The goal of the design of a wastewater disposal system is to minimize detrimental effects of the discharge on the environment. Submerged multiport diffusers have been found to be an efficient way of maximizing initial dilution and meeting regulatory requirements. This chapter presents a review of all facets of the design of a multiport diffuser, and gives details of two examples.

10.1 THE DESIGN PROCESS

Prior to initiation of a design the basic variables of quantity and quality of the effluent, the general location of the facilities, and sometimes even a preliminary recommendation for the outfall design have usually been determined from a planning study. The design engineer must then choose the best particular design to meet the basic objectives of achieving sufficient dispersion and mixing of the effluent with the receiving water so that quality standards are not violated. The design process is basically iterative; the final chosen design is usually the culmination of many trials. In the process not only engineering factors but also economics, aesthetics, and other social and intangible factors are of concern. Sometimes the latter factors may dictate the choice among

390

several technically viable alternatives. In this chapter only the technical aspects are considered.

The primary goal of an outfall diffuser system is to accomplish rapid initial mixing of the effluent with the ambient water. Indeed, only the initial jet mixing is under the control of the designer since the subsequent turbulent diffusion and transport processes are dependent primarily on the flow in the ambient and are hence controlled by nature. The rapid initial mixing is usually accomplished by means of a diffusion structure which consists of a manifold containing many discharge ports along a line. For example, the 3.05-m diameter, 8350-m long outfall built by the Orange County Sanitation Districts has a 1829-m long diffuser at the end with 500 ports spaced at 7.32 m on each side of the pipe. The ports range in size from 7.52 to 10.49 cm and discharge at an average velocity of 4 m/sec at the design flow of 12.7 m^3/sec (Carollo Engineers, 1970).

The initial dilution obtained in a large outfall diffusion system for sewage effluent depends primarily on the discharge, the length of the diffuser, the depth of discharge, and the ambient currents and density stratification. Among these, only the length and depth of discharge are under the control of the engineer. Thus one of the first steps in the design process is to determine these fundamental diffuser characteristics. The secondary design variables such as port size and spacing are of less importance in determining the initial dilutions obtained and their specification may be deferred to a later stage in the design process.

The existence of an ambient density stratification due usually to temperature variations often plays a vital role in the determination of effluent plume disposition. It has long been recognized that the diluted buoyant effluent, when mixed sufficiently with the cold bottom water may stay submerged and not reach the surface. On the other hand, the dilution is an increasing function of the distance traveled by the buoyant jet. It is possible for a waste field to be deeply submerged resulting in less dilution. In the design process, it is important to have knowledge of the expected range of ambient density profiles so that the dilution and submergence characteristics of candidate designs can be evaluated.

Following the initial dilution, the waste field is subjected to further mixing which is usually of much less importance. While the initial dilution obtained in a typical well-designed diffuser is of the order of 100 or more, the further mixing obtainable in the environment in a matter of hours is usually only a factor of approximately five or ten. Of more importance than the further dispersion is the transport by the currents. This aspect can have a strong bearing on the choice of the location of the discharge structure.

An equally important factor in the design is the internal system hydraulics of the outfall and diffuser. The diffuser is basically a manifold and for proper operation the flow distribution along it must be reasonably uniform. In addition, the overall hydraulics of the outfall system also includes possibly a pumping station or energy dissipator. These must be considered to assure proper operation of the system. The structural design must ensure that the outfall will not be damaged by catastrophic events (e.g., extreme waves or earthquakes) which may occur once or twice in the life of the structure. We do not discuss the structural design here, except to note that at times structural requirements may affect the hydraulic design.

10.2   MIXING PHENOMENA

Consider a typical large ocean outfall discharging sewage effluent through a long diffusion structure as shown schematically in Fig. 10.1. For purposes of discussion, it is convenient to divide the mixing and dispersion processes into three zones. Near the discharge, the mixing is accomplished in the buoyant jets, a phenomenon governed by the momentum and buoyancy of the discharge. Far from the discharge, the transport and mixing are accomplished by the ocean currents and turbulence and are relatively insensitive to the exact discharge conditions or the fact that the discharge had been buoyant. In between these two regions, there is a zone where the sewage field is established. The dynamics in this region depends on both the momentum and buoyancy in the discharge and the ocean currents. Evaluation of candidate diffuser designs must include analysis in all three regions even though only the first is under the control of the designer.

10.2.1   Initial Mixing

10.2.1.1   Dilution in the Rising Plume

The theory of buoyant jets and plumes, presented in Chapter 9, is generally limited to simple geometries and ambient conditions such as single round jets and infinitely long slot jets. Currents are considered mostly for ambients of uniform density, and are steady and uniform. In practice the ocean is often density stratified in a nonlinear and dynamic manner, and the currents are usually neither steady nor uniform. No predictive capability yet exists which can accurately forecast the mixing of wastewater discharges in the general case. Fortunately, effective design can be made without such general tools. In the remainder of this section we review some simple dilution formulas which have generally been found to serve as adequate guides in making design choices. The formulas given here differ somewhat from those presented in the body of Chapter 9 (see Section 9.2), but they are the ones usually used for practical outfall design.

A large ocean outfall diffuser is usually many times longer than the discharge depth. The ratio of length of discharge manifold to depth of discharge is of the order of 10 or more. For preliminary estimation, the discharge may therefore

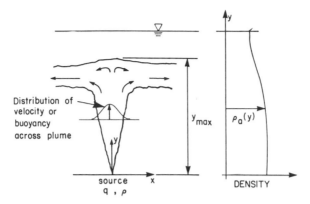

Figure 10.1 Schematic drawing of a rising plume from a submerged multiport diffuser. [After Koh and Brooks (1975).]

be considered two dimensional. [Most large diffuser systems contain many small ports spaced at a distance on the order of a few meters resulting in many small jets. However, they tend to merge together to form a curtain fairly rapidly (see Fig. 10.1).] It has also been shown in Chapter 9 that all buoyant discharges tend to the plume solution. One of the most useful results for estimating the dilution is therefore the solution for a two-dimensional buoyant plume. When the ambient is uniform and motionless, the plume solution is often written (see Table 9.3, p. 332)

$$S = 0.38 \, g'^{1/3} d / q^{2/3}, \tag{10.1}$$

where S is the centerline dilution, $g' = g \, \Delta\rho/\rho$, ρ the density of the discharge, $\Delta\rho$ the density difference between the ambient fluid and the discharged fluid, g the gravitational acceleration, d the vertical distance above the source, and q the initial discharge per unit length.

Since the vertical distance available for dilution is limited by the depth of discharge while q is inversely proportional to the length of the diffuser, it is seen that the dilution obtainable from a line diffuser is improved by increasing either the discharge depth or the length of the diffuser. In other words, from the point of view of initial dilution, a shorter deeper diffuser is essentially equivalent to a longer shallower one if there is no density stratification. It is interesting to inquire as to which might be the more economical. Assume that an outfall is to be aligned perpendicular to shore with the diffuser parallel to shore. Let L_0 be the length of outfall pipe (excluding diffuser), L the length of diffuser, and α the bottom slope at the site.

It can be deduced from Eq. (10.1) that

$$r \equiv \frac{dS/dL_0}{dS/dL} = \frac{3}{2} \alpha \frac{L}{d}. \tag{10.2}$$

If $r > 1$, then it would take less pipe to achieve the same incremental improvement in dilution by increasing L_0 rather than L. If α is 0.01 (typical in Southern California) and d is 60 m, the critical value L_e is 4050 m. For L larger than L_e, it is less costly to increase L_0. This value of L_e is larger than the length of any diffuser in use. This implies that for purposes of achieving a given dilution it is almost always cheaper to increase the length of the diffuser. However, dilution is not the only parameter of concern as will be discussed shortly.

Example 10.1.   Using Eq. (10.1) estimate the initial dilution for $q = 0.01$ m²/sec, $\Delta\rho/\rho = 0.025$, $d = 45$ m. From Eq. (10.1),

$$S = 0.38 \times \frac{(9.81 \times 0.025)^{1/3}(45)}{(0.01)^{2/3}} = 230. \quad \blacksquare$$

It may be noted that the value $d = 45$ m is typical of large sewage outfalls, while $\Delta\rho/\rho = 0.025$ is the difference between sewage effluent and sea water. The above example shows that to achieve a dilution of about 230 in approximately 45 m of water, the length of the diffuser needs to be such that the discharge per unit length q be about 0.01 m²/sec or that there should be 100 m of diffuser for every cubic meter per second of waste flow. Table 10.1 shows that this is largely true for almost all of the recent major Pacific Ocean outfalls.

The simple result in Eq. (10.1) is for a two-dimensional line plume (source of buoyancy only) in a uniform still environment. When the ambient is stratified with a linear density profile, the corresponding plume equation is

$$S = 0.31 \, g'^{1/3} y_{max}/q^{2/3} \tag{10.3}$$

where y_{max}, the maximum height of rise of the jet, is given by

$$y_{max} = 2.84(g'q)^{1/3}\left\{\frac{-g}{\rho}\frac{d\rho_a}{dy}\right\}^{-1/2}, \tag{10.4}$$

where $\rho_a(y)$ is the ambient density, $g' = g \, \Delta\rho/\rho$, and $\Delta\rho$ is the density difference between the discharge fluid and the ambient fluid at the level of the discharge.

The coefficients in these equations have not been confirmed by experiments but are based on assuming that the entrainment process is unaffected by the stratification.

Example 10.2.   Apply Eqs. (10.3) and (10.4) to the case $q = 0.01$ m²/sec, $\Delta\rho/\rho = 0.025$, $d = 60$ m, and where the ocean temperature varies linearly with depth. Assume that the temperature is 20°C on the surface and 17°C on the bottom. The salinity is constant at 34 ‰.

The densities of seawater are obtained from Appendix A. The density gradient is

$$-\frac{1}{\rho}\frac{d\rho_a}{dy} = \frac{(24.767 - 24.020) \times 10^{-3}}{60} \, m^{-1} = 1.25 \times 10^{-5} \, m^{-1}.$$

TABLE 10.1

Summary of Characteristics of Major Pacific Ocean Outfalls (U.S.A.)

	Year operation began	Pipe diameter (inside) (inches)	Length of main outfall (excl. diff.) (ft)	Length of diffuser, L (ft)	Depth of discharge (nominal) (ft)	Design average flow, Q (ft³/sec)	Port diameters[a] (inches)	Port spacing (average)[b] (ft)	Velocity of disch. (nominal) for av. flow (fps)	Q/L (ft²/sec)	Area factor (total port area/pipe area)
Sanitation Districts of Los Angeles County White Point No. 3	1956	90	7,900	2400	200–210	232	6.5–7.5	24	8	0.097	0.63
City of Los Angeles at Hyperion	1960	144	27,525	7920	195	651	6.75–8.13	48	13	0.082	0.44
San Diego	1963	108	11,500	2688	200–210	363	8.0–9.0[c]	48	15	0.135	0.39
Sanitation Districts of Los Angeles County Whites Point No. 4	1965	120	7,440	4440	165–190	341	2.0–3.6	6	9	0.077	0.51
Metrop. Seattle (West Point)	1965	96	3,050	600	210–240	194	4.5–5.75	3	6	0.323	0.60
Sanitation Districts of Orange County, Calif.	1971	120	21,400	6000	175–195	450	2.96–4.13	12	13	0.075	0.45
Honolulu (Sand Island)	1975	84	9,120	3384	220–235	164	3.00–3.53	12	10	0.048	0.44

[a] Exclusive of end ports, which are usually somewhat larger.

[b] Length of diffuser divided by number of ports; real spacings on each side of the pipe are twice the values indicated.

[c] Blocked by orifice plates with openings of 6.5–7 in. for early years' low flow.

Equations (10.4) and (10.3) then gives

$$y_{max} = 2.84(9.81 \times 0.025 \times 0.01)^{1/3}[9.81 \times 1.25 \times 10^{-5}]^{-1/2}$$

$$= 35 \text{ m}$$

$$S = 0.31 \frac{(9.81 \times 0.025)^{1/3} \times 35}{(0.01)^{2/3}} = 150.$$

Hence the waste field is submerged at approximately 25 m depth with a dilution of 150. To achieve this submergence required only a 3°C temperature difference between the surface and bottom waters in the ambient ocean. ■

Equations (10.3) and (10.4) are for the case when the discharge is oriented vertically. For horizontal discharge, using typical discharge conditions in a sewage outfall, the coefficients 0.31 and 2.84 in Eqs. (10.3) and (10.4) are modified to 0.36 and 2.5, respectively. For design calculations, these latter coefficients are probably more appropriate. It should be stressed, however, that the entire process of dilution estimation is approximate since not only are the results not verified experimentally, but also the analyses have not included all aspects of the overall phenomenon such as the effect of an ocean current and the non-linear character of the density profiles.

The equations for a point plume (pertaining to a single outlet discharge) corresponding to Eqs. (10.1), (10.3), and (10.4) are

$$S = 0.089 \frac{g'^{1/3}y^{5/3}}{Q^{2/3}} \tag{10.5}$$

$$S = 0.071 \frac{(g')^{1/3}y_{max}^{5/3}}{Q^{2/3}} \tag{10.6}$$

$$y_{max} = 3.98(Qg')^{1/4}\left(-\frac{g}{\rho}\frac{d\rho_a}{dy}\right)^{-3/8}. \tag{10.7}$$

Example 10.3. Apply Eqs. (10.6) and (10.7) to the case when $d = 60$ m, $\Delta\rho/\rho = 0.025$, $Q = 6 \text{ m}^3/\text{sec}$, $-(1/\rho) d\rho_a/dy = 1.25 \times 10^{-5} \text{ m}^{-1}$ (same as previous example but no diffuser is used).

Equation (10.7) gives

$$y_{max} = 3.98(6 \times 9.81 \times 0.025)^{1/4}(9.81 \times 1.25 \times 10^{-5})^{-3/8} = 128 \text{ m}.$$

Since y_{max} is larger than the depth, the waste field will surface. The dilution obtained is estimated by Eq. (10.6) using 60 as y_{max} to be

$$S = 0.071 (9.81 \times 0.025)^{1/3}(60)^{5/3}/(6)^{2/3} = 12 \quad ■$$

Comparison of these results with those of the previous example shows clearly the advantage of a diffuser. Since Q is 6 m³/sec and q is 0.01 m²/sec in these examples, the diffuser length is 600 m.

It must be pointed out that, strictly speaking, Eqs. (10.1), (10.3), (10.4), (10.5), (10.6), and (10.7) *cannot* be applied directly to the evaluation of dilution and submergence of candidate design even if the ambient is motionless and the plume approximation is valid. The example calculations above are made merely to show the typical values obtainable and to delineate the influence of the diffuser. To appreciate why these results cannot be used it is first necessary to discuss the implications of the second phase of the dispersion process, viz., the establishment of the waste water field, since it affects the dilution obtainable.

10.2.1.2 *Establishment of the Wastewater Field*

Referring to Fig. 10.1, which shows schematically the rise of the buoyant effluent, it may be observed that application of the theories of buoyant jets and plumes to obtain dilution estimates must take into account the presence of the waste field above. In other words, even though mixing is presumably occurring from the discharge all the way to y_{max}, dilution of the effluent with clean ocean water ceases when the plume reaches the bottom of the waste field. The additional mixing occurring above that level does not contribute to the dilution in the sense that it does not lower the concentration of pollutants, since the mixing is with the previously discharged diluted effluent. Rather, the mixing which occurs for $y > y_b$ (see Fig. 10.2) only serve the purpose of evening out the differences in concentration between the various cross-sectional portions of the rising plume.

At present, only approximate estimates can be made as to this effect of blocking due to the finite thickness of the waste water field. Referring to Fig. 10.2, let h be the thickness of waste field, b the width of waste field (normal to current), u the current speed, S_{aw} the average dilution in waste field, Q_0

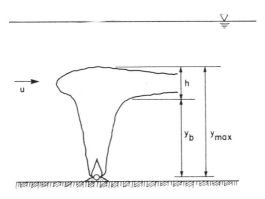

Figure 10.2 Definition sketch for establishment of the wastewater field. [After Koh and Brooks (1975).]

the discharge from diffuser, y_b the y coordinate at bottom of waste field, y_{max} the y coordinate at top of waste field. By continuity,

$$Q_0 S_{aw} = ubh = ub(y_{max} - y_b). \tag{10.8}$$

It will now be assumed that the average dilution at elevation y is proportional to y. This is nearly true in all buoyant line plume and jet cases for y sufficiently large.

Thus, assuming S_{aw} is the same as the average dilution in the plume at $y = y_b$,

$$S_{aw}/S_a = y_b/y_{max}, \tag{10.9}$$

where S_a is the calculated value at the top of the plume ($y = y_{max}$) disregarding the presence of blocking of the finite thickness of the waste field. Hence Eq. (10.8) becomes

$$Q_0 S_a = ub(y_{max}/y_b)(y_{max} - y_b). \tag{10.10}$$

Equation (10.10) may be solved for y_b/y_{max} to give

$$y_b/y_{max} = 1/(1 + Q_0 S_a/uby_{max}) \tag{10.11}$$

or substituting into (10.9), the dilution S_{aw} is given by

$$S_{aw} = S_a(1 + Q_0 S_a/uby_{max})^{-1}. \tag{10.12}$$

The thickness of the waste field is $h = y_{max} - y_b = y_{max}(1 - y_b/y_{max})$ and is therefore

$$h = y_{max}\{(Q_0 S_a/uby_{max})/(1 + Q_0 S_a/uby_{max})\}. \tag{10.13}$$

Before this concept can be applied in practice, it is necessary to determine u, b, and S_a. It should further be noted that if the product ub is very small, then the analysis fails to be valid because the assumption in Eq. (10.9) breaks down.

For a current which is perpendicular to the diffuser, b may simply be taken as the length of the diffuser. For a current parallel to the diffuser, the width b must be estimated differently. The next two subsections describe the procedures for a uniform and a stratified ambient.

10.2.1.2.1 Uniform ambient. Consider first the case when the ambient is uniform in density so that the waste plume rises to the surface. It will be assumed that a line source of buoyancy of strength $g'q$ and length L is operating under a parallel current u. The quantities of import are, therefore,

$$g'q, \quad u, \quad L, \quad b.$$

It will now be assumed that u and L occur as a pair in the combination L/u. This is equivalent to saying that the problem may be approximated by the two-dimensional problem of the time dependent release of a buoyant fluid at rate $g'q$ starting at time 0. (This may be visualized by imagining an observer traveling with the current and encountering the diffuser at time $t = 0$.)

Applying dimensional analysis to the three variables

$$g'q, \qquad L/u, \qquad b$$

it may be seen that there is only one dimensionless group

$$(g'q)^{1/3}(L/u)/b$$

which must therefore be a constant, say, C. Hence, subject to the approximations above,

$$b = C(g'q)^{1/3}(L/u). \tag{10.14}$$

Limited laboratory experiments (R. M. Towill Corp., 1972; Koh and Fan, 1970; Koh, 1976; Roberts, 1977) show that this is a good approximation and that the constant C is about 1.2.

Substituting Eq. (10.14) into Eq. (10.12) gives

$$\frac{S_{aw}}{S_a} = 1/(1 + Q_0 S_a/1.2(g'q)^{1/3} L y_{max}). \tag{10.15}$$

Equation (10.15) expresses the ratio of S_{aw}, the actual average dilution to S_a, the average dilution in the absence of blocking. Unfortunately, no analysis is available to obtain an estimate of the latter which is also affected by the current. Until such is available, we shall assume that S_a is still given by Eq. (10.3) except for the conversion from centerline dilution to average dilution

$$S_a = \sqrt{2}S = 0.54((g'q)^{1/3} y_{max}/q).$$

This assumption is equivalent to assuming that there is no current below y_b. With this assumption,

$$\frac{S_{aw}}{S_a} = \frac{1}{1 + \dfrac{0.54}{1.2}} = 0.7, \tag{10.16}$$

so that

$$S_{aw} = \frac{0.38(g'q)^{1/3} y_{max}}{q}. \tag{10.17}$$

Comparing Eq. (10.16) with (10.9) shows that the thickness of the sewage field in this case is about 30% of the depth.

It should be noted that Eq. (10.17) does not depend on the magnitude of u the current speed. Thus it infers that the dilution obtainable is not sensitive to the current speed for a parallel current. Moreover, Eqs. (10.12) and (10.16) in combination show that for a current flowing perpendicular to the diffuser, the dilution including the effect of blocking is at least as given by Eq. (10.17), even for very low currents, and S_{aw} increases for higher perpendicular currents.

Example 10.4.   Evaluate the average dilution including the effect of blocking obtained in a water depth of 45 m for a 600 m diffuser discharging 6 m³/sec of sewage effluent with $\Delta\rho/\rho = 0.025$. Use a perpendicular current of 0.25 m/sec and a parallel current. Assume uniform ambient density.

We first evaluate S_a, the average dilution without the effect of blocking

$$S_a = 0.54 \frac{(9.81 \times 0.025)^{1/3}(45)}{(0.01)^{2/3}} = 325.$$

Using Eq. (10.12), the average dilution for a perpendicular current of 0.25 m/sec is

$$S_{aw} = 325/(1 + 6 \times 325/0.25 \times 600 \times 45) = 255.$$

For a parallel current, Eq. (10.17) gives

$$S_{aw} = 0.38 \frac{(9.81 \times 0.025)^{1/3}(45)}{(0.01)^{2/3}} = 230. \quad \blacksquare$$

10.2.1.2.2   Submerged sewage field in stratified ambient.   We now consider the case when the sewage field is submerged so that there is no net buoyancy in the waste field. In this case, however, there must be some ambient density stratification. The horizontal spreading process in this case is very complex. However, simple dimensional analysis can also be applied. It should first be noted that in the previous case of a buoyant discharge in a uniform ambient, the quantity $g'q$ can also be interpreted as the rate of discharge of kinematic potential energy. In the present case of a neutrally buoyant waste field in a density stratification with $\varepsilon = -(1/\rho)\,d\rho/dy$, the rate of discharge of kinematic potential energy is $g\varepsilon q_1{}^2$, where q_1 is the rate of discharge into the waste field. The variables of import now are, therefore

$$g\varepsilon q_1{}^2, \qquad L/u, \qquad b$$

so that, from dimensional analysis,

$$b = C_1(g\varepsilon q_1{}^2)^{1/4}\, L/u. \tag{10.18}$$

Limited laboratory experiments (Wu, 1965) indicate that the coefficient C_1 is 0.8. Substituting Eq. (10.18) into (10.12) gives

$$S_{aw} = \frac{S_a}{1 + Q_0 S_a/0.8(g\varepsilon q_1{}^2)^{1/4} L y_{max}}. \tag{10.19}$$

But

$$q_1 = S_{aw}\, Q_0/L. \tag{10.20}$$

Hence

$$S_{aw} = \frac{S_a}{1 + Q_0 S_a/0.8L(\varepsilon g S_{aw}^2(Q_0{}^2/L^2))^{1/4} y_{max}}. \tag{10.21}$$

Solving for S_{aw} gives

$$S_{aw} = S_a - A\sqrt{\frac{A^2}{4} + S_a} + \frac{A^2}{2}, \qquad (10.22)$$

where

$$A = \frac{S_a(Q_0/L)^{1/2}}{y_{max} 0.8(\varepsilon g)^{1/4}}. \qquad (10.23)$$

It may be noted that this result is again independent of current speed.

It must be cautioned that the above analyses on the effect of blocking by a finite waste field thickness are only approximate and based on sparse experimental results. However, until more research results are available on the subject, these approximate results can be used with caution to estimate the range of dilutions obtainable in candidate diffuser designs.

Example 10.5. Calculate the dilution and submergence characteristics including the effect of blocking by a parallel current for the following case: $Q = 6$ m³/sec, $d = 60$ m, $\Delta\rho/\rho = 0.025$, diffuser length $= 600$ m, $-(1/\rho)\,d\rho_a/dy = 1.25 \times 10^{-5}$ m^{-1}.

We first obtain $y_{max} = 2.84(9.81 \times 0.01 \times 0.025)^{1/3}[9.81 \times 1.25 \times 10^{-5}]^{-1/2} = 35$ m. The sewage field is therefore submerged. Next we obtain S_a

$$S_a = \sqrt{2} \times 0.31 \frac{(9.81 \times 0.025)^{1/3} \times 35}{(0.01)^{2/3}} = 207.$$

Equation (10.23) then gives

$$A = \frac{207 \times 0.01^{1/2}}{35 \times 0.8 \times (9.81 \times 1.25 \times 10^{-5})^{1/4}} = 7$$

Finally Eq. (10.22) gives

$$S_{aw} = 207 - 7\sqrt{\frac{7^2}{4} + 207} + \frac{7^2}{2} = 128. \quad \blacksquare$$

10.2.1.3 Effect of Currents on Plume Dilution

All the above analysis techniques are for the case where the buoyant rise phase of the effluent plume occurs in a motionless ambient. In practice the sea is seldom motionless. Moreover, the formulations for estimating the initial dilution presented previously assumes a two-dimensional line plume. In practice, while the lengths of typical major diffusers are usually much larger than the depth so that one expects the line plume to be a good approximation for the initial dilution, nonetheless, the length is still finite. To obtain the effluent plume characteristics at a distance commensurate or larger than the diffuser length, the two-dimensional assumption cannot be made. The combination of a current and finite length diffuser makes the problem very complex. Roberts (1977, 1979) has performed laboratory experiments on this problem for the case where the ambient water is of uniform density. His research result indicates that the minimum surface dilution S_m which usually occurs near the discharge, can be expressed as

$$S_m q/ud = f(F, \theta), \tag{10.24}$$

where q is unit discharge, u the ambient current speed, d the water depth, θ the angle of current with respect to diffuser orientation, and F the Froude number defined as

$$F = u^3 \left/ \frac{\Delta\rho}{\rho} \, qg, \right. \tag{10.25}$$

where $\Delta\rho/\rho$ is relative density difference between effluent and ambient and g is gravitational acceleration. It can readily be shown that the line plume solution including blocking [Eq. (10.17)] reduces to

$$S_m q/ud = 0.27 F^{-1/3}, \tag{10.26}$$

where $0.27 = 0.38/\sqrt{2}$ and the factor $\sqrt{2}$ is to convert from average to minimum dilution S_m. The results of Roberts' experiments are summarized in Fig. 9.40 in Chapter 9. It should be cautioned that the laboratory experiments were performed at a small scale where the Reynolds numbers are quite small. Their applicability to the field is, therefore, still an unresolved question. For design calculations, the procedure outlined above [which resulted in Eqs. (10.17) and (10.22)] and in subsequent sections should be pursued since estimates based on them will be conservative. Moreover, ocean currents are frequently nonuniform with depth with shear occurring at the thermocline. The assumption of low or no current for $y < y_b$ (see Fig. 10.2) may in fact be a reasonable one.

Roberts' data further confirmed that for low values of F the orientation of the diffuser with respect to the current does not play a major role in determining the dilution at the surface.

10.2.1.4 Effects of Stratification in the Receiving Water

Before any analyses of buoyant jets can be made, it is first necessary to have a knowledge of the ambient conditions. Typical results of field investigations on the oceanographic conditions existing at the site usually contain measurements of the stratification, in the form of temperature and salinity as functions of depth. Examination of these data frequently reveals that the density stratification is neither linear nor steady. Figure 10.3 shows some typical density profiles obtained off Sand Island near Honolulu during the field investigations effort for the purpose of designing the Sand Island Outfall for the City and County of Honolulu.

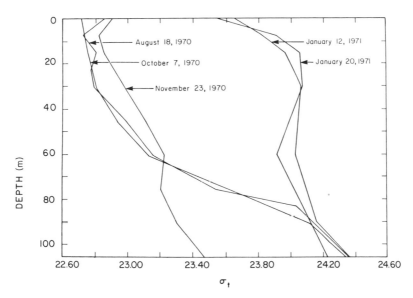

Figure 10.3 Density profiles measured off Sand Island, Hawaii. [After R. M. Towill Corp. (1972).]

Examination of Fig. 10.3 reveals that the density stratification is not uniform. Two approaches may be envisioned to overcome this problem. First, it is noted that the basic technique applied to a uniform stratification can also be applied to a nonlinear stratification. However, this means that a numerical solution must be obtained for each case on an individual basis. This can be readily done by a computer solution. Computer programs for this purpose for a single round jet, a single slot jet, and even a row of equally spaced round jets can be found in Ditmars (1969), Sotil (1971), and Koh and Fan (1970), respectively.

Another approach is to assume that the result may be approximated by assuming that the stratification is uniform between the discharge depth and the maximum height of rise (Brooks, 1973). This latter approach lends itself readily to a graphical solution. Since the basic analyses of dilution and submergence are approximate in nature, this simpler approach is quite useful for preliminary design purposes. The chosen design can then be evaluated using a computer program during the final design phase.

Return now to the expression for $y_{\max}$

$$y_{\max} = 2.5(g'q)^{1/3}[-(g/\rho)\, d\rho_a/dy]^{-1/2}. \qquad (10.4a)$$

The replacement of the actual density profile by an equivalent linear one between $y = 0$ and $y_{\max}$ implies

$$-\frac{g}{\rho}\frac{d\rho_a}{dy} = \frac{g\Delta\rho_a}{\rho y_{\max}}, \qquad (10.27)$$

where $\Delta\rho_a$ is the density difference between $y = 0$ and $y = y_{max}$. Substituting into Eq. (10.4a) gives

$$y_{max} = 6.25(g'q)^{2/3}(\rho/g\Delta\rho_a). \tag{10.28}$$

Since for a given design, $g'q$, is known, Eq. (10.28) may be interpreted as a relation between y and $\Delta\rho_a$ to be solved simultaneously with the ambient density stratification profile in the form $\Delta\rho_a(y)$. This may be done graphically by plotting the ambient profile and the hyperbola given in Eq. (10.28) and finding the intersection. In practice, there would be several ambient profiles and they may all be plotted on the same sheet. Moreover, change in the diffuser length simply implies a change in the hyperbola. Thus by plotting two sets of curves, with one on transparent paper, many candidate designs can be evaluated simultaneously. It may further be noted that for a horizontal discharge

$$S = 0.36\frac{(g'q)^{1/3}y_{max}}{q} = 0.36\frac{g'y_{max}}{(g'q)^{2/3}} \tag{10.29}$$

and using Eq. (10.28),

$$S = 2.25(\Delta\rho_d/\Delta\rho_a), \tag{10.30}$$

where $\Delta\rho_d$ is the density difference between the ambient and the discharge. Thus an additional scale may be placed on the density profile graph to allow the direct reading of the dilution. This process of graphical analysis is illustrated in Fig. 10.4.

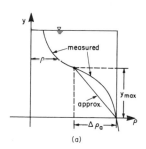

(a)

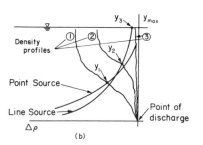

(b)

Figure 10.4 (a) Approximation of non-linear density profile by a linear one. (b) Diagram for solving equations with measured density profiles. [After Brooks (1973).]

The same process may also be carried out for the case of a point source (rather than a line source). The equations for a point source of buoyancy corresponding to (10.28) and (10.30) are readily shown to be

$$y_{max} = 9.1(Qg')^{2/5}(g \, \Delta\rho_a/\rho)^{-3/5} \tag{10.31}$$

and

$$S = 2.8(\Delta\rho_d/\Delta\rho_a). \tag{10.32}$$

10.2.1.5 Selection of Design Density Profiles

Since the dilution and submergence characteristics of a diffuser depend on the ambient density stratification, it is essential that the designer evaluate the diffuser performance for a range of observed density profiles. Daily variations in density stratification may be caused by tidal effects, internal waves, and variations of currents. Usually stratification will be more pronounced during the warmer months when insolation raises the temperature in the surface layer. Stratification may also increase during the wet season if runoff depresses the surface salinity. A typical annual regime of stratification may be obtained by averaging many density profiles for each month such as was done in the design study for the Orange County Outfall (Fig. 10.5). If insufficient data are available to define the annual regime the designer must exercise care in interpreting his results and realize the shortcomings in the data. Evaluations of dilution and submergence are at most as good as the data on density stratification.

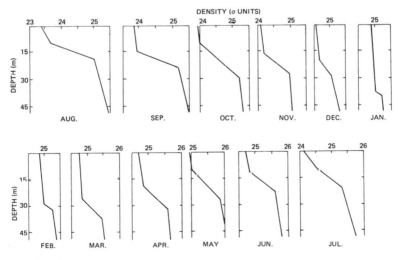

Figure 10.5 Monthly average density profiles used for design of Orange County Outfall, California. [After John Carollo Engineers (1970).]

Before leaving the subject of dilution calculations, three points should be reemphasized. First, all the techniques are approximate and results are no more accurate than perhaps $\pm 20\%$. Second, the initial dilution obtained in a multiple port diffusion structure is the only part of the overall dispersion and mixing process under the control of the designer. The main variables under his control are primarily the length of the diffuser and the discharge depth. Secondary parameters such as port size and spacing, angle of discharge, often modify the result only in a minor way. Third, the entire analysis assumes that the discharge does not influence the dynamics in the ambient. While the last assumption is in general quite good for typical multiple port diffusion structures used for sewage effluents, it is not true for thermal outfalls.

10.2.2   Further Transport and Dispersion Processes

After the initial dilution by jet mixing, the discharged effluent is subject to further diffusion and transport in the environment. These transport and dispersion processes are generally not under the control of design, but they are of importance in determining whether a waste discharge would be detrimental in a specific area. Transport of even diluted sewage to a beach should be avoided as much as practicable. If it cannot be avoided entirely the diffusion structure should be placed as far offshore as possible to increase the travel time. Longer travel times are important for bacterial decay. It was pointed out previously that except for cases with steep bottom slope, it always takes less pipe to increase the length of the diffuser to achieve a given dilution. One of the reasons for discharging further offshore is to increase travel time to the beaches if onshore currents are of concern. Figure 10.6 shows the cluster of wastewater outfalls which serve Los Angeles County. Currently only two of them [the 90 in. (2.3 m) and the 120 in. (3.03 m)] are operating. Both of them discharge at approximately a depth of 60 m. It can be seen that they are somewhat more than a nautical mile from the nearest land. Clearly one of the design considerations should be an evaluation of the potential that the wastewater might reach the beaches and the resulting public health hazards.

A waste field can reach the beach by one of two ways. If the discharge rises immediately to the surface it can be transported onshore by surface currents. If the initial field is submerged it can be brought towards the beach by subsurface currents and then surfaced by upwelling. The designer should be particularly wary of the possibilities of upwelling, especially if a waste field is initially formed at large depth, resulting in poor dilutions, and later upwelled to bring poorly diluted effluent to the surface. Normally a monitoring program will have been carried out before the final design, to obtain adequate estimates of the density stratification and ocean transport. To establish meaningful estimates of the behavior of the wastewater field, it is essential that the

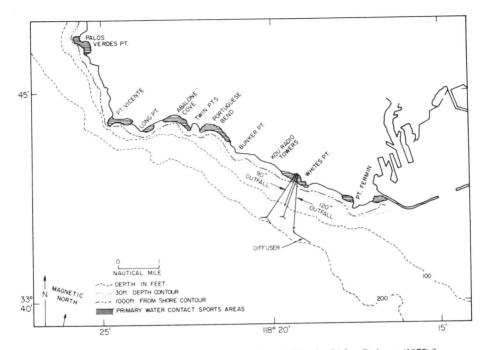

Figure 10.6 Ocean outfalls near Whites Point, California. [After Roberts (1977).]

monitoring program encompass all the typical conditions which might occur. In particular, to obtain seasonal variations, measurements should extend at least for one year. To assess diurnal changes, measurements should be made at sufficiently small sampling intervals. In general, the planning of a monitoring program to obtain the required information subject to limited resources is a difficult task but is beyond the scope of this book. In the following, some of the tools available to extract the needed information on ocean transport from current meter data will be briefly described and illustrated.

A typical observed time series of ocean currents off Whites Point is shown in Fig. 10.7. From data like these one can obtain the basic statistical estimates such as the mean current. A histogram in terms of speed and direction can also be obtained. Another frequently employed method of examining ocean transport data is the construction of a progressive vector diagram. This is a plot of the current vector in sequence and is intended to provide a visual appreciation of the movement of water particles. Clearly, the progressive vector diagram is not the same as a drogue track since the latter takes into account the spatial variations in the current while the former does not. Figure 10.8 illustrates a progressive vector diagram obtained using a portion of the data shown in Fig. 10.7.

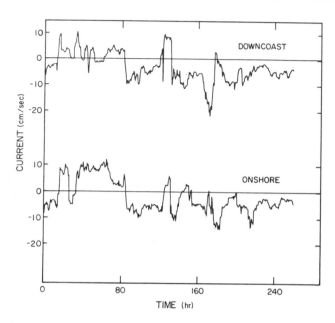

Figure 10.7   Example measured ocean currents off southern California.

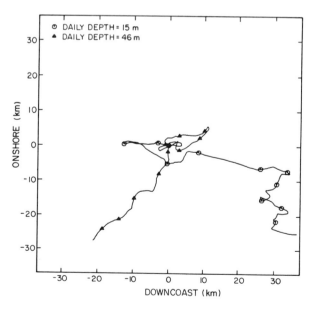

Figure 10.8   Progressive vector diagram based on currents in Fig. 10.7.

For estimating the advection of wastewater plumes a more meaningful construct than the progressive vector diagram is the streakline. To illustrate this concept, we shall assume that the ocean current is spatially homogeneous but temporally variable. Let $\mathbf{u}(t)$, $0 < t < T$ be the measured current velocity. Then at time T the location of the waste particles released at time $\tau < T$ is now at

$$\mathbf{x}(\tau, T) = \int_{\tau}^{T} \mathbf{u}(t)\, dt,$$

where $\mathbf{x} = 0$ is the diffuser location. The quantity $\mathbf{x}(\tau, T)$ when considered as a function of τ $(0 < \tau < T)$ then represents the locus of particles released between $t = 0$ and $t = T$ when observed at time T. Now let $\zeta = T - t$ so that $d\zeta = -dt$. Then we have

$$\mathbf{x}(\tau, T) = \int_{0}^{T-\tau} \mathbf{u}(T - \zeta)\, d\zeta.$$

This integral can be obtained from the data by reversing time and integrating backwards. In practice, if $\mathbf{u}(t)$ is sampled by a current meter at interval Δt resulting in a time series $\{\mathbf{u}_i\}$ $i = 1, 2, \ldots, N$, the integral can be obtained by reversing the order of the sequence and simply integrating. Under the same assumption of spatial homogeneity, an estimate of the transport probabilities to various points such as the nearby beaches can be estimated by first calculating $\mathbf{x}(t, T)$. For each T, and t, the value of $\mathbf{x}(t, T)$ can be tested for whether or not it is shoreward of some imaginary line (such as the shoreline). By varying t and T, this would provide some measure of the probability of transport to shore within various times. This type of analysis is illustrated in Figs. 10.9 and 10.10 for the Sand Island Outfall. Figure 10.9 shows typical traces of $\mathbf{x}(t, T)$. In that design effort, two lines were constructed termed "reef line" and "recreation line." The former marks the outer edge of the coral reef while the latter is displaced 1000 ft (305 m) seaward. Figure 10.10 shows the estimated probabilities of transport to the reef line using four different sets of current meter measurements. It must be pointed out that this method of analysis is subject to the rather gross assumption that the ocean current is spatially homogeneous and hence the results should be regarded only as a guide. The estimates of the probabilities can be expected to be progressively worse for longer travel times.

Sometimes it is also useful to obtain an estimate of further diffusion by lateral mixing using the analysis of Brooks (1960). Using the $\frac{4}{3}$rds power law for eddy diffusivity (see Section 3.4), Brooks's analysis of dilution in a uniform current gives

$$\frac{C_0}{C_{max}} = \left\{ \mathrm{erf}\left[\frac{3/2}{(1 + 8\varepsilon_0 t/w^2)^3 - 1} \right] \right\}^{-1/2}, \tag{10.33}$$

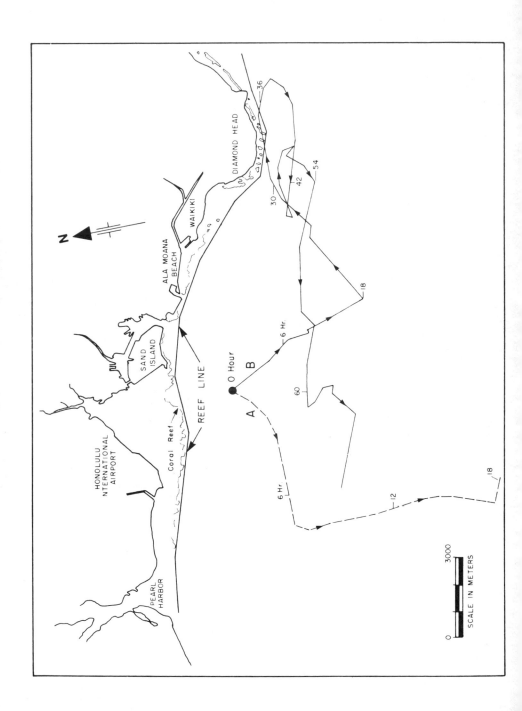

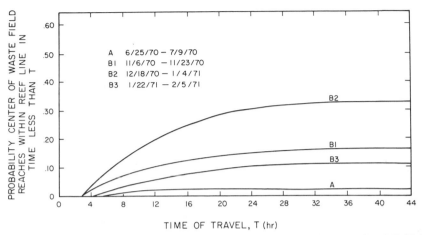

TIME OF TRAVEL, T (hr)

Figure 10.10 Cumulative probability of shoreward transport to reef line (see Fig. 10.9). [After R. M. Towill Corp. (1972).]

where ε_0 is the initial value of the horizontal turbulent diffusion coefficient corresponding to width w, C_0 is the waste concentration after the establishment of the waste field but before further diffusion, and C_{max} is the centerline concentration after travel time t. The rate of increase of further dilution with time t is thus tempered by both ε_0 and w. For w large, i.e., a large diffuser, the rate is small. For typical values of the parameters in the case of long diffusers, C_0/C_{max} increases only slowly with time of travel.

Example 10.6. Evaluate C_0/C_{max}, the further dilution factor at $t = 1, 3, 10$ hours for $w = 30$ m and 600 m using Eq. (10.33); assume $\varepsilon_0 = 0.01w^{4/3}$ in c.g.s. units.

	C_0/C_{max} ($\varepsilon_0 = 0.01\ w^{4/3}$)	
t (hr)	$w = 30$ m	$w = 600$ m
1	2.6	1.03
3	8.1	1.41
10	40	3.4

From this example, it is clear that the further dilution by ocean turbulence is relatively minor for large diffusers while it can be quite effective for small diffusers. ∎

Figure 10.9 Example streaklines based on measured currents for visualization of ocean transport of wastewater plume. [After R. M. Towill Corp. (1972).]

10.3   OUTFALL AND DIFFUSER HYDRAULICS

10.3.1   Manifold Hydraulics

The primary purpose of a multiple port diffuser is to distribute the flow evenly along the entire length of the structure. Thus proper design demands that the discharge per port should be nearly uniform from one end of the diffuser to the other. To achieve this goal involves the solution of a manifold problem which will be discussed in more detail in this section.

The achievement of a uniform flow distribution is by no means the only requirement for a well designed diffuser. Other basic hydraulic requirements for sewage effluent include

(i)   maintaining adequate velocities in the diffuser pipe to prevent deposition,
(ii)   providing for means of cleaning or flushing the system,
(iii)   ensuring that no seawater instrusion occurs or that all ports flow full,
(iv)   keeping the head loss reasonably small to minimize pumping.

These requirements have resulted in typical diffusers having the following features:

(i)   the diffuser pipe diameters are reduced in steps towards the far end,
(ii)   a flap gate is usually provided at the end which can be removed for flushing,
(iii)   ports are relatively small and so that the total port area downstream of any section is less than the pipe area at that section,
(iv)   ports have generally rounded corners.

A multiple port diffuser is basically a manifold. For the case of sewage effluent discharged into the ocean, the manifold flow problem is complicated by two factors. First, there is a density difference between the seawater outside of the diffuser and the sewage effluent inside (the density difference is approximately 0.025 gm/cc in the ocean). Second, the friction along the pipe changes the hydraulic head inside. The discharge through the port is dependent on the pressure difference across the port. Thus both the density difference and the friction play a role in determining the flow distribution. (The density difference is of no consequence if the ports are all on a level. However, this is almost never the case.)

The flow from a single port can be adequately represented by an equation of the form

$$Q_p = C_D A_p \sqrt{2gE}$$

where Q_p is the port discharge, A_p the port area, E the difference in total head across the port, g the gravitational acceleration, and C_D the discharge coefficient.

This equation is a semiempirical representation of the Bernoulli equation where the discharge coefficient C_D is to account for various losses, contractions,

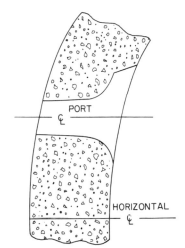

Figure 10.11   Schematic section of port
configuration cast in wall of diffuser pipe.

and flow nonuniformities. In general, the value of C_D must be found experimentally and depends not only on the geometrical characteristics of the port but also on the ratio of the velocity head in the diffuser to the total head E. For a port with rounded entrance cast directly into the wall of the diffuser pipe (see Fig. 10.11) it has been found that

$$C_D = 0.975(1 - V_d^2/2gE)^{3/8} \tag{10.34}$$

while if the entrance is not rounded (i.e., a sharp edged port), then

$$C_D = 0.63 - 0.58(V_d^2/2gE). \tag{10.35}$$

These are based on laboratory experiments performed on ports which are small compared to the diffuser pipe (port diameter less than one-tenth of the pipe diameter). Here V_d is the velocity in the diffuser pipe.

In some applications, the discharge ports actually consist of riser-nozzle assemblies. Risers are necessary when the main diffuser pipe is completely buried under the ocean bottom. In that event, the discharge coefficient is dependent on the entire geometrical characteristics of the assembly. Experimental data exist on various special configurations. One configuration is shown in Fig. 10.12. This is the configuration used in the discharge structures for the new Units 2 and 3 at the San Onofre Nuclear Generating Station near San Clemente, California. It may be noticed that

(i)   the riser diameter is larger than the port diameter,

(ii)   the entrance to the riser is rounded,

(iii)   an elbow is used to direct the discharge in a more nearly horizontal direction,

(iv)   an expansion and a bell mouth is used beyond the exit plane of the nozzle.

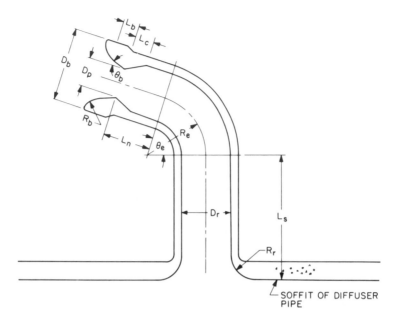

Figure 10.12   Schematic of nozzle-riser assemblies. [After Koh (1973).]

(i) and (ii) are primarily to reduce energy loss. (iii) is to achieve better mixing of the discharge with the surrounding sea water and (iv) is for reducing the head loss during reverse flow when the discharge nozzles are actually used as intake ports.

While experimental tests are the only reliable way to obtain C_D, estimates can be made for the discharge coefficient in many cases. These estimates may be used for preliminary calculations. For a riser-nozzle assembly as shown in Fig. 10.12, an energy equation may be written between the diffuser and the ambient at the location of the vena contracta of the jet

$$E = \frac{V_j^2}{2g} + x_{en}\frac{V_r^2}{2g} + f_r\frac{L_r}{D_r}\frac{V_r^2}{2g} + x_1\frac{V_r^2}{2g} + x_c\frac{V_r^2}{2g}, \qquad (10.36)$$

where x_{en}, x_1, x_c are the head loss coefficients for the entrance from diffuser pipe to riser; elbow; and contraction, respectively. V_r is the velocity in the riser; L_r, D_r are the length and diameter of the riser; f_r is the friction factor for the riser; V_j the jet velocity; and g the gravitational acceleration. Equation (10.36) may be written, using the port velocity V_p, as

$$E = \left\{\left(x_{en} + f_r\frac{L_r}{D_r} + x_1 + x_c\right)\left(\frac{D_p}{D_r}\right)^4 + \frac{1}{C_c^2}\right\}\frac{V_p^2}{2g} \equiv X\frac{V_p^2}{2g}, \qquad (10.37)$$

where D_p is the port diameter, $C_c = V_p/V_j$ the jet contraction coefficient. Based on experiments performed by McNown (1954) on branching and

TABLE 10.2
Coefficients of Jet Contraction

D_p/D_r	$\alpha = 45°$ C_c	$\alpha = 90°$ C_c	$\alpha = 135°$ C_c	$\alpha = 180°$ C_c
0.0	0.746	0.611	0.537	0.500
0.1	0.747	0.612	0.546	0.513
0.2	0.747	0.616	0.555	0.528
0.3	0.748	0.622	0.566	0.544
0.4	0.749	0.631	0.580	0.564
0.5	0.752	0.644	0.599	0.586
0.6	0.758	0.662	0.620	0.613
0.7	0.768	0.687	0.652	0.646
0.8	0.789	0.722	0.698	0.691
0.9	0.829	0.781	0.761	0.760
1.0	1.000	1.000	1.000	1.000

combining pipes it can be shown that, for D_r/D_d (D_d is the diffuser pipe diameter) less than $\frac{1}{4}$,

$$x_{en} \cong 0.406 + (V_d/V_r)^2 \equiv x_c + (V_d/V_r)^2, \tag{10.38}$$

where V_d is the velocity in the diffuser. The coefficient 0.406 is for a sharp-edged entrance and is confirmed by both free streamline theory and laboratory experiments. For rounded entrances, its value should be smaller, perhaps 0.1 or 0.2.

The value of the contraction coefficient C_c can be deduced based on free streamline theory and is a function of the diameter ratio D_p/D_r and α defined by

$$\alpha = \tan^{-1}(D_r - D_p)/2L_c \tag{10.39}$$

(where L_c is the length of contraction, see Fig. 10.12). Table 10.2 shows the values for various D_p/D_r and α values.

In Eq. (10.38), the quantity V_d is the velocity in the diffuser pipe upstream of the riser. For hydraulic calculations it is more convenient to express the discharge coefficient in terms of the velocity V downstream of the riser.

$$V = V_d - r^2 V_p, \tag{10.40}$$

where

$$r = D_p/D_d. \tag{10.41}$$

Combining these equations, it can be shown that

$$C_D = \frac{-r^2(V/\sqrt{2gE}) + X(1 - V^2/2gE) + r^4}{X + r^4}. \tag{10.42}$$

Experimental data (Koh, 1973) shows that the agreement is quite good for $V = 0$ for a variety of discharges and port configurations with a single choice of the various loss coefficients.

For small ports cast into the walls of the pipe, $r = 0$ and $X = x_e + 1/C_c^2$. Hence

$$C_D = \left[1 - \frac{V^2}{2gE}\right]^{1/2} \frac{1}{\sqrt{x_e + 1/C_c^2}}. \qquad (10.43)$$

This may be compared with Eqs. (10.34) and (10.35) obtained experimentally for a small port with rounded and sharp edged entrances respectively.

10.3.1.1   Calculation Procedure

The calculation procedure used in the design of a multiport diffuser may be formulated mathematically as follows (see Rawn *et al.* 1961).

Let D be the diameter of the pipe; d_n the diameter of the nth port, counting from the offshore end; a_n the area of the nth port; V_n the mean pipe velocity between the nth port and the $(n + 1)$th port; $\Delta V_n = V_n - V_{n-1}$ the increment of velocity due to discharge from the nth port (or group of ports); $h_n = \Delta p_n/\gamma$ the difference in pressure head between the inside and the outside of the diffuser pipe just upstream of the nth port (expressed in meters of effluent such as sewage); $E_n = h_n + (V_n^2/2g)$ the total head at the nth port (same either side by assumption); C_D the discharge coefficient for ports; q_n the discharge from the nth port; h_{f_n} the head loss due to friction between $(n + 1)$th and the nth port; L_n the distance between the $(n + 1)$th and the nth port; f the Darcy friction factor; Δz_n the change in elevation between the $(n + 1)$th and the nth port [measured to center of port; positive when $(n + 1)$th port is not as deep as the nth port]; $\Delta s/s$ the relative difference in specific gravity between discharging fluid and ambient fluid.

First it is necessary to select E_1; then q_1 for the first port is

$$q_1 = C_D a_1 \sqrt{2gE_1} = C_D(\pi/4)d_1^2 \sqrt{2gE_1}. \qquad (10.44)$$

Next, one finds the velocity in the pipe

$$V_1 = \Delta V_1 = \frac{q_1}{(\pi/4)D^2}$$

and velocity head $V_1^2/2g$.

Proceeding to port No. 2, one finds E_2 by

$$E_2 = E_1 + h_{f1} + (\Delta s/s)\Delta z_1. \qquad (10.45)$$

The ratio $(V_1^2/2g)/E_2$ is calculated for use in Eq. (10.34), (10.35), or (10.42) to find C_D. Then

$$q_2 = C_D a_2 \sqrt{2gE_2}$$

and

$$V_2 = V_1 + \Delta V_2 = V_1 + \frac{q_2}{(\pi/4)D^2}.$$

This procedure is continued step by step back up the diffuser using the general relations

$$C_D = \text{function of } \left(\frac{V_{n-1}^2}{2g}\bigg/E_n\right), \tag{10.46}$$

$$q_n = C_D a_n \sqrt{2gE_n}, \tag{10.47}$$

$$\Delta V_n = \frac{q_n}{(\pi/4)D^2}, \tag{10.48}$$

$$V_n = V_{n-1} + \Delta V_n, \tag{10.49}$$

$$h_{f_n} = f\,\frac{L_n}{D}\frac{V_n^2}{2g}. \tag{10.50}$$

and

$$E_{n+1} = E_n + h_{f_n} + (\Delta s/s)\Delta z_n, \tag{10.51}$$

The procedure is readily carried out with a digital computer and a number of trial designs can be easily investigated.

If the port discharges and pipe velocities change slowly, it is expedient to make the stepwise calculations for small groups of ports. In this case, Eq. (10.48) is changed to read

$$\Delta V_n = m\,\frac{q_n}{(\pi/4)D^2},$$

where m is the number of ports considered in a group.

By the nature of the calculations, it is apparent that one cannot decide on a particular total flow before starting the calculations. It is necessary to estimate the flow from the end port (q_1) which will correspond to the desired total flow.

10.3.1.2 Selection of Port Sizes and Pipe Sizes

During the process of the calculation, the designer is at liberty to change the pipe size, the port size, and/or the port spacing. To keep the velocity high enough at the end of the diffuser, it is sometimes necessary to reduce the size of the pipe in one or more steps from the beginning to the end of the diffuser. The size of the discharge ports may be varied in order to keep the discharge uniform from port to port. The spacing between ports is rather inflexible, inasmuch as practical considerations usually give preference to a spacing which is either equivalent to the length of a pipe section or multiple or simple fraction thereof. The entire design process inevitably requires some trial and error arrangements in order to get one arrangement which is satisfactory at various total rates of flow.

For a diffuser which is placed at zero slope, the relative distribution of flow would be the same at all rates of discharge. This is because all the head terms

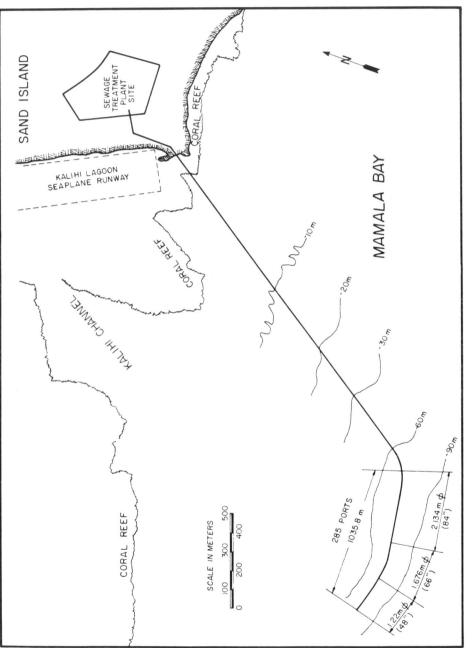

Figure 10.13a Schematic plan of Sand Island Outfall, Hawaii. [After R. M. Towill Corp. (1972).]

SAND ISLAND

SEWAGE TREATMENT PLANT SITE

KALIHI LAGOON SEAPLANE RUNWAY

CORAL REEF

CORAL REEF

KALIHI CHANNEL

CORAL REEF

MAMALA BAY

-10 m

-20 m

-30 m

-60 m

-90 m

285 PORTS
1035.8 m

2.134 m φ
(84")

1.676 m φ
(66")

1.22 m φ
(48")

SCALE IN METERS
0   100   200   300   400   500

N

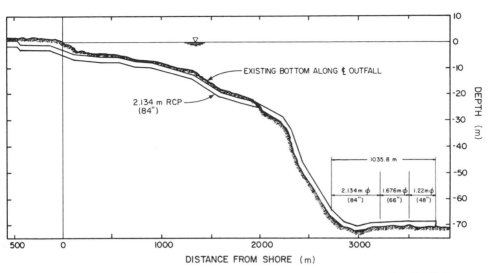

Figure 10.13b Schematic profile of Sand Island Outfall, Hawaii. [After R. M. Towill Corp. (1972).]

are proportional to the square of the velocity. In that case, where there are no differential elevations, one calculation would suffice for all rates of flow. For example, to double the rate of flow, one would need only to quadruple all the heads and double all the velocities and discharges.

It is essential that the end of the diffuser pipes be bulkheaded, otherwise the flow will not be forced out of the discharge ports near the end of the diffuser, and an excess of flow will be discharged through the open-ended pipe. The bulkheads should be removable for flushing the line.

In the process of making the hydraulic calculations it was found that a good rule of thumb was to assure that the sum of all the port areas is less than the cross-sectional area of the outfall pipe. It is impossible to make a diffuser flow full if the aggregate jet area exceeds the pipe cross-section area, since that would mean that the average velocity of discharge would have to be *less* than the velocity of flow in the pipe. Experience indicates that the best area ratio (total port area/pipe area) is usually between $\frac{1}{3}$ and $\frac{2}{3}$; the ports should be small enough to get good flow distribution among the ports, but not so small as to increase the total head unduly. Another good rule to keep in mind is that the effects of pipe friction and pressure recovery on flow distribution tend to cancel each other if $fL/D = 3$, where f is Darcy friction factor, L the length of diffuser, and D the diameter of diffuser (see Camp and Graber, 1968, 1970).

As an example of a multiple port diffuser, the Sand Island outfall diffuser for the City and County of Honolulu is shown illustrated in Fig. 10.13. The last

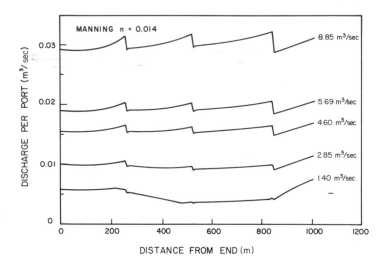

Figure 10.14   Distribution of discharge from diffuser ports for Sand Island Outfall, Hawaii. [After R. M. Towill Corp. (1972).]

1030 m of the outfall contains 282 ports ranging in size from 7.62 to 8.97 cm. The variation of port discharge and velocity in the diffusers are shown in Fig. 10.14 for the case of a value of Manning's n of 0.014. The end of that diffuser is equipped with a flap gate. The flushing velocity in the diffuser with the flap gate open is shown in Fig. 10.15 for $n = 0.016$.

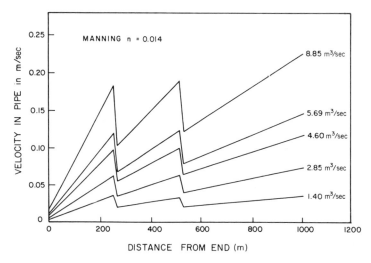

Figure 10.15   Variation of velocity in diffuser pipe, Sand Island Outfall, Hawaii. [After R. M. Towill Corp. (1972).]

10.4   AN EXAMPLE DESIGN: THE SAND ISLAND OUTFALL
IN HONOLULU, HAWAII

In this section, a specific outfall diffuser system, viz., the Sand Island Outfall in Honolulu, Hawaii will be examined in some detail to illustrate the applications of the analyses developed in previous sections to an actual case.

Sand Island is located on the south coast of the island of Oahu in the Hawaiian archipelago and has served as the beginning point of an old ocean outfall discharging untreated sewage in 12 m of water with essentially no diffuser structure. In the late 1960s, realizing the necessity to improve the general waste disposal facilities around the islands, a special study program (which has come to be known as the Oahu Water Quality Program) was undertaken by the City and County of Honolulu. The results of the study included the specific recommendation that improved waste treatment and disposal facilities be constructed at three locations to serve the expanding needs of the rapidly growing island population both residents and transients. These are (i) at Sand Island to serve the Southeastern portion of the island (the most heavily populated areas), (ii) at Barbers Point to serve the Southwestern portion, and (iii) at Kailua to serve the Northeastern portion. The Sand Island waste treatment and disposal system is the largest among the three since it serves the most people. Its design was undertaken in the period 1970–1972; construction of the outfall was initiated in 1974 and completed in 1975. The waste treatment facilities are to be built in stages at a later date.

The design period of the Sand Island Outfall was 1970–2020, a period of 50 years. The waste flows estimated based on population and water use projections are summarized in Table 10.3 which gives the range for the design period and for wet and dry weather.

The anticipated effluent characteristics (after primary treatment) are summarized in Table 10.4.

TABLE 10.3

Estimated Pollutant Loads[a]

Description	1970 Total	1990 Total	2020 Total
Population (thousands)	332	458	676
Average sewage flow (m³/sec)	1.22	1.78	2.85
Dry weather infiltration (m³/sec)	1.18	1.18	1.18
Design average flow (m³/sec)	2.75	3.56	4.64
Design maximum hourly flow (m³/sec)	3.46	4.43	5.7
Wet weather infiltration (m³/sec)	3.15	3.15	3.15
Design peak flow (m³/sec)	6.62	7.58	8.85

[a] Source: OWQP Report (1972).

TABLE 10.4

Parameter	Anticipated effluent characteristics	Parameter	Anticipated effluent characteristics
(a) Settleable solids	0.1–1.0 mg/liter ($\leqq 90\%$ removal)	(g) Total dissolved solids	2500 mg/liter
(b) Floatables	3 ppm (65% removal)	(h) Turbidity	20–30 JTU
(c) BOD, median noncanning	100 mg/liter	(i) Suspended solids, median	50 mg/liter (70% removal)
median canning	200 mg/liter (45% removal)	(j) Total P (average)	5 mg/liter (5–20% removal)
(d) Coliforms (total), median	Estimated as 500,000/mliter	(k) Total N (average)	22 mg/liter (5–20% removal)
(e) Coliforms (fecal), median	Estimated as 100,000/mliter	(l) Temperature	71–79°F
(f) pH	6.7–7.5		

It is interesting to point out that specified in the contract document from the City and County of Honolulu is that the effluent field be submerged. To comply with this requirement would imply that the discharge be located deeper than the lowest point the surface mixed layer in the ocean ever reaches. This turns out not to be practicable. In fact, it is at times undesirable as will be discussed subsequently.

The water quality standards which must be met are specified by the Public Health Regulations of the Department of Health, State of Hawaii and are summarized in Table 10.5. Guidelines and objectives as recommended by the Oahu Water Quality Program are also included in that table. The latter represent a more reasonable approach although the former are the legal standards.

10.4.1 Preliminary Overall Considerations

After examining the available information on waste flow characteristics, receiving water characteristics, ocean density stratification, bathymetric and topographic features, it became obvious that

(i) some of the water quality standards are virtually impossible to meet. This is notably true for phosphorus where the natural ocean background level already exceed the standards.

(ii) ensuring sewage field submergence would necessitate the placement of the diffuser at a depth in excess of 90 m.

(iii) the bathymetry is such that the bottom slope becomes progressively steeper as the depth increases (on the order of 10–20% or more).

(iv) much of the area in the offshore region at the site displays rocky outcrops (either volcanic or coral origin) on the sea bottom.

TABLE 10.5

Listing of Water Quality Requirements for Sand Island Offshore Area

	Criteria	Existing state standards	OWQP recommendations
(a)	Settleables	None	Quantitative objectives not recommended
(b)	Floatables	None	As a guide, oily matter less than 20 mg/m², sewage solids less than 1.5 mg/m²
(c)	Color, odor	None	Odor—As a guide, less than that detectable by threshold odor determinations (110:1 dilution for present S.I. effluent). Color—No recommendations in platinum–cobalt scale. As a guide, extinction coefficients for red, blue, green light not increased more than 10%
(d)	Total coliforms	Median no greater than 1000/100 mliter, no more than 10% greater than 2400/100 mliter	Median no greater than 1000/100 mliter, 90% less than 2400/100 mliter
(e)	Fecal coliforms	Arithmetic mean no greater than 200/100 mliter, no more than 10% greater than 400/100 mliter in 30-day period	Log mean no greater than 200/100 mliter, 90% less than 400/100 mliter
(f)	pH	Less than ½ unit variation from natural conditions, not outside range 7.0–8.5 from other than natural conditions	95% less than 8.2, no more than 5% less than 7.8
(g)	Dissolved oxygen	Not less than 5.0 mg/liter	Mean no less than 6.0, no more than 5% less than 5.0 mg/liter
(h)	Salinity	Not applicable to Class A waters	95% less than 35.00, no more than 5% less than 33.15 ppt
(i)	Secchi disk depth or equivalent light extinction coefficient	Less than 10% variation from natural conditions	No more than 5% less than 60 ft (Secchi disk)
(j)	Total phosphorus	Not greater than 0.025 mg/liter	Log mean no greater than 0.015 mg/liter, 95% less than 0.055 mg/liter
(k)	Total nitrogen	Not greater than 0.15 mg/liter	Log mean no greater than 0.1 mg/liter, 95% less than 0.300 mg/liter
(l)	Temperature	Not more than 1.5°F variation from natural conditions	Summer / Fall / Winter / Spring: 26.5 (27.0) / 26.5 (27.0) / 24.5 25.0 / — (26.0). Upper figs ≦ arithmetic mean. Lower figs ≦ 95%ile. ( ) = estimated values
(m)	Heavy metals, pesticides	Not applicable	Studies recommended

The combination of (ii) and (iii) leads to the question of whether it would be practicable, or even feasible, to construct a large pipeline and securely anchor it at the great depth and steep side slope. Item (iv) implies that the alignment of the outfall and the location of the diffuser must be chosen in such a way as to avoid encountering these rocky outcrops and the associated added costs of underwater grading of the sea bottom.

10.4.2   Design Philosophy

In developing the diffuser design, high dilution was considered to be of overriding importance while submergence of the sewage field was given a lower priority. This was because (i) the objection to a surfacing sewage field is much removed if the dilution is sufficiently high and (ii) ensuring submergence in the winter when the density stratification is slight requires a long diffuser at great depths which leads to very deep submergence in the summer, which in turn results actually in low dilution (since dilution is proportional to the height of rise of the plume). It is deemed undesirable to have a poorly diluted sewage field even when it is deeply submerged since not only is the natural biodegradation capability less at depth, there is also the potential of later upwelling of this poorly diluted field. All factors considered, the decision was made to seek no more depth (after 60 m) than necessary to (1) achieve submergence most of the time while avoiding excessive submergence with attendant low dilutions in the summer months and (2) locate the diffuser at a suitable topographic site with reasonable bottom slope and smooth bottom texture. The final choice of the diffuser site was also influenced by consideration of the prevailing ocean currents, the shoreward transport probabilities, and their relation to public health. (See Figs. 10.9 and 10.10 in Section 10.2.2.)

10.4.3   Final Design

The plan and profile of the final chosen design are shown schematically in Fig. 10.13. The westerly outfall alignment was necessary not only to reach the diffuser location which is the best topographic site but also to lengthen the path and decrease the likelihood of shoreward transport to the heavily populated and much utilized Waikiki area. The diffuser is located at approximately 72 m depth and is 1030 m long. The pipe size chosen is 84-in. (2.13 m) inside diameter based on a balance of initial cost, operating cost (pumping), and maintaining adequate velocities for flushing. The resulting dilution and submergence characteristics are shown in Fig. 10.16 for the design density profiles in Fig. 10.17. From Fig. 10.16 it is readily observed that dilution and submergence are inversely related. During the times in summer when there is a relatively strong density stratification, the sewage field would be submerged

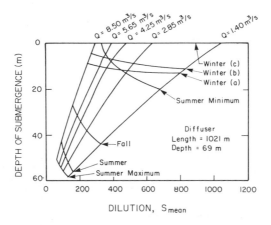

Figure 10.16   Submergence and dilution characteristics for Sand Island Outfall, Honolulu, Hawaii. [After R. M. Towill Corp. (1972).]

quite deeply with dilution on the order of 100, whereas when the sewage field surfaces, the range of dilution is 300–1000 depending on the flow rate. In the design process, a large combination of diffuser lengths and depths were evaluated. For each candidate, a graph such as shown in Fig. 10.16 was prepared. It was found that for a given diffuser depth, the dilution increases monotonically with increasing diffuser length. However, for a given diffuser length, the dilution obtained at times decreases with increasing diffuser depth. This is because

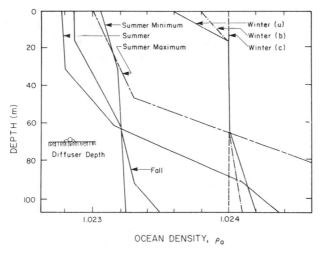

Figure 10.17   Density profiles used in design of Sand Island Outfall, Honolulu, Hawaii. [After R. M. Towill Corp. (1972).]

increasing the diffuser depth placed it in a region of stronger density stratification. The chosen design represents a compromise to achieve high dilution
and submergence most of the time.

As shown in Fig. 10.13, the outfall is buried to a depth of about 23 m after
which it gradually emerges onto the ocean floor at a depth of 26 m. Beyond
this point, the pipe is placed on the ocean bottom using rock ballast appro-
priately sized to ensure pipe stability. The deep burial and ballast sections
are necessary to protect the physical integrity of the pipeline against catastrophic
wave attack. The design wave has a significant wave height of 8.2 m, significant
wave period of 11.7 sec and maximum wave height of 14 m. This was based on
hindcasting of past hurricanes taking into consideration refraction and effect
of land masses in transforming the hurricane generated waves.

After the length and profile of the diffuser was selected, the detailed design of
the ports was performed using the calculation procedure described in Section
10.3.1.

10.5 DESIGN OF STRUCTURES FOR THERMAL DISCHARGES

The discharge of once-through cooling water from coastal power plants
is a common method of achieving the cooling necessary in the generation of
electrical power. One of the reasons why many power plants are located along
the coast of some large body of water, such as the ocean, is the availability of
cooling water. It should be noted that waste heat from power plants, unlike
other waste residues is not subject to treatment and must be disposed. The
second law of thermodynamics precludes the use of the energy contained in
the cooling water except for direct use such as space heating, which though
attractive in principal, has serious logistic and economic shortcomings.

Many existing coastal power plants discharge their cooling water via canals
to the sea, a practice commonly referred to as "across the beach" discharge.
Others build a submerged pipeline and discharge the warmed effluent from a
large single point outlet structure at some distance from shore. More recently,
the establishment of discharge regulations resulted in the building of diffusion
structures to effect rapid mixing of the warm effluent with the ambient receiving
water. Furthermore, the implementation of the Federal Water Pollution
Control Act Amendments of 1972 has completely altered the philosophy
of thermal discharges. At present, discharge to the navigable waters is forbidden
except under variances, and closed cycle cooling (e.g., cooling towers or ponds) is
universally required regardless of circumstances. Such rigid federal law has
received much criticism from various interested parties, particularly under
the sometimes conflicting interests of energy and environmental protection.
It is not possible to predict exactly what will come to pass in the future, although
we hope that some rational approach will be adopted.

10.5.1 Similarity and Differences between Thermal and Wastwater Discharges

At first glance, thermal and wastewater discharges are physically quite similar. Both involve buoyant effluents in a large body of water. Further examination, however, reveals that there is much difference between the two: thermal discharges are larger in quantity (of the order of 10 times more), less in buoyancy (of the order of $\frac{1}{10}$), are usually made in shallower water (a few meters rather than a few tens of meters), and require less dilution (factor of 10 rather than 10^2). Because of the large quantities of flow, the momentum in the discharge has at times been utilized as an added control to impart a drift in the plume away from the intake and discharge structures to avoid recirculation. Another difference lies in the fact that thermal discharges are always associated with an intake structure. These are either shoreline structures in an embayment or offshore intake structures connected to the pump wells via submerged pipeline. The former has the advantage of a lower suction head requirement whereas the latter tends to withdraw the cooler bottom water. Finally, the construction costs associated with the larger pipelines in a thermal discharge are also higher.

The combination of large flow, shallow discharge, and large momentum implies that thermal discharges from multiport diffusers tend to be active rather than passive resulting in a modification of the nearshore circulation. Analysis using the available theories on buoyant jets are deficient in that no good method has yet been found to incorporate the effects of such modifications into the analyses. While it is possible to estimate the dilutions which would result in candidate systems using the results of these theories, they should be looked upon as merely rough estimates useful only in preliminary calculations. Frequently, a physical model of the proposed structures is used to determine the expected thermal plume behavior. Other complications in thermal discharge systems are the possibility of recirculation between the discharge and intake and reentrainment of old diluted effluent into the discharge plume. The latter phenomena, though possible in wastewater systems, are much less likely due to the usually much larger depths of discharge which in turn imply a much greater availability of diluting water.

10.5.2 Hydraulic Modeling of Buoyant Discharge Systems

Current practice in the design of thermal discharge systems often requires the modeling of candidate designs in the laboratory. Hydraulic modeling requires not only geometric and kinematic similitude but also dynamic similitude so that model observations are representative of prototype behavior. The discharge of thermal waste into the ocean involves a problem in density-stratified flow, and, for dynamic similitude, the most important phenomenon which must be modeled is the interplay between momentum and inertia of the flow and the internal gravity forces due to buoyancy. These play dominant roles both near and to some distance from the discharge point. Far away from the discharge, when mixing has occurred to a sufficient degree, the dominant mechanisms change to those due to ambient turbulence and currents and surface heat exchange. In general, it is not possible to preserve similarity on all counts. Some phenomena are usually sacrificed and not modeled in favor of the more important ones. The following lists the phenomena or mechanisms which are of some importance during some phase of the mixing process resulting from a buoyant ocean discharge:

geometrical configuration	bottom and interfacial friction
source mass flux	energy dissipation
source momentum flux	surface heat exchange
source buoyancy flux	ocean turbulence
ocean density stratification	ocean currents
ocean temperature stratification	

Laboratory modeling which provides dynamic similitude can be obtained by maintaining the numerical values of various dimensionless numbers the same between model and prototype using either physical arguments, as described in Section 1.5, or by nondimensionalization of the basic conservation equations and boundary conditions. The dimensionless groups which are important to the thermal discharge problem are found to be:

Froude number $\qquad F_s = u/\sqrt{gd}$

Internal Froude number $\qquad F = \dfrac{u}{\sqrt{g(\Delta\rho/\rho)d}}$

Reynolds number $\qquad R = ud/v$

Friction factor $\qquad f$

Surface heat exchange $\qquad k = K/\rho c_p u.$
coefficient number

Here u is a characteristic velocity, d a characteristic length, v the kinematic viscosity, g the gravitational acceleration, ρ the density, c_p the specific heat, $\Delta\rho$ the density difference between discharge and ambient, and K the surface heat exchange coefficient (defined as the rate of transfer of heat through the water surface per unit surface area and per unit temperature difference between the actual and "equilibrium" water temperature; K is used frequently in thermal plume analyses in place of the more complex but more accurate analysis of surface heat transfer presented in Section 6.2).

To achieve dynamic similitude, the values of these numbers in the model and the prototype must be the same. It is readily apparent that this cannot be achieved unless the model is a full scale one. If we use the subscripts p, m, and r to denote prototype, model, and ratio of prototype to model, then similitude requires $F_{s_r} = F_r = R_r = f_r = k_r = 1$. Furthermore, $d_r = \lambda$, some chosen length scale, then $F_{s_r} = 1$ requires

$$u_r = \sqrt{\lambda}$$

since $g_r = 1$. $F_r = 1$ further requires $(\Delta\rho/\rho)_r = 1$. The ratio of Reynolds numbers is thus $R_r = u_r d_r/v_r = \lambda^{3/2}/v_r$. If water is used in the model then $v_r = 1$ so that $R_r = \lambda^{3/2}$ a number usually much larger than unity. Therefore, one must give up either the Reynolds number or the Froude number. In thermal discharges, the Reynolds number is always delegated to secondary importance with the proviso that its value in the model be sufficient to achieve turbulent flow. This means that the discharge jet Reynolds number based on jet exit velocity and diameter should be larger than approximately 1000–2000. This would result in a turbulent jet in the model. When there is an ocean current, the Reynolds number based on the current velocity and total water depth is also of importance. While the flow will be turbulent in the prototype, it is usually not possible to maintain a sufficiently large Reynolds number to provide turbulent flow in the model.

Related to the Reynolds number of the current, and of more importance is the friction factor ratio f_r. The value for f in the prototype is usually quite small for sandy bottom since the Reynolds number is also quite large. In the model, however, due to a much smaller Reynolds number, the friction factor would be larger. In many instances, it is impossible to decrease the friction factor in the model since the boundary may already be hydrodynamically smooth. The only recourse to properly model friction is to use a distorted model which will be discussed subsequently. Of even more importance than the bottom friction is the interfacial friction between the warm buoyant surface plume and the cooler seawater beneath. At present, there is not sufficient information available to properly determine the value of the interfacial friction. If it is assumed to be some constant fraction of the bottom friction, then distortion is also needed to correctly model this phenomena.

Proper modeling of the surface heat exchange mechanism requires $k_r = 1$. Since $(\rho c_p)_r = 1$, and $u_r = \sqrt{\lambda}$, we have $k_r = K_r/\sqrt{\lambda} = 1$ requiring $K_r = \sqrt{\lambda}$. Thus the model value of K, the surface heat exchange coefficient, should be made smaller than the prototype value by the ratio $\sqrt{\lambda}$. This can be achieved by controlling the atmosphere in the laboratory. If the model is built in a laboratory without control of the room atmosphere, then it is likely that k_r will not be unity and distortion may be required. The phenomenon of surface heat exchange, however, is usually not of importance in the vicinity of the discharge so that unless the model covers a very large area, it may be ignored.

When the discharge structure being modeled is a multiple port diffuser where the effluent from the many ports intermix, it is possible to properly model the friction effects by means of a distorted model. Locally near the jets, the model must not be distorted. Thus the jet diameter, spacing, and depth should all be scaled according to the vertical scale. Far from the discharge, due to the individual jets being quickly intermixed, the characteristics of the diffusion structure would be governed by the mass, momentum and buoyancy fluxes per unit length of the diffuser. The total length of the diffuser should be modeled by the horizontal length scale. Since the horizontal dimension is foreshortened, this has the effect of decreasing the importance of friction. It should be cautioned, however, that large distortion factors should be avoided, since in this case the end effects become too important.

10.5.3   An Example Design: The San Onofre Units 2 and 3 Thermal Discharge System

As an example of a thermal discharge, the design of outfalls and intakes of the San Onofre Nuclear Generating Station, Units 2 and 3 near San Clemente, California, will be described and discussed (see Koh, *et al.*, (1974)).

The San Onofre Generating Station is located near San Clemente, California, and is operated jointly by the Southern California Edison Company and the San Diego Gas and Electric Company. The first unit was put into operation in the early sixties. It is a nuclear generator with an output of 450 Mw(e). The utilities decided to install two more nuclear units (Units 2 and 3) at the same site where each unit will have the capacity of 1140 Mw(e). The original Unit 1 uses a once through cooling system where 22 m^3/sec of ocean water is pumped through the condensers. The intake and discharge structures (single point structures) are both offshore in depths of approximately 9 m. For the new units, it is not possible to duplicate the basic design of the Unit 1 discharge structure and meet the new thermal discharge requirements in California.

Table 10.6 shows the characteristics of the cooling water requirements for the operation of the 1140 Mw(e) units. Table 10.7 describes the California Thermal Plan which defines the requirements which the discharge must meet.

Before embarking on the discussion of the design, it is of interest to delineate the differences between the San Onofre thermal discharge and the Sand Island wastewater discharge in the context of design considerations:

TABLE 10.6
Cooling Water Requirements for
San Onofre Units 2 and 3

Flow rate	52.5 m^3/sec per unit
Temperature rise	20°F (11.1°C)
Power generation	1140 Mw(e) per unit

TABLE 10.7
*Summary of California Thermal Plan in Effect During
Design of San Onofre Units 2 and 3*

Effluent $\Delta T \leq 20°F$ (11.1°C)
   $\Delta T <$ 4°F at shoreline (2.22°C)
   $\Delta T <$ 4°F at bottom
   $\Delta T <$ 4°F beyond 1000 ft from discharge
               structure more than 50% of any
               tidal cycle
   ΔT is with reference to "natural" temperature

(1) The flow in the San Onofre discharge is 52.5 m³/sec per unit resulting in 105 m³/sec for the two new units. The flow is not expected to vary during periods of normal operation. In contrast, the Sand Island discharge will have variable flow ranging from about 1 to 9 m³/sec.

(2)   The density difference between the effluent and the receiving water is 0.025 gm/cc for the Sand Island sewage effluent (seawater versus fresh water) and that for San Onofre is 0.003 gm/cc (warm seawater versus cooler seawater, $\Delta T_0 = 20°F$).

(3)   The dilution to be obtained in San Onofre is on the order of 5–10. That at Sand Island is 100–1000.

(4)   The depth of discharge at Sand Island is 72 m. That at San Onofre is on the order of 12 m.

It is readily observed that while both are problems of buoyant jets dispersing in an ocean environment, there are very large differences in scale. Of great importance in the thermal discharge is the large flow rate and large momentum flux. In the design chosen for the San Onofre discharge system, these have been taken advantage of as an integral part of the design concept.

Examination of the bathymetry at the site reveals that the ocean bottom is primarily sandy with a very gentle slope (on the order of $\frac{1}{200}$). To reach a depth of 15 m requires a distance of approximately 2450 m offshore. To meet the requirement of the California Thermal Plan [basically $\Delta T \leq 20°F$ (11.1°C) for the effluent, and $\Delta T \leq 4°F$ (2.2°C) on the surface beyond 1000 ft (305 m) of the discharge] necessitates an initial dilution of five. The total volume rate of flow of the two new units is 105 m³/sec. The total volume rate of fluid involved after a dilution of five is 525 m³/sec. At a flowing depth of say 10 m, a total diffuser length of L, and a current speed of say 6 cm/sec, the volume flux of ocean water would be $0.6L$. Equating 525 m³/sec to $0.6L$ reveals that L would be 875 m. Thus, one expects that the diffusers would need to be almost 1000 m long. Examination of the ocean currents at San Onofre reveals that while there are usually currents on the order of 10 or 15 cm/sec, there exist times lasting on the order of a day when the currents are very low (less than 2.5 cm/sec). Since the $\Delta T \leq 2.2°C$ requirement must be met 50% of each tidal

cycle, the design must be based on the "worst" day. It is also apparent from
the above that operation of such a discharge structure would modify the ocean
circulation near the intake and discharge structures. Techniques developed
for the mathematical analysis of buoyant jets and plumes as presented in
Chapter 9, assume that the jets do not modify conditions in the ambient. This
is one important reason why, in determining the design of the discharge structures at San Onofre, a physical model study was undertaken. The hydraulic
modeling technique discussed in general terms in Section 10.5.2 will now be
illustrated using the San Onofre study as the example.

For a thermal discharge from a large multiport diffuser of length L, such
as for the San Onofre Units 2 and 3, it is convenient to distinguish three parts
of the flow field: (1) near field, (2) intermediate field, and (3) far field.

In the near field, the dominant dynamic features are the jet discharge and
entrainment of ambient fluid producing the initial dilution. The extent of this
zone is only a few multiples of the depth. The individual jets are characterized
by the following (see Fig. 10.18):

q port discharge,
d jet diameter $= \sqrt{C_c}D$,
D port diameter,
C_c contraction coefficient,
u_j $q/(\pi d^2/4)$,
ϕ jet inclination to horizontal,
S_0 initial dilution at the surface (centerline of jet),
g' $g\Delta\rho/\rho$,
$\Delta\rho/\rho$ relative density difference (ambient less discharge),
b port spacing,
u_c current velocity
θ angle of current direction to diffuser alignment,
α angle of current direction to jet direction,
a height above bottom to center of nozzle,
y_0 depth over center of nozzle.

The jet dilution at the surface is

$$S_0 = f(y_0/d, F, b/y_0, a/d, \phi, \theta, \alpha, u_j/u_c),$$

where

$$F = u_j/\sqrt{g'd}.$$

It is presumed that the port spacing is close enough so that there is interference
between adjacent jets, as desired for approximating a line source. In this
situation the Reynolds number is not modeled, but it should be kept high
enough to ensure turbulent flow from the model nozzles.

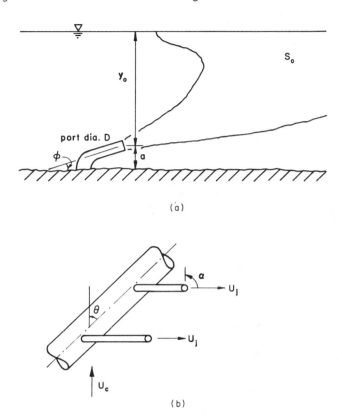

Figure 10.18 (a) Exit port orientation and nomenclature. (b) Diffuser orientation and nomenclature. [After Koh *et al.* (1974).]

To model the flow in this zone it is essential to have undistorted geometry and to follow the Froude law ($F_r = 1$). In the notation of ratios (the subscripts r, p, and m, refer to ratio, prototype, and model, respectively),

$$y_r = y_p/y_m = d_p/d_m.$$

This length ratio applies to all lengths, and angles are preserved. The Froude law is

$$F_r = 1 = u_r/(g'_r y_r)^{1/2}$$

or

$$u_r = \sqrt{g'_r y_r}.$$

The velocity ratio applies to both the jet velocity and ambient current velocity.

Note that the Froude number is based on g', rather than g; the Froude number thus defined is called the densimetric Froude number. In the San

Onofre model, free surface effects are deemed unimportant so that there is no need to model for the ordinary Froude number, based on g. Therefore the density ratio $(\Delta\rho/\rho)_r$ or g'_r need not be unity.

The Reynolds number R of any of the jets is

$$R = u_j d/\nu,$$

where ν is the kinematic viscosity of discharge fluid. The ratio of Reynolds numbers in a Froude model is

$$R_r = u_r d_r/\nu_r$$

$$R_r = g'_r{}^{1/2} y_r^{3/2}/\nu_r.$$

Typical prototype jet Reynolds numbers for the San Onofre discharge will be

$$R_p = u_j d/\nu,$$

$$= \frac{(4)(0.5)}{9.3 \times 10^{-7}},$$

$$= 2.2 \times 10^6.$$

$$\nu \doteq 9.3 \times 50^{-7} \text{ m}^2/\text{sec (seawater at 25°C)},$$

$$u_j \doteq 4 \text{ m/sec†}, \qquad d \doteq 0.5 \text{ m†}$$

To obtain fully turbulent jets, the smallest tolerable model Reynolds number R_m is about 10^3. Therefore

$$R_r < 2.2 \times 10^6/10^3 = 2200$$

and

$$g'_r{}^{1/2} y_r^{3/2}/\nu_r < 2200.$$

In the laboratory the discharge temperatures used were about 35°C, giving $\nu_m = 7.25 \times 10^{-7}$ m^2/sec, $\nu_r = 1.28$, and

$$g'_r = \frac{(\Delta\rho/\rho)_p}{(\Delta\rho/\rho)_m} = \frac{0.0028}{0.0041} = 0.68.$$

The restriction on the vertical scale ratio then becomes

$$y_r < \left\{ \frac{(2200)(1.28)}{(0.68)^{1/2}} \right\}^{2/3} = 226.$$

Thus if $y_r = y_p/y_m$ is larger than 226, the jet Reynolds number will be too small for adequately turbulent flow. For the tests performed, the investigators chose a scale of $y_r = 200$.

---

† Exact values are not given because they vary slightly from one end of the diffuser to the other.

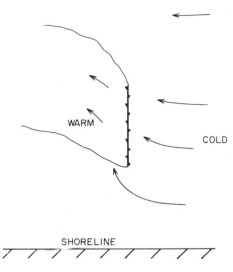

Figure 10.19 Schematic of typical flow pattern in intermediate field for diffuser perpendicular to shore with jets aimed offshore. [After Koh *et al.* (1974).]

Beyond the initial mixing zone, in the intermediate field, the flow becomes essentially horizontal in two layers—an upper warm water layer (the diluted discharge) overriding the ambient seawater. Currents are induced in the ambient seawater by the entrainment of the jets. The overall patterns of these induced currents, toward and away from the diffuser, were studied in the model basin. The distances of concern are of the order of several times the total diffuser length L, as depicted in Fig. 10.19.

The induced flow pattern at this scale is driven by the momentum and buoyancy in the diffuser discharge. Resistance to flow at this scale arises from bottom friction and interfacial friction between warm and cool layers.

To reproduce the correct buoyancy effects the Froude Law must again be followed:

$$F_r = u_r / \sqrt{g'_r y_r} = 1, \qquad \text{or,} \qquad u_r = \sqrt{g'_r y_r},$$

where y_r is the vertical length ratio, g'_r the effective gravity ratio $(g \Delta \rho / \rho)_r$. This scaling also ensures the equivalence of all Richardson numbers; a typical Richardson number is defined as

$$\text{Ri} = -\frac{g}{\rho}\frac{d\rho}{dy} \bigg/ \left(\frac{du}{dy}\right)^?.$$

The ratio of Richardson numbers is

$$\text{Ri}_r = \frac{g'_r y_r}{u_r^2} = F_r^{-2} = 1.$$

The equivalence of Richardson numbers ensures that internal stratified flow phenomena such as generation and breaking of interfacial waves are correctly modeled dynamically.

The frictional effects should also be properly modeled. The bottom friction of the current depends on the friction factor f which is

$$f = f(R_c, 4y_1/k),$$

where y_1 is the total local depth, k the bottom roughness, and R_c is the Reynolds number for the current, given as

$$R_c = 4u_c y_1/v.$$

Typical values for the prototype at San Onofre are

$$u_c = 5 \text{ cm/sec},$$

$$y_1 = 12 \text{ m},$$

$$v = 12 \times 10^{-7} \text{ m}^2/\text{sec (seawater at 15°C)},$$

$$R_c = \frac{4 \times 0.05 \times 12}{12 \times 10^{-7}} = 2 \times 10^6.$$

The surface roughness is quite low for the sandy bottom off San Onofre and might be taken as $k = 15$ mm. Then $4y_1/k = 4(12)/0.015 = 3200$, and by the Moody friction factor diagram we find for $R_c = 2 \times 10^6$ and $4y_1/k = 3200$,

$$f_p = 0.015.$$

At the chosen model scales of $y_r = 200$ and $g'_r = 0.68$, $u_r = \sqrt{200 \times 0.68} = 11.7$, giving a model current $u_{cm} = 5/11.7 = 0.427$ cm/sec. The model depth is 0.06 m and the viscosity (fresh water at 19.4°C) is $v_m = 10.2 \times 10^{-7}$ m²/sec. Therefore in the model, the Reynolds number for the ambient current is

$$R_{cm} = \frac{4u_{cm}u_m}{v_m} = \frac{4(0.00427)(0.06)}{10.2 \times 10^{-7}} = 1 \times 10^3.$$

This is still in the laminar range; the model friction factor is estimated to be $f_m = 0.06$. (Note that predictions of friction factors are somewhat uncertain at Reynolds numbers near critical.) The model friction factor would still be 0.06 even if the sand in the bed of the basin were finer, because of the low Reynolds number.

The ratio of friction factors is

$$f_r = f_p/f_m = 0.015/0.06 = 0.25.$$

Thus to counteract the resulting excessive friction which would arise in an undistorted model, the horizontal dimensions (L) should be foreshortened by distortion; namely

$$f_r = y_r/L_r = 0.25,$$

or $L_r/y_r =$ distortion factor (DF) $= 4$. Intuitively, this result may be understood by remembering that

$$f \propto \text{friction slope} \propto \frac{\text{vertical distance}}{\text{horizontal distance}}.$$

For interfacial friction between warm and cold layers, a similar argument applies; the model interfacial friction is proportionately too large, and is counteracted by reducing horizontal dimensions. Interfacial friction factors depend on both Richardson and Reynolds numbers in a complex way. A distortion factor of four may be taken as a first estimate, pending further research on the matter. Large distortion factors of the order of 10 should be avoided in models of this type, but on the other hand, it is believed that model distortion is necessary for models of this size.

If the model is to be distorted, then the length of the diffuser must be scaled by the horizontal length, not the vertical. For n ports at spacing b,

$$L = nb,$$

$$L_r = n_r b_r.$$

The near-field scaling necessitates taking $b_r = y_r$ to preserve the correct ratio of b/y_0; otherwise the initial mixing would be incorrectly modeled.

The number of ports must thus be scaled as

$$n_r = L_r/y_r = \text{distortion factor}.$$

In other words if the distortion factor is four, the number of ports is reduced also by a factor of four, as the length L is reduced by factor four (from what it would be undistorted). The diffuser is *locally undistorted* in terms of ports *per unit length*, and momentum, volume, and buoyancy fluxes *per unit length*. The ocean current was similarly reproduced: momentum and volume fluxes per unit width are scaled by Froude Law, but with overall flow field widths and lengths reduced relatively by the distortion factor.

This approach is justified *only* when the ratio of model diffuser length L_m to model depth y_m is still large ($\gtrsim 10$) and the number of ports in the model is still large ($n_m \gtrsim 10$).

There is still one conflict of scaling, namely the length of the initial mixing zone in front of the diffuser. This distance should really be scaled in an *undistorted* way, a certain number of flow depths, or by the vertical length ratio y_r. Thus if the mixing zone extends for 1 m in the model when $y_r = 200$ and $L_r = 800$, it should be interpreted to be 200 m rather than 800 m in the prototype. The initial mixing zone is thus somewhat *too large* in a distorted model but this is not believed to be a serious problem. To be conservative, the horizontal distance may be scaled according to L_r,

The time-scale ratio in a distorted model to produce the correct horizontal displacements in unsteady flow must be

$$t_r = L_r/u_r = L_r/\sqrt{g'_r y_r}.$$

For example, given $L_r = 800$, $g'_r = 0.68$, $y_r = 200$, then $t_r = 68.6$. In other words, one day in the prototype becomes 21 minutes in the model.

The far field, (the zone of drift flow far away from the diffuser) is not represented by the hydraulic model used for the San Onofre study. Most of the heat loss is in the far field. For a unidirectional flow the rate of temperature decay is

$$\Delta T/\Delta T_s = \exp[-(K/\rho c_p h)t],$$

where h is the depth of heated layer, $t = x/u$ the time of travel, ΔT the excess temperature above the equilibrium temperature, and ΔT_s is the initial value of ΔT at the end of the initial mixing zone. For San Onofre, $K/\rho c_p \approx 0.5$ m/day. If $h \sim 9.5$ m, then the exponent in the equation becomes $-0.053t$, for t in days. The half-life is then found from $e^{-0.053t} = 0.5$ or $t = 13$ days. For one day the heat loss is only of the order of 5 %.

If heat loss were to be scaled in the model, then

$$\left(\frac{K}{\rho c_p h}t\right)_r = \frac{K_r L_r}{(\rho c_p)_r y_r u_r} = 1,$$

or taking $(\rho c_p)_r = 1$,

$$L_r/y_r = \mathrm{DF} = \sqrt{g'_r y_r}/K_r.$$

The order of magnitude of the distortion factor would be found as follows:

$$y_r = 200,$$

$$g'_r = 0.68$$

$$K_r = 1.33$$

$$(K_{\mathrm{lab}} \approx \tfrac{3}{4}K_{\mathrm{field}}),$$

$$\mathrm{DF} = \frac{(0.68 \times 200)^{1/2}}{1.33} = 8.8.$$

In other words, the model loses heat so fast that a distortion of almost nine is required to counteract it; or put another way, if the model is distorted only by a factor of four, then the heat loss in the model will be about 2.2 times too

MODEL COORDINATES (m)

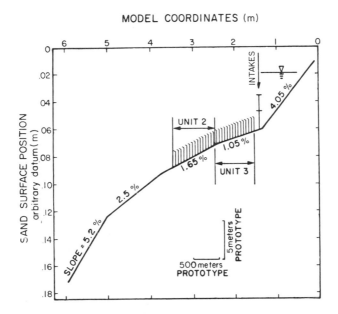

Figure 10.20 Schematic profile of offshore circulating water system, San Onofre Nuclear Generating Station Units 2 and 3, California. [After Koh *et al.* (1974).]

fast. In one simulated prototype day the model would lose 11% of the heat instead of 5%. Some adjustment of the results is necessary for tests which reproduce time periods of the order of prototype days.

In summary, the hydraulic modeling is concerned with the *near* and *intermediate* fields which are dominated by hydrodynamic effects, rather than the *far* field which is dominated by heat loss and long-time advection.

Using the chosen scales, many laboratory tests were conducted for the San Onofre Nuclear Generating Station discharges, including various configurations of the structures, lengths and distances offshore. The tests also encompassed a variety of different ocean currents. The ultimate design was based on these investigations.

The final chosen design for the San Onofre discharge system is shown schematically in Fig. 10.20. The diffusers are oriented perpendicular to shore to ensure the intercepting of the ocean current which is primarily alongshore. There are 63 discharge nozzles along each of the diffusers space 12.2 m apart. The nozzles are oriented so that they discharge at an angle of 20° up from the horizontal direction to avoid the jet impinging the bottom. They are also oriented at ±25° with respect to the centerline of the pipe in an offshore direction. In this manner, during periods of low ocean currents, an offshore drift flow is induced by the discharge jets to avoid reentrainment and recirculation.

TABLE 10.8
Summary of San Onofre Unit 2 and 3 Diffusers

Length	768 m per unit
Number of nozzles	63 per unit
Port spacing	12.2 m
Jet diameter at vena contracta	0.52 m (approximate)
Jet velocity at vena contracta	4 m/sec (approximate)
Nozzle angle up from horizontal	20°
Nozzle angle with axis of diffuser (in horizontal plane)	±25° (alternating)
Elevation of center of nozzle above the sea floor	1.74 m
Orientation	perpendicular to shore
Distance from shore for most inshore nozzle	1065 m for Unit 3 (approximate)
	1830 m for Unit 2 (approximate)

Table 10.8 summarizes the characteristics of the final adopted design and Figs. 10.21–10.23 show the surface isotherms measured in the laboratory under several ocean current conditions. The effect of the offshore momentum can be observed in these figures and can be seen to be of importance in affecting the plume under low current conditions. There are as yet no prototype verifications possible for this outfall since the plant is not yet in operation.

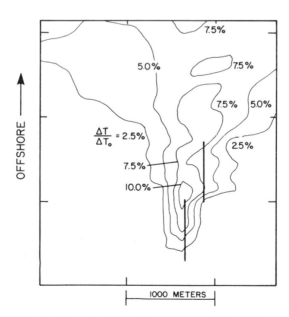

Figure 10.21 Surface isotherms based on laboratory model simulation of San Onofre Units 2 and 3 for zero ambient (longshore) current ($u = 0$). [After Koh *et al.* (1974).]

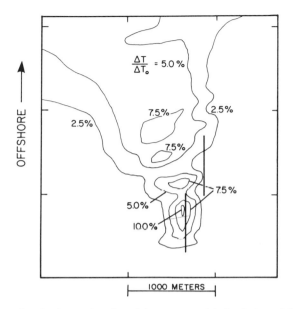

Figure 10.22 Surface isotherms based on laboratory model simulation of San Onofre Units 2 and 3 for steady ambient (longshore) current ($u = 0.1$ knots (5 cm/sec)). [After Koh *et al.* (1974).]

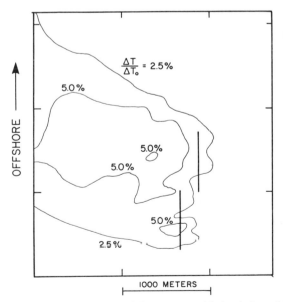

Figure 10.23 Surface isotherms based on laboratory model simulation of San Onofre Units 2 and 3 for steady ambient (longshore) current ($u = 0.25$ knots (12.5 cm/sec)). [After Koh *et al.* (1974).]

Appendix A

An Estimator for the Density of Seawater

Engineering practice involving density stratified flow frequently requires accurate knowledge of the density of seawater. Density is a function of temperature, salinity, and pressure and is usually expressed in "σ-units" defined by

$$\sigma_{S,T,p} = (\rho_{S,T,p} - 1)1000 \qquad (1)$$

where $\rho_{S,T,p}$ is the density in gm/cc. At atmospheric pressure, the notation σ_t is generally used for $\sigma_{S,T,p}$.

Empirical equations have been developed based on work by Forch *et al.* (1902) to calculate σ_t as function of temperature T and salinity S and form the basis of hydrographic tables, the first such being by Knudsen (1901). A recent frequently used table is the one from U.S. Navy H. O. Publication 615 (1952). Figure A.1 can be used to obtain an approximate density over the whole range of normally encountered temperatures and salinities, and Table A.1 gives some densities of fresh water. For more accurate values the following procedure can be used.

DEVELOPMENT OF THE ESTIMATOR

Since most seawater falls in a narrow range of salinity, the detailed tables in H. O. Publication 615 are not necessary in engineering practice for almost all cases. The dependence of σ_t on S is nearly linear while the dependence on T

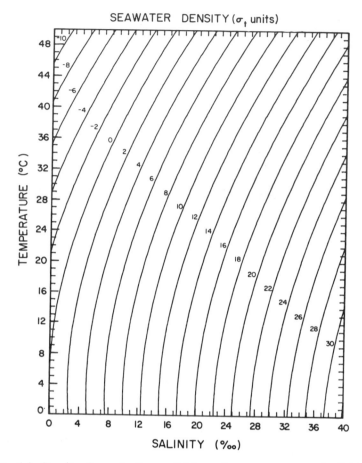

Figure A.1   Density of seawater (in σ_t units) for various temperatures and salinities.

is much more nonlinear. For a small range of S, linear interpolation suffices to a high degree of accuracy. Thus, write

$$\sigma_t(T, S) \cong \sigma_t(T, S_0) + \frac{\partial \sigma_t}{\partial S}(T, S_0)[S - S_0],$$

where S_0 is some reference salinity, say in the middle of the range of interest. We choose S_0 to be 34‰ since the salinity of ocean water is typically in the range 33‰–35‰. Table A.2 is constructed giving values of $\sigma_t(T, 34)$ and $\partial \sigma_t/\partial S \, (T, 34)$ for T in the range 0–44.9°C in increments of 0.1°C. While the values of $\sigma_t(T, 34)$ and $\partial \sigma_t/\partial S \, (T, 34)$ in Table A.2 are given to four and five figures after the decimal, respectively, the result should be rounded to three since it is only accurate to ± 0.001 σ-units in the range 33‰ $\leq S \leq$ 35‰.

TABLE A.1
Density of Fresh Water (in σ_t Units)

Temp. (°C)		Temp. (°C)		Temp. (°C)		Temp. (°C)		Temp. (°C)	
0.0	−0.132	10.0	−0.274	20.0	−1.772	30.0	−4.329	40.0	−7.762
0.2	−0.118	10.2	−0.292	20.2	−1.813	30.3	−4.390	40.2	−7.839
0.4	−0.106	10.4	−0.310	20.4	−1.855	30.4	−4.451	40.4	−7.916
0.6	−0.094	10.6	−0.329	20.6	−1.897	30.6	−4.512	40.6	−7.993
0.8	−0.083	10.8	−0.349	20.8	−1.940	30.8	−4.574	40.8	−8.071
1.0	−0.073	11.0	−0.369	21.0	−1.983	31.0	−4.636	41.0	−8.148
1.2	−0.063	11.2	−0.389	21.2	−2.027	31.2	−4.698	41.2	−8.227
1.4	−0.054	11.4	−0.410	21.4	−2.071	31.4	−4.761	41.4	−8.305
1.6	−0.046	11.6	−0.432	21.6	−2.115	31.6	−4.824	41.6	−8.384
1.8	−0.039	11.8	−0.454	21.8	−2.160	31.8	−4.887	41.8	−8.463
2.0	−0.032	12.0	−0.476	22.0	−2.205	32.0	−4.951	42.0	−8.542
2.2	−0.026	12.2	−0.500	22.2	−2.251	32.2	−5.015	42.2	−8.622
2.4	−0.020	12.4	−0.523	22.4	−2.297	32.4	−5.079	42.4	−8.702
2.6	−0.015	12.6	−0.547	22.6	−2.344	32.6	−5.144	42.6	−8.782
2.8	−0.011	12.8	−0.572	22.8	−2.390	32.8	−5.209	42.8	−8.862
3.0	−0.008	13.0	−0.597	23.0	−2.438	33.0	−5.274	43.0	−8.943
3.2	−0.005	13.2	−0.623	23.2	−2.485	33.2	−5.339	43.2	9.024
3.4	−0.003	13.4	−0.649	23.4	−2.533	33.4	−5.405	43.4	−9.106
3.6	−0.001	13.6	−0.675	23.6	−2.582	33.6	−5.472	43.6	−9.187
3.8	−0.000	13.8	−0.702	23.8	−2.630	33.8	−5.538	43.8	−9.269
4.0	−0.000	14.0	−0.730	24.0	−2.680	34.0	−5.605	44.0	−9.351
4.2	−0.000	14.2	−0.758	24.2	−2.729	34.2	−5.673	44.2	−9.434
4.4	−0.001	14.4	−0.787	24.4	−2.779	34.4	−5.740	44.4	−9.517
4.6	−0.003	14.6	−0.816	24.6	−2.830	34.6	−5.808	44.6	−9.600
4.8	−0.005	14.8	−0.845	24.8	−2.880	34.8	−5.877	44.8	−9.683
5.0	−0.008	15.0	−0.875	25.0	−2.931	35.0	−5.945	45.0	−9.767
5.2	−0.012	15.2	−0.905	25.2	−2.983	35.2	−6.014	45.2	−9.851
5.4	−0.016	15.4	−0.936	25.4	−3.035	35.4	−6.083	45.4	−9.935
5.6	−0.021	15.6	−0.968	25.6	−3.087	35.6	−6.153	45.6	−10.019
5.8	−0.026	15.8	−1.000	25.8	−3.140	35.8	−6.223	45.8	−10.104
6.0	−0.032	16.0	−1.032	26.0	−3.193	36.0	−6.293	46.0	−10.189
6.2	−0.039	16.2	−1.065	26.2	−3.246	36.2	−6.363	46.2	−10.274
6.4	−0.046	16.4	−1.098	26.4	−3.300	36.4	−6.434	46.4	−10.360
6.6	−0.054	16.6	−1.131	26.6	−3.354	36.6	−6.505	46.6	−10.446
6.8	−0.062	16.8	−1.166	26.8	−3.408	36.8	−6.577	46.8	−10.532
7.0	−0.071	17.0	−1.200	27.0	−3.463	37.0	−6.649	47.0	−10.619
7.2	−0.081	17.2	−1.235	27.2	−3.518	37.2	−6.721	47.2	−10.705
7.4	−0.091	17.4	−1.271	27.4	−3.574	37.4	−6.793	47.4	−10.792
7.6	−0.102	17.6	−1.307	27.6	−3.630	37.6	−6.866	47.6	−10.879
7.8	−0.113	17.8	−1.343	27.8	−3.686	37.8	−6.939	47.8	−10.967
8.0	−0.125	18.0	−1.380	28.0	−3.743	38.0	−7.012	48.0	−11.055
8.2	−0.137	18.2	−1.417	28.2	−3.800	38.2	−7.086	48.2	−11.143
8.4	−0.150	18.4	−1.455	28.4	−3.857	38.4	−7.160	48.4	−11.231
8.6	−0.164	18.6	−1.493	28.6	−3.915	38.6	−7.234	48.6	−11.320
8.8	−0.178	18.8	−1.531	28.8	−3.973	38.8	−7.308	48.8	−11.409
9.0	−0.192	19.0	−1.570	29.0	−4.032	39.0	−7.383	49.0	−11.498
9.2	−0.208	19.2	−1.610	29.2	−4.091	39.2	−7.458	49.2	−11.587
9.4	−0.223	19.4	−1.649	29.4	−4.150	39.4	−7.534	49.4	−11.677
9.6	−0.240	19.6	−1.690	29.6	−4.209	39.6	−7.610	49.6	−11.767
9.8	−0.256	19.8	−1.730	29.8	−4.269	39.8	−7.686	49.8	−11.857
10.0	−0.274	20.0	−1.772	30.0	−4.329	40.0	−7.762	50.0	−11.948

TABLE A.2

σ_t and $\partial\sigma_t/\partial S$ at $S = 34\%_{00}$ over a temperature range of 0–44.9°C[a]

ΔT	$T = \Delta T$	$T = 5 + \Delta T$	$T = 10 + \Delta T$	$T = 15 + \Delta T$	$T = 20 + \Delta T$	$T = 25 + \Delta T$	$T = 30 + \Delta T$	$T = 35 + \Delta T$	$T = 40 + \Delta T$
0.0	27.3206.80554	26.9050.79125	26.1918.77910	25.2210.76884	24.0201.76044	22.6082.75378	20.9974.74873	19.1952.74524	17.2052.74316
0.1	27.3156.80524	26.8935.79103	26.1748.77887	25.1991.76865	23.9939.76028	22.5779.75366	20.9632.74866	19.1572.74518	17.1635.74313
0.2	27.3104.80493	26.8819.79076	26.1577.77864	25.1772.76848	23.9676.76015	22.5475.75354	20.9290.74858	19.1192.74513	17.1217.74310
0.3	27.3050.80461	26.8701.79050	26.1405.77841	25.1552.76830	23.9412.75999	22.5170.75343	20.8946.74849	19.0811.74509	17.0798.74309
0.4	27.2995.80432	26.8583.79024	26.1232.77820	25.1331.76811	23.9147.75984	22.4865.75331	20.8602.74841	19.0429.74503	17.0379.74304
0.5	27.2939.80402	26.8463.78999	26.1058.77798	25.1109.76793	23.8882.75969	22.4559.75320	20.8257.74834	19.0046.74496	16.9959.74303
0.6	27.2881.80371	26.8342.78972	26.0883.77777	25.0886.76775	23.8615.75955	22.4252.75308	20.7911.74825	18.9663.74492	16.9538.74300
0.7	27.2823.80342	26.8220.78946	26.0707.77756	25.0662.76756	23.8348.75941	22.3944.75298	20.7565.74817	18.9279.74486	16.9116.74297
0.8	27.2762.80312	26.8097.78920	26.0530.77733	25.0437.76738	23.8080.75926	22.3636.75285	20.7218.74808	18.8894.74481	16.8694.74295
0.9	27.2701.80281	26.7972.78896	26.0352.77711	25.0211.76720	23.7811.75911	22.3326.75276	20.6870.74800	18.8508.74477	16.8271.74292
1.0	27.2637.80252	26.7847.78868	26.0172.77639	24.9985.76701	23.7542.75897	22.3016.75264	20.6521.74792	18.8122.74471	16.7847.74290
1.1	27.2573.80223	26.7720.78844	25.9992.77657	24.9758.76685	23.7271.75882	22.2705.75253	20.6172.74785	18.7735.74466	16.7422.74287
1.2	27.2507.80193	26.7592.78818	25.9811.77646	24.9529.76668	23.7000.75868	22.2394.75243	20.5821.74777	18.7346.74461	16.6997.74286
1.3	27.2440.80164	26.7463.78792	25.9629.77625	24.9300.76648	23.6727.75853	22.2081.75230	20.5470.74768	18.6958.74457	16.6571.74284
1.4	27.2372.80135	26.7333.78767	25.9446.77603	24.9070.76631	23.6454.75839	22.1768.75220	20.5119.74760	18.6568.74451	16.6144.74281
1.5	27.2302.80104	26.7202.78741	25.9262.77582	24.8839.76613	23.6180.75827	22.1454.75209	20.4766.74754	18.6178.74448	16.5717.74280
1.6	27.2230.80075	26.7070.78717	25.9077.77562	24.8607.76596	23.5906.75812	22.1139.75198	20.4413.74745	18.5788.74442	16.5289.74278
1.7	27.2158.80046	26.6936.78691	25.8891.77539	24.8374.76578	23.5630.75798	22.0824.75188	20.4059.74738	18.5396.74437	16.4859.74275
1.8	27.2084.80016	26.6801.78667	25.8703.77518	24.8140.76563	23.5354.75784	22.0507.75177	20.3704.74731	18.5003.74432	16.4430.74274
1.9	27.2009.79987	26.6666.78642	25.8515.77498	24.7906.76543	23.5077.75771	22.0190.75168	20.3348.74724	18.4610.74428	16.3999.74272
2.0	27.1933.79959	26.6529.78618	25.8326.77477	24.7670.76527	23.4799.75757	21.9872.75156	20.2992.74716	18.4216.74425	16.3568.74269
2.1	27.1855.79930	26.6391.78592	25.8136.77457	24.7434.76509	23.4520.75743	21.9553.75148	20.2635.74709	18.3822.74419	16.3136.74267
2.2	27.1776.79901	26.6252.78569	25.7945.77435	24.7197.76494	23.4240.75729	21.9234.75136	20.2277.74702	18.3426.74416	16.2703.74266
2.3	27.1695.79872	26.6111.78543	25.7753.77414	24.6958.76476	23.3960.75716	21.8914.75127	20.1919.74695	18.3030.74411	16.2269.74265

2.4	27.1614.79845	26.5970.78519	25.7560.77394	24.6719.76459	23.3678.75703	21.8593.75116	20.1559.74689	18.2633.74408	16.1835.74263
2.5	27.1531.79816	26.5828.78494	25.7366.77374	24.6479.76442	23.3396.75690	21.8271.75105	20.1199.74680	18.2236.74402	16.1400.74262
2.6	27.1447.79787	26.5684.78470	25.7172.77353	24.6239.76425	23.3113.75676	21.7948.75096	20.0838.74673	18.1837.74397	16.0964.74259
2.7	27.1361.79759	26.5540.78445	25.6976.77333	24.5997.76408	23.2830.75664	21.7625.75085	20.0477.74666	18.1438.74394	16.0527.74259
2.8	27.1274.79730	26.5394.78423	25.6779.77312	24.5755.76393	23.2545.75650	21.7301.75075	20.0114.74660	18.1038.74390	16.0090.74258
2.9	27.1186.79701	26.5247.78398	25.6581.77292	24.5511.76375	23.2260.75636	21.6976.75066	19.9751.74652	18.0638.74387	15.9652.74256
3.0	27.1097.79674	26.5099.78374	25.6382.77271	24.5267.76358	23.1974.75624	21.6650.75056	19.9388.74646	18.0236.74384	15.9213.74254
3.1	27.1006.79645	26.4950.78349	25.6183.77251	24.5022.76343	23.1687.75612	21.6323.75047	19.9023.74640	17.9834.74379	15.8774.74253
3.2	27.0914.79617	26.4800.78326	25.5982.77232	24.4776.76326	23.1399.75598	21.5996.75038	19.8658.74632	17.9431.74376	15.8333.74253
3.3	27.0821.79590	26.4649.78302	25.5780.77211	24.4529.76311	23.1110.75584	21.5668.75027	19.8291.74626	17.9028.74371	15.7892.74251
3.4	27.0727.79562	26.4497.78278	25.5578.77191	24.4281.76294	23.0821.75572	21.5339.75018	19.7924.74620	17.8623.74368	15.7451.74250
3.5	27.0631.79533	26.4344.78253	25.5374.77171	24.4033.76279	23.0531.75560	21.5010.75009	19.7557.74612	17.8218.74364	15.7008.74250
3.6	27.0534.79507	26.4189.78230	25.5170.77153	24.3783.76260	23.0240.75546	21.4679.74998	19.7188.74606	17.7812.74361	15.6565.74248
3.7	27.0436.79480	26.4034.78207	25.4965.77133	24.3533.76245	22.9948.75534	21.4348.74989	19.6819.74600	17.7406.74358	15.6121.74248
3.8	27.0337.79453	25.3878.78183	25.4758.77113	24.3282.76230	22.9655.75522	21.4016.74982	19.6449.74596	17.6998.74353	15.5676.74247
3.9	27.0236.79424	25.3720.78160	25.4551.77094	24.3030.76213	22.9362.75510	21.3684.74971	19.6079.74588	17.6590.74350	15.5230.74246
4.0	27.0134.79396	26.3562.78139	25.4343.77075	24.2777.76198	22.9068.75497	21.3350.74963	19.5707.74583.	17.6181.74347	15.4784.74245
4.1	27.0031.79370	26.3402.78114	25.4134.77055	24.2523.76181	22.8773.75485	21.3016.74954	19.5335.74576	17.5772.74344	15.4337.74244
4.2	26.9927.79343	26.3241.78090	25.3924.77036	24.2269.76167	22.8477.75473	21.2681.74944	19.4962.74570	17.5361.74339	15.3889.74244
4.3	26.9822.79315	26.3080.78069	25.3713.77017	24.2013.76152	22.8180.75461	21.2345.74934	19.4589.74564	17.4950.74336	15.3440.74243
4.4	26.9715.79288	26.2917.78046	25.3501.76997	24.1757.76135	22.7883.75449	21.2009.74925	19.4214.74559	17.4538.74333	15.2991.74242
4.5	26.9607.79263	26.2753.78023	25.3288.76978	24.1500.76120	22.7585.75436	21.1672.74918	19.3839.74553	17.4126.74332	15.2541.74242
4.6	26.9498.79236	26.2588.78000	25.3074.76959	24.1242.76105	22.7286.75426	21.1334.74908	19.3463.74547	17.3712.74327	15.2090.74242
4.7	26.9388.79208	26.2422.77977	25.2859.76941	24.0983.76089	22.6986.75412	21.0995.74901	19.3087.74541	17.3298.74324	15.1638.74241
4.8	26.9276.79182	26.2255.77954	25.2644.76921	24.0723.76074	22.6685.75401	21.0655.74890	19.2709.74535	17.2884.74323	15.1186.74241
4.9	26.9164.79155	26.2087.77933	25.2427.76903	24.0463.76059	22.6384.75389	21.0315.74883	19.2231.74530	17.2468.74318	15.0733.74240

[a] Example: at $T = 21.3°C$, $\sigma_t = 23.6727$, $\partial\sigma_t/\partial S = 0.75853 \ (‰)^{-1} \Rightarrow \sigma_t(S = 34.5‰) = 23.6727 + 0.75853 \times 0.5 = 24.052$.

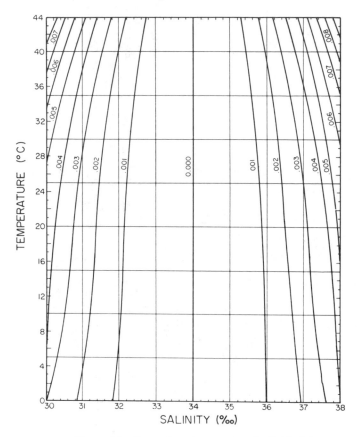

Figure A.2   Corrections to σ_t as given by Table A.2.

The range of applicability of Table A.2 can be extended. Let $\sigma_t(T, S)$ be the correct value and $\sigma_t'(T, S)$ be the value calculated by use of Table A.2. Define $\Delta\sigma_t$ by

$$\sigma_t(T, S) = \sigma_t'(T, S) + \Delta\sigma_t(T, S).$$

In other words, $\Delta\sigma_t$ is the correction which must be added to the value obtained from Table A.2 to get the "correct" σ_t. Since the correction is relatively small, it can be conveniently and accurately obtained from a graph as shown in Fig. A.2 which covers the range $30\%_{oo} \leq S \leq 38\%_{oo}$.

The procedure to obtain $\sigma_t(T, S)$ is thus as follows:

(1)   Obtain $\sigma_t(T, 34)$ and $\partial\sigma_t/\partial S\ (T, 34)$ from Table A.2.
(2)   Obtain $\sigma_t'(T, S) = \sigma_t(T, 34) + \partial\sigma_t/\partial S\ (T, 34) \times (S - 34)$
(3)   Obtain $\Delta\sigma_t(T, S)$ from Fig. A.2.
(4)   Obtain $\sigma_t(T, S) = \sigma_t'(T, S) + \Delta\sigma_t(T, S)$
(5)   Round result to 0.001 σ unit.

The accuracy of this procedure is ± 0.001 σ units. For engineering practice an accuracy of ± 0.01 σ units is often sufficient. In that event, steps 3 and 4 can often be ignored.

Example A.1.   Find the density of seawater at a salinity of 30.8 ‰ and a temperature of 24°C.

Solution.   From Table A.2 we find that at $S = 34‰, T = 24°C, \sigma_t = 22.9068$, and $\partial \sigma_t / \partial S = 0.75497$. Therefore

$$\sigma_t' = 22.9068 + 0.75497 (30.8 - 34)$$
$$= 20.4909.$$

From Fig. A.2 $\Delta \sigma_t = 0.003$. Therefore the correct value of the density is

$$\sigma_t = 20.491 + 0.003 = 20.494$$

or

$$\rho = 1.020494 \text{ gm/cc} = 1020.494 \text{ kg/m}^3.$$

Appendix B

Fluid Properties

TABLE B.1

Some properties of Pure Water

Temperature, T (°C)	Density, ρ (kg/m³)	Thermal[a] expansivity, $10^6\alpha$ (°C⁻¹)	Vapor pressure, $10^{-3}v_p$ (N/m²)	Latent heat of vaporization, $10^6 L$ (J/kg)	Specific heat, c_p (J/kg °C)	Thermal conductivity, k (W/m °C)	Thermal diffusivity, $10^8\kappa$ (m²/sec)	Viscosity, $10^3\mu$ (kg/msec)	Kinematic viscosity, $10^6\nu$ (m²/sec)
0	999.868	−68	0.61	2.501	4217.4	0.564	13.4	1.787	1.787
5	999.992	16	0.87	2.489				1.519	1.519
10	999.726	88	1.23	2.477	4191.9	0.578	13.8	1.307	1.307
15	999.125	151	1.70	2.465				1.139	1.140
20	998.228	207	2.33	2.454	4181.6	0.598	14.2	1.002	1.004
25	997.069	257	3.16	2.442				0.8904	0.893
30	995.671	303	4.23	2.430	4178.2	0.607	14.6	0.7975	0.801
35	994.055	346	5.62	2.415				0.7194	0.724
40	992.233	385	7.38	2.406	4178.3	0.628	15.2	0.6529	0.658
45	990.233	423	9.58	2.391				0.5960	0.602
50	988.052	458	12.3	2.382	4180.4			0.5468	0.553
60	983.20	523	19.9	2.357	4184.1	0.652	15.8	0.4665	0.475
70	977.77	584	31.1	2.333	4189.3			0.4042	0.413
80	971.80	641	47.3	2.308	4196.1	0.669	16.4	0.3547	0.365
90	965.31	696	70.1	2.283	4204.8			0.3147	0.326
100	958.36	750	101.3	2.257	4215.7	0.671	16.6	0.2818	0.294

[a] $\alpha = -d(\ln \rho)/dT$.

451

TABLE B.2
Some Properties of Seawater at 34‰ Salinity[a]

Temperature, T (°C)	Density, ρ (kg/m³)	Specific heat, c_p (J/kg °C)	Kinematic viscosity, $10^6 \nu$ (m²/sec)
0	1027.32	3989	1.8
5	1026.91	3992	1.6
10	1026.19	3995	1.4
15	1025.22	3997	1.2
20	1024.02	4000	1.1
25	1022.61	4002	0.94

[a] See also Table A.2.

TABLE B.3
Some Properties of Pure Ice[a]

Latent heat of melting	L	334×10^3	J/kg
Density	ρ	917	kgm/m³
Specific heat	c_p	2075	J/kg °C
Thermal conductivity	k	2.3	W/m °C
Thermal diffusivity	κ	1.2×10^{-6}	m²/sec

[a] $\sim -5°C$.

TABLE B.4
Some Properties of Dry Air[a]

Density	ρ	1.225	kg/m³
Specific heat	c_p	1012	J/kg °C
Thermal conductivity	k	0.0253	W/m °C
Thermal diffusivity	κ	0.204×10^{-4}	m²/sec
Viscosity	μ	1.78×10^{-6}	kg/msec
Kinematic viscosity	ν	14.5×10^{-6}	m²/sec

[a] 15°C and 1 atm (101.3×10^3 N/m²).

TABLE B.5
Some Properties of Aqueous Solutions of Sodium Chloride

Kilograms of NaCl per cubic meter of solution (kg/m³ = gm/liter)	0	10	20	30	40	50	60	70
Percentage salt by wt. at 20°C	0	0.995	1.976	2.943	3.898	4.841	5.772	6.690
Density of solution at 15°C (kg/m³)	999.13	1006.30	1013.39	1020.41	1027.35	1034.25	1041.05	1047.83
Density of solution at 20°C (kg/m³)	998.23	1005.30	1012.29	1019.22	1026.07	1032.88	1039.60	1046.32
Specific heat of solution at constant pressure (J/kg °C) at 20°C	4182	4127	4075	4024	3975	3929	3884	3841

Table of Conversion Factors

To obtain	From	Multiply by
(1) Length [L]		
meters (m)	feet (ft)	0.3048
meters (m)	yards (yd)	0.9144
meters (m)	miles (miles)	1609.34
meters (m)	nautical miles	1852
(2) Area [L^2]		
square meters (m^2)	square feet (ft^2)	0.09290
square meters (m^2)	acres (acre)	4047
square meters (m^2)	square miles (sq. mile)	2.59×10^6
(3) Volume [L^3]		
cubic meters (m^3)	U.S. gallons (gal)	0.0037854
cubic meters (m^3)	Imperial gallons (Imp. gal)	0.0045460
cubic meters (m^3)	acre-feet (acre-ft)	1233.49
cubic meters (m^3)	cubic feet (ft^3)	0.028317
(4) Velocity [LT^{-1}]		
meters per second (m/sec)	feet per second (ft/sec)	0.3048
meters per second (m/sec)	miles per hour (mph)	0.44704
meters per second (m/sec)	knots (kt)	0.5148
(5) Discharge [L^3T^{-1}]		
cubic meters per second (m^3/sec)	cubic feet per second (cfs)	0.02832
cubic meters per second (m^3/sec)	gallons per minute (gpm)	63.08×10^{-6}
cubic meters per second (m^3/sec)	million gallons per day (Mgd)	0.04382
cubic meters per second (m^3/sec)	acre-ft per year (Af/yr)	39.1×10^{-6}
(6) Mass [M]		
kilograms (kg)	slugs	14.594
(7) Density [ML^{-3}]		
kilograms per cubic meter (kg/m^3)	slugs per cubic foot (sl/ft^3)	515.4
(8) Force [MLT^{-2}]		
newtons (N)	pounds force (lb. f)	4.4482
(9) Pressure [$ML^{-1}T^{-2}$]	(pascal = newton/m^2)	
pascals (P)	pounds per square inch (psi)	6895
pascals (P)	pounds per square foot (psf)	47.88
pascals (P)	atmospheres (atm)	101.3×10^3
pascals (P)	feet of water (ft)	2988.2
(10) Energy [ML^2T^{-2}]	(joule = newton meter)	
joules (J)	British thermal units (BTU)	1054.8
joules (J)	gram calories (20°C) (cal)	4.181
joules (J)	kilowatt hours (kwh)	3.6×10^6
joules (J)	foot pounds (ft lb)	1.356
(11) Power [ML^2T^{-3}]	(watt = joule per second)	
watt (W)	British thermal unit per hour (BTU/hr)	0.2931
watt (W)	horsepower (hp)	745.7
watt (W)	foot pounds per minute (ft lb/m)	0.0226

Notation for Chapters 1–9

Symbols used frequently in Chapters 1–9 are defined in the following. Symbols used in Chapter 10 and all other symbols are defined where they first appear. Numbers in parentheses indicate the equation in which the symbol first appears with the listed definition; where a symbol has more than one meaning the definitions are listed in the order in which they appear.

A cross-sectional area   (4.27)

A_s surface area of a lake   (6.35)

b a measure of the width of a jet   (9.8)

B initial specific buoyancy flux   (9.6)

c_j jet coefficient   (9.14)

c_p plume coefficient   (9.31)

C mass concentration of diffusing solute   (2.1)

C_m concentration on the axis of a jet   (9.9)

C_p specific heat of water   (6.35)

C' deviation from the cross-sectional mean concentration   (4.5)

$\bar{C}$ cross-sectional mean concentration   (4.4)

d depth of an open channel flow   (5.1)

D molecular diffusion coefficient   (2.1)

f Darcy–Weisbach friction factor   (1.2)

      functional relationship   (2.10)

      freshness   (7.47)

F internal Froude number U/NH (6.10)

        diffuser Froude number U^3/b (9.148)

F_i internal Froude number $U(\Delta\rho gH/\rho_0)^{-1/2}$ (6.12)

g acceleration of gravity

g' effective acceleration of gravity $g\Delta\rho/\rho$ (9.7)

Gr   Grashof number $N^2H^4\nu^{-2}$ (6.13)

h characteristic transverse dimension of a shear flow   (4.2)

        thickness of the epilimnion   (6.43)

h_B terminal height of rise of a simple plume   (9.58)

h_M terminal height of rise of a simple momentum jet   (9.57)

H depth of a reservoir   (6.8)

$\tilde{H}$ rate of transfer of heat from a reservoir to the atmosphere through the water surface, per unit surface area   (6.35)

I the value of a dimensionless integral   (4.47)

k first order decay coefficient   (5.31)

K longitudinal dispersion coefficient   (4.16)

l_L Lagrangian length scale   (3.35)

l_M characteristic length scale for a buoyant jet ($M^{3/4}/B^{1/2}$ for a round jet)   (9.36)

l_Q characteristic length scale for a jet ($Q/M^{1/2}$ for a round jet)   (9.10)

L distance required for complete cross-sectional mixing   (5.10)

        length of a reservoir   (6.8)

m specific momentum flux   (9.2)

M mass   (2.10)

        initial specific momentum flux of a jet   (9.5)

$\dot{M}$ input of mass per unit time   (2.41)

        mass transport through a cross section per unit time   (4.15)

n concentration   (3.41)

N buoyancy frequency $(\varepsilon'g)^{1/2}$ (6.6)

        ratio of momentum to buoyancy length scales $(h_M/h_B)^8$ (9.59)

p probability   (2.15)

P potential energy of water in a reservoir   (6.29)

        tidal prism   (7.35)

Pr   Prandtl number $\nu\kappa^{-1}$ (6.14)

q solute mass flux   (2.1)

        partial discharge   (5.13)

        turbulent velocity in the epilimnion   (6.47)

        volume flux per unit length of a diffuser   (9.148)

Q total discharge of a river   (5.29)

        initial volume flux of a jet   (9.4)

Q_d dilution discharge   (7.32)

Q_e effluent discharge   (7.30)

Q_f freshwater discharge   (7.5)

r radial coordinate   (4.23)

R Lagrangian autocorrelation function   (3.22)
        radius of a curve   (5.5)
        tidal exchange ratio   (7.25)

R_i^* Richardson number $\alpha g \Delta T h / u_*^2$ (6.55)

R_i^q Richardson number $\alpha \Delta T g h / q^2$ (6.53)

R_0 jet Richardson number ($l_Q/l_M = Q B^{1/2}/M^{5/4}$ for a round jet)   (9.40)

R_p plume Richardson number ($\mu B^{1/2}/m^{5/4}$ for a round plume)   (9.32)

s slope (5.1)

S salinity   (7.4)

S_0 ocean salinity   (7.4)

t time (1.6)

T period of oscillation   (4.49)
        water temperature   (6.35)
        top width of underflowing water   (6.95)
        tidal period   (7.1)

T_c time scale for cross-sectional mixing h^2/D (4.53)

T_L Lagrangian time scale   (3.35)

T' dimensionless time scale for cross-sectional mixing T/T_c (7.1)

u velocity in the x direction   (2.55)

u_f velocity associated with penetrative convection   (6.40)

u' deviation of velocity from the cross-sectional mean   (4.2)

$\bar{u}$ cross-sectional mean velocity   (4.2)

$\bar{u}^z$ depth averaged velocity   (5.13)

u^* shear velocity at a wall boundary or channel bottom $(\tau_0/\rho)^{1/2}$ (4.38)
        shear velocity at the surface of a reservoir $[C_D(\rho_A/\rho_w)U^2]^{1/2}$ (6.47)

U velocity of a single particle   (3.20)
        peak or characteristic velocity in a shear flow   (4.20)
        wind speed 10 m above the water surface   (6.17)
        velocity of a uniform cross flow   (9.62)

U_f freshwater discharge velocity Q_f/A (7.13)

U_t mean or rms tidal velocity   (7.10)

v velocity in the y direction   (4.60)

w_m velocity on the centerline of a jet   (9.8)

w time-averaged jet velocity   (9.8)

W width of an open channel flow   (5.8)

x Cartesian coordinate direction in the streamwise direction   (1.4)
        transverse or radial distance from the jet axis   (9.8)

x' dimensionless longitudinal distance $x\varepsilon_t/\bar{u}W^2$ (5.8)
        dimensionless longitudinal distance $U_f x/2K$ (7.40)

y Cartesian coordinate direction

y' dimensionless transverse distance y/W (5.8)

Y rate of supply of tracer mass to a jet   (9.19)

z Cartesian coordinate vertically upward   (5.14)

        Cartesian coordinate direction distance along the axis of a jet   (9.11)

z_B length scale for a plume in a crossflow $(B/U^3$ for a round plume)   (9.98)

z_M length scale for a jet in a crossflow $(M^{1/2}/U$ for a round jet)   (9.62)

z_t terminal height of rise of a negatively buoyant jet   (9.52)

$\bar{z}$ z coordinate of the jet axis   (9.74)

α dimensionless decay coefficient $4Kk/\bar{u}^2$ (5.34 and Chap. 7)

        volume coefficient of thermal expansion of water $-(1/\rho_0)(d\rho/dT)$ (6.36 and Chap. 9)

        entrainment coefficient   (9.107)

β specific buoyancy flux   (9.3)

$\delta(x)$ Dirac delta function   (2.18)

ϵ rate of energy dissipation per unit mass   (3.59)

ε turbulent mixing coefficient   (3.30a)

ε_t transverse mixing coefficient   (5.4)

$\varepsilon_v, \varepsilon_z$ vertical mixing coefficient   (5.2), (6.41)

ε' length scale established by the ambient density gradient $-(1/\rho_0)(d\rho/dz)$ (6.2)

ζ dimensionless distance along the axis of a jet   (9.42)

κ von Karman constant   (4.41)

        thermal diffusivity   (6.15)

λ length scale for a jet in a stratified crossflow $U/(g\varepsilon')^{1/2}$ (9.104)

        ratio of jet widths for tracer and velocity b_T/b_w (9.132)

μ mean of a distribution   (2.19)

        specific mass flux (volume flux)   (9.1)

$\bar{\mu}$ dimensionless volume flux   (9.41)

ν kinematic viscosity   (6.14)

ξ Cartesian coordinate in the same direction as x (2.27)

        Cartesian coordinate in a moving coordinate system defined by $\xi = x - \bar{u}t$ (4.6)

ρ density of the fluid   (4.37)

        density change induced by any motion   (6.1)

ρ_a ambient water density   (6.1)

ρ_e density above ρ_0 when motion is absent   (6.1)

ρ_0 density of water at the mean temperature of a reservoir   (6.1)

        reference density   (9.55)

$\Delta\rho$ difference in density between the surrounding fluid and the fluid in a jet (9.3)

σ^2 variance of a distribution   (2.20)

τ a timelike variable   (2.39)

τ_0 shear stress at a wall or boundary   (4.37)

ψ concentration   (3.49)

References

Abbott, M. B., and Rasmussen, C. H. (1977). On the numerical modelling of rapid expansions and contractions in models that are two-dimensional in plan. *Proc. Congr. Int. Assoc. Hydraul. Res.* *17th* **2**, 229–238.

Abraham, G. (1963). Jet Diffusion in Stagnant Ambient Fluid. Delft Hydraulics Laboratory Publ. No. 29.

Abraham, G. (1967). Jet with negative buoyancy in homogeneous fluid. *J. Hydraul. Res.* **5**, 235–248.

Abraham, G., Karelse, M., and Lases, W. B. P. M. (1975). Data requirement for one-dimensional mathematical modelling of salinity intrusion in estuaries. *Proc. Congr. Int. Assoc. Hydraul. Res.* *16th* **3**, 275–283.

Adams, A. (1965). "Natural Light Photography," Basic Photo 4. Morgan and Morgan, Inc., Hastings on Hudson, New York.

Adams, E. E., Stolzenbach, K. D., and Harleman, D. R. F. (1975). Near and Far Field Analysis of Buoyant Surface Discharges into Large Bodies of Water. Ralph M. Parsons Laboratory for Water Resources and Hydrodynamics, Massachusetts Institute of Technology Rep. No. 205.

Albertson, M. L., Dai, Y. B., Jensen, R. A., and Rouse, H. (1950). Diffusion of submerged jets. *Trans. Am. Soc. Civ. Eng.* **115**, 639–664.

Aris, R. (1956). On the dispersion of a solute in a fluid flowing through a tube. *Proc. R. Soc. London Ser. A* **235**, 67–77.

Arons, A. B., and Stommel, H. (1951) A mixing-length theory of tidal flushing. *Trans. Am. Geophys. Un.* **32**, 419.

Bardey, P. R. (1977). Height of Rise of a Momentum Jet in a Stagnant Linearly Stratified Fluid. Unpublished Term Paper, California Institute of Technology, Pasadena, California.

Batchelor, G. K. (1949). Diffusion in a field of homogeneous turbulence. I. Eulerian analysis. *Aust. J. Sci. Res.* **2**, 437–450.

Batchelor, G. K. (1952). Diffusion in a field of homogeneous turbulence. II. The relative motion of particles. *Proc. Cambridge Philos. Soc.* **48**, 345–362.

459

Bean, B. R., Emmanuel, C. B., Gilmer, R. O., and McGavin, R. E. (1975). The spatial and temporal variations of the turbulent fluxes of heat, momentum and water vapor over Lake Ontario. *J. Phys. Oceanogr.* **5**, 532–540.

Becker, H. A., Hottel, H. C., and Williams, G. C. (1967). The nozzle-fluid concentration field of the round, turbulent, free jet. *J. Fluid Mech.* **30**, 285.

Bella, D. A., and Grenny, W. J. (1970). Finite difference convection errors. *J. Sanit. Eng. Div. Proc. Am. Soc. Civ. Eng.* **96**, 1361–1375.

Bendat, J. S., and Piersol, A. G. (1971). "Random Data: Analysis and Measurement Procedures." Wiley (Interscience), New York.

Bennet, J. P. (1971). Convolution approach to the solution for the dissolved oxygen balance equation in a stream. *Water Resour. Res.* **7**, 580–590.

Blumberg, A. F. (1975). A Numerical Investigation into the Dynamics of Estuarine Circulation. Technical Rep. 91, Chesapeake Bay Institute, The Johns Hopkins Univ. Baltimore, Maryland.

Bobb, W. H., Boland, A., Jr., and Banchetti, A. J. (1973). Houston Ship Channel Galveston Bay, Texas, Rep. 1, Hydraulic and Salinity Verification. U.S. Army Waterways Experiment Station Technical Rep. H-73-12.

Bowden, K. F. (1963). The mixing processes in a tidal estuary. *Int. J. Air Water Pollut.* **7**, 343–356.

Bowden, K. F. (1967a). Circulation and diffusion. *In* "Estuaries" (G. H. Lauff, ed.), pp. 15–36. AAAS Publ. No. 85, Washington, D.C.

Bowden, K. F. (1967b). Stability effects on mixing in tidal currents. *Phys. Fluids Suppl.* **10**, S278–S280.

Bowden, K. F. (1970). Turbulence II. *Oceanogr. Mar. Biol. Ann. Rev.* **8**, 11–32.

Bowden, K. F. (1978). Mixing processes in estuaries. *In* "Estuarine Transport Processes" (B. Kjerfve, ed.), Belle W. Baruch Library in Marine Science Number 7. Univ. of South Carolina Press, Columbia, South Carolina.

Bowden, K. F., and Gilligan, R. M. (1971). Characteristic features of estuarine circulation as represented in the Mersey Estuary. *Limnol. Oceanogr.* **16**, 490–502.

Briggs, G. A. (1965). A plume rise model compared with observations. *J. Air Pollut. Contr. Assoc.* **15**, 433–438.

Briggs, G. A. (1975). Plume rise predictions. *In* "Lectures on Air Pollution and Environmental Impact Analysis," Chapter 3, sponsored by American Meteorological Society, September 29–October 3, Boston, Massachusetts.

Brogdon, J. J. Jr. (1972). Grays Harbor Estuary, Washington, Report 1, Verification and Base Tests. U.S. Army Waterways Experiment Station, Vicksburg, Mississippi Technical Rep. H-72-2.

Brooks, N. H. (1960). Diffusion of sewage effluent in an ocean current. *Proc. Int. Conf. Waste Disposal Mar. Environ., 1st* 246–267. Pergamon, Oxford.

Brooks, N. H. (1973). Dispersion in Hydrologic and Coastal Environments. Environmental Protection Agency Rep. 660/3-73-010; also W. M. Keck, Laboratory Rep. KH-R-29, California Institute of Technology, Pasadena, California.

Cacchione, D., and Wunsch, C. (1974). Experimental study of internal waves over a slope. *J. Fluid Mech.* **66**, 223–239.

Caldwell, D. R., Brubaker, J. M., and Neal, V. T. (1978). Thermal microstructure on a lake slope. *J. Limnol. Oceanogr.* **23**, 372–374.

Camp, T. R., and Graber, S.D. (1968). Dispersion conduits. *J. Sanit. Eng. Div. Proc. Am. Soc. Civ. Eng.* **94**, 31–39.

Camp, T. R. , and Graber, S. D. (1970). Discussion of Vigander, Elder, and Brooks. *J. Hydraul. Div. Proc. Am. Soc. Civ. Eng.* **96**, 2631–2635.

Cannon, G. A. (1969). Observations of Motion at Intermediate and Large Scales in a Coastal Plain Estuary. Technical Rep. 52, Chesapeake Bay Institute, Johns Hopkins Univ., Baltimore, Maryland.

Caro-Cordero, R., and Sayre, W. W. (1977). Mixing of Power-Plant Heated Effluents with the Missouri River. Iowa Institute of Hydraulic Research Rep. No. 203.

Carslaw, H. S., and Jaeger, J. C. (1959). "Conduction of Heat in Solids." 2nd ed. Oxford Univ. Press (Clarendon), London and New York.

Chandrasekhar, S. (1943). Stochastic problems in physics and astronomy. *Rev. Mod. Phys.* **15** (1), 1–89 (see Chapter II, p. 20).

Chang, Y. C. (1971). Lateral Mixing in Meandering Channels. PhD. dissertation, Univ. of Iowa, Iowa City, Iowa.

Chassaing, P., George, J., Claria, A., and Sananes, F. (1974). Physical characteristics of subsonic jets in a cross stream. *J. Fluid Mech.* **62**, 41–64.

Chatwin, P. C. (1970). The approach to normality of the concentration distribution of a solute in a solvent flowing along a straight pipe. *J. Fluid Mech.* **43**, 321–352.

Chatwin, P. C. (1975). On the longitudinal dispersion of passive containment in oscillatory flows in tubes. *J. Fluid Mech.* **71**, 513–527.

Chen, C. J., and Rodi, W. (1976). A Review of Experimental Data of Vertical Turbulent Buoyant Jets. Iowa Institute of Hydraulic Research Rep. No. 193.

Cheng, R. T. (1978). Modeling of hydraulic systems by finite-element methods. *Adv. Hydrosci.* **11**, 208–284.

Chow, V. T. (1959). "Open Channel Hydraulics," 1st ed. McGraw-Hill, New York.

Chu, V. H., and Goldberg, M. B. (1974). Buoyant forced plumes in a cross flow. *J. Hydraul. Div. Am. Soc. Civ. Eng.* **100** (HY9), 1203–1214.

Collar, R. H. F., and Mackay, D. W. (1973). The Clyde model: circulation and pollution. *In* "Mathematical and Hydraulic Modelling of Estuarine Pollution" (A. L. H. Gameson, ed.), pp. 201–210. Water Pollution Research Technical Paper No. 13, Department of the Environment, London.

Corcos, G. M., and Sherman, F. S. (1976). Vorticity concentrations and the dynamics of unstable shear layers. *J. Fluid Mech.* **73**, 241–264.

Corrsin, S. (1943). Investigation of Flow in an Axially Symmetric Heated Jet of Air. NACA Wartime Rep. W-94.

Corrsin, S., and Uberoi, M. S. (1950). Further Experiments on the Flow and Heat Transfer in a Heated Turbulent Air Jet. NACA Rep. 998.

Cox, G. C., and Macola, A. M. (1967). Predicting Salinity in an Estuary. *Am. Soc. Civ. Eng. Environm. Eng. Conf.*, Dallas, Texas. Conf. Preprint 433.

Cramer, H., and Leadbetter, M. R. (1967). "Stationary and Related Processes; Sample Function Properties and Their Application." Wiley (Interscience), New York.

Crank, J. (1956). "The Mathematics of Diffusion." Oxford Univ. Press (Clarendon), London and New York.

Crank, J. (1975). "The Mathematics of Diffusion," 2nd ed. Oxford Univ. Press, (Clarendon), London and New York.

Crawford, T. V., and Leonard, A. S. (1962). Observations of buoyant plumes, in calm stably stratified air. *J. Appl. Meteorol.* **1**, 251–256.

Crickmore, M. J. (1972). Tracer tests of eddy diffusion in field and model. *J. Hydraul. Div. Proc. Am. Soc. Civ. Eng.* **98**, 1737–1752.

Crow, S. C., and Champagne, F. H. (1971). Orderly structure in jet turbulence. *J. Fluid Mech.* **48**, 547–596.

Csanady, G. T. (1963). Turbulent diffusion in Lake Huron. *J. Fluid Mech.* **17**, 360–384.

Csanady, G. T. (1966). Accelerated diffusion in the skewed shear flow of lake currents. *J. Geophys Res.* **71**, 411–420.

Csanady, G. T. (1972). Frictional currents in the mixed layer at the sea surface. *J. Phys. Oceanogr.* **2**, 498–508.

Csanady, G. T. (1976). Mean circulation in shallow seas. *J. Geophys. Res.* **71**, 5389–5399.

Day, T. J. (1975). Longitudinal dispersion in natural channels. *Water Resour. Res.* **11**, 909–918.

Denton, R. A. (1978). Entrainment by Penetrative Convection at Low Peclet Number. Ph.D. Thesis, Univ. of Canterbury, Christchurch, New Zealand.

Ditmars, J. D. (1969). Computer Program for Round Buoyant Jets Into Stratified Ambient Environments. Tech. Memo. 69-1, W. M. Keck Laboratory of Hydraulics and Water Resources, California Institute of Technology, Pasadena, California.

Dyer, K. R. (1974). The salt balance in stratified estuaries. *Estuarine Coastal Mar. Sci.* **2**, 273–281.

Einstein, A. (1927). "Investigations on the Theory of Brownian Movement. . . . Assisted with Notes by R. Fürth" (Transl. by A. D. Cowper). Methuen, London (Reprinted by Dover, New York, 1956).

Elder, J. W. (1959). The dispersion of marked fluid in turbulent shear flow. *J. Fluid Mech.* **5**, 544–560.

Elder, R. A., and Wunderlich, W. O. (1972). Inflow Density Currents in TVA Reservoirs. *Int. Symp. Stratified Flows, Novosibirsk.*

Ellison, T. H., and Turner, J. S. (1959). Turbulent entrainment in stratified flows. *J. Fluid Mech.* **6**, 423–448.

Ellison, T. H., and Turner, J. S. (1960). Mixing of dense fluid in a turbulent pipe flow. *J. Fluid Mech.* **8**, 514–544.

Fair, G. M., Geyer, C., and Okun, D. A. (1971). "Elements of Water Supply and Waste Water Disposal." Wiley, New York.

Fan, L.-N. (1967). Turbulent Buoyant Jets into Stratified or Flowing Ambient Fluids. Technical Rep. KH-R-15, W. M. Keck Laboratory of Hydraulics and Water Resources, California Institute of Technology, Pasadena, California

Fan, L.-N., and Brooks, N. H. (1969). Numerical Solutions of Turbulent Buoyant Jet Problems. Technical Rep. KH-R-18, W. M. Keck Laboratory of Hydraulics and Water Resources, California Institute of Technology, Pasadena, California.

Feller, W. (1950). "An Introduction to Probability Theory and its Applications." Wiley, New York.

Fick, A. (1855). On liquid diffusion. *Philos. Mag.* **4**(10), 30–39.

Fischer, H. B. (1966). Longitudinal Dispersion in Laboratory and Natural Streams. Technical Rep. KH-R-12, California Institute of Technology, Pasadena, California.

Fischer, H. B. (1967a). The mechanics of dispersion in natural streams. *J. Hydraul. Div. Proc. Am. Soc. Civ. Eng.* **93**, 187–216.

Fischer, H. B. (1967b). Transverse Mixing in a Sand-Bed Channel. U.S. Geological Survey Professional Paper 575-D, D267-D272.

Fischer, H. B. (1968a). Dispersion predictions in natural streams. *J. Sanit. Eng. Div. Proc. Am. Soc. Civ. Eng.* **94**, 927–944.

Fischer, H. B. (1968b). Methods for Predicting Dispersion Coefficients in Natural Streams with Applications to Lower Reaches of the Green and Duwamish Rivers, Washington. U.S. Geological Survey Professional Paper 582-A.

Fischer, H. B. (1969). The effect of bends on dispersion in streams. *Water Resour. Res.* **5**, 496–506.

Fischer, H. B. (1970). A Method for Predicting Pollutant Transport in Tidal Waters. Contribution 132, Water Resources Center, Univ. of California.

Fischer, H. B. (1971). The dilution of an undersea sewage cloud by salt fingers. *Water Res.* **5**, 909–915.

Fischer, H. B. (1972a). Mass transport mechanisms in partially stratified estuaries. *J. Fluid Mech.* **53**, 671–687.

Fischer, H. B. (1972b). A Lagrangian Method for Predicting Pollutant Dispersion in Bolinas Lagoon, Marin County, California, U.S. Geological Survey Professional Paper 582-B.

Fischer, H. B. (1974). Numerical modelling of dispersion in estuaries. *Int. Symp. Discharge Sewage from Sea Outfalls, London* Paper No. 37, pp. 1–8. (Reprinted in "Discharge of Sewage from Sea Outfalls," Pergamon, Oxford, 1975.)

Fischer, H. B. (1975). Discussion of "Simple method for predicting dispersion in streams" by R. S. McQuivey and T. N. Keefer. *J. Environ. Eng. Div. Proc. Am. Soc. Civ. Eng.* **101**, 453–455.

Fischer, H. B. (1976a). Mixing and Dispersion in Estuaries. *Ann. Rev. Fluid Mech.* **8**, 107–133.

Fischer, H. B. (1976b). Some remarks on computer modeling of coastal flows. *J. Waterways Harbors Coastal Eng. Div. Proc. ASCE* **102**, 395–406.

Fischer, H. B. (1977). A numerical model of the Suisun Marsh, California, *Proc. Congr. Int. Assoc. Hydraul. Res. 17th, Baden-Baden, Germany* **3**, 425–432.

Fischer, H. B. (1978). On the tensor form of the bulk dispersion coefficient in a bounded skewed shear flow. *J. Geophys. Res.* **83**, 2373–2375.

Fischer, H. B., and Dudley, E. (1975). Salinity intrusion mechanisms in San Francisco Bay, California. *Proc. Congr. Int. Assoc. Hydraul. Res., 16th, Sao Paulo, Brazil* **1**, 124–133.

Fischer, H. B., and Hanamura, T. (1975). The effect of roughness strips on transverse mixing in hydraulic models. *Water Resour. Res.* **2**, 362–364.

Fischer, H. B., and Holley, E. R. (1971). Analysis of the use of distorted hydraulic models for dispersion studies. *Water Resour. Res.* **7**, 46–51.

Fischer, H. B., and Kirkland, W. B. Jr. (1978). Flushing of South San Francisco Bay; Results of Dye Concentration Measurements in the Bay-Delta Physical Model. Hugo B. Fischer, Inc. Rep. HBF-78/01.

Florkowski, T , Davis, T. G., Wallander, B., and Prabhakar, D. R. L. (1969). The measurement of high discharges in turbulent rivers using tritium tracer. *J. Hydrol.* **8**, 249–264.

Forch, C., Knudsen, M., and Sorensen, S. P. L. (1902). Beriche uber die Konstantenbestimmungen zur Aufstellung der hydrographischen Tabellen. *D. Kgl. Danske Vidensk. Selsk. Skrifter, 6. Raekke, naturvidensk. og mathem., Afd.*, XII.1, 151 pp.

Ford, D. (1978). Unpublished data, Environmental Laboratory, Waterways Experiment Station, U.S. Army Corps of Engineers, Vicksbury, Mississippi.

Forstall, W., and Gaylord, E. W. (1955). Momentum and mass transfer in a submerged water jet. *J. Appl. Mech.* **22**, 161–164.

Fourier, J. B. J. (1822). "Theorie Analytique de la Chaleur." Didot, Paris (see also translation by Freeman, Dover, New York, 1955).

Fox D. G. (1970). Forced plume in a stratified fluid. *J. Geophys. Res.* **75**, 6818–6835.

Friehe, C. A., and Schmitt, K. F. (1976). Parameterization of air-sea interface fluxes of sensible heat and moisture by the bulk aerodynamic formulas. *J. Phys. Oceanogr.* **6**, 801–809.

Fukuoka, S. (1974). A Laboratory Study on Longitudinal Dispersion in Alternating Shear Flows. Research Bulletin C12, Department of Engineering, James Cook Univ. of North Queensland.

Fukuoka, S., and Sayre, W. W. (1973). Longitudinal dispersion in sinuous channels. *J. Hydraul. Div. Proc. Am. Soc. Civ. Eng.* **99**, 195–218.

George, W. R., Jr., Alpert, R. L., and Tamanini, F. (1977). Turbulence measurements in an axisymmetric buoyant plume. *Int. J. Heat Mass Transfer* **20**, 1145–1154.

Glenne, B., and Selleck, R. E. (1969). Longitudinal estuarine diffusion in San Francisco Bay, California. *Water Res.* **3**, 1–20.

Glover, R. E. (1964). Dispersion of dissolved or suspended materials in flowing streams. U.S. Geological Survey Professional Paper 433-B.

Godfrey, R. G., and Frederick, B. J. (1970). Dispersion in Natural Streams. U.S. Geological Survey Professional Paper 433-K.

Graf, W. H. (1971). "Hydraulics of Sediment Transport." McGraw-Hill. New York.

Grant, A. J. (1974). A numerical model of instability in axisymmetric jets. *J. Fluid Mech.* **66**, 707–724.

Gregg, M. C. (1976). Finestructure and microstructure observations during the passage of a mild storm. *J. Phys. Oceanogr.* **6**, 528–555.

Halpern, D. (1974). Observations of the deepening of the wind-mixed layer in the northeastern Pacific Ocean. *J. Phys. Oceanogr.* **4**, 454–466.

Hamblin, P. F., and Carmack, E. C. (1978). River induced currents in a Fjord Lake. *J. Geophys. Res.* **83**, 885–899.

Hansen, D. V. (1965). Currents and mixing in the Columbia River estuary. Ocean Science and Ocean Engineering. *Trans. Joint Conf. Mar. Tech. Soc. Am. Soc. Limnol. Oceanogr., Washington, D. C.* pp. 943–955.

Hansen, D. V., and Rattray, M. (1965). Gravitational circulation in straits and estuaries. *J. Mar. Res.* **23**, 104–122.

Hansen, D. V., and Rattray, M. (1966). New dimensions in estuary classification. *Limnol. Oceanogr.* **11**, 319–325.

Harden, T. O., and Shen, H. T. (1979). Numerical simulation of mixing in natural rivers. *J. Hyd. Div. Proc. Am. Soc. Civ. Eng.* **105**, 393–408.

Heaps, N. S., and Ramsbottom, A. E. (1966). Wind effects on the water in a narrow two-layered lake. *Philos. Trans. R. Soc. London A* **259**, 391–430.

Hebbert, B., Imberger, J., Loh, I., and Patterson, J. (1979). Collie River underflow into the Wellington Reservoir. *J. Hydraul. Div. Am. Soc. Civ. Eng.* **105**, 533–546.

Heisenberg, W. (1948). Zur Statistichen Theorie der Turbulenz. *Z. Phys.* **124**, 628–657.

Henderson, F. M. (1966). "Open Channel Flow." Macmillan, New York.

Hetling, L. J., and O'Connell, R. L. (1966). A study of tidal dispersion in the Potomac River. *Water Resour. Res.* **2**, 825–841.

Hicks, B. B. (1972). Some evaluations of drag and bulk transfer coefficients over water bodies of different sizes. *Boundary-Layer Meteorol.* **3**, 201–213.

Hicks, B. B. (1975). A procedure for the formulation of bulk transfer coefficients over water. *Boundary-Layer Meteorol.* **8**, 515–524.

Hicks, B. B., Drinklow, R. L., and Grauze, G. (1974). Drag and bulk transfer coefficients associated with a shallow water surface. *Boundary-Layer Meteorol.* **6**, 287–297.

Hill, B. J. (1972). Measurement of local entrainment rate in the initial region of axisymmetric turbulent air jets. *J. Fluid Mech.* **51**, 773–779.

Hinstrup, P., Kej, A., and Kroszynski, U. (1977). A high accuracy two-dimensional transport-dispersion model for environmental applications. *Proc. Congr. Int. Assoc. Hydraul. Res. 17th*, **3**, 129–137.

Hinze, J. O., and van der Hegge Zijnen (1949). Transfer of heat and matter in the turbulent mixing zone of an axially symmetric jet. *Appl. Sci. Res.* **A1**, 435–461.

Hirst, E. (1971). Buoyant jets discharged to quiescent stratified ambients. *J. Geophys. Res.* **76**, 7375–7384.

Hirst, E. (1972). Zone of flow establishment for round buoyant jets. *Water Resour. Res.* **8**(5), 1234–1246.

Holley, E. R., and Abraham, G. (1973a). Laboratory studies on transverse mixing in rivers. *J. Hydraul. Res.* **11**, 219–253.

Holley, E. R., and Abraham, G. (1973b). Field tests on transverse mixing in rivers. *J. Hydraul. Div. Proc. Am. Soc. Civ. Eng.* **99**, 2313–2331.

Holley, E. R., Siemons, J., and Abraham, G. (1972). Some aspects of analyzing transverse diffusion in rivers. *J. Hydraul. Res.* **10**, 27–57.

Holly, F. M., Jr., and Preissmann, A. (1977). Accurate calculation of transport in two dimensions. *J. Hydraul. Div. Proc. Am. Soc. Civ. Eng.*, **103**, 1259–1277.

Hoult, D. P., and Weil, J. C. (1972). Turbulent plume in a laminar cross flow. *Atmos. Environ.* **6**, 513–531.

Hoult, D. P., Fay, J. A., and Forney, L. J. (1969). A theory of plume rise compared with field observations. *J. Air Pollut. Control Assoc.* **19**, 585–590.

Imberger, J. (1976). Dynamics of a horizontally stratified estuary. *Proc. Coastal Eng. Conf. Am. Soc. Civ. Eng. Honolulu, Hawaii, July 11–17*, Vol. 4, pp. 3108–3123.

Imberger, J., Thompson, R. T., and Fandry, C. (1976). Selective withdrawal from a finite rectangular tank. *J. Fluid Mech.* **78**, 489–512.

Imberger, J., Kirkland, W. B., Jr., and Fischer, H. B. (1977). The Effect of Delta Outflow on the Density Stratification in San Francisco Bay. Hugo B. Fischer, Inc., and Waterfront Design Associates Rep. HBF-77/02.

Imberger, J., Patterson, J., Hebbert, B., and Loh, I. (1978). Dynamics of reservoir of medium size. *J. Hydraul. Div. Am. Soc. Civ. Eng.* **104**, 725–743.

Inglis, C. C., and Allen, F. H. (1957). The regimen of the Thames estuary as affected by currents, salinities, and river flow. *Proc. Inst. Civ. Eng.* **7**, 827–868.

Ippen, A. T., and Harleman, D. R. F. (1961). One Dimensional Analysis of Salinity Intrusion in Estuaries. U. S. Army Corps of Engineers, Waterways Experiment Station, Vicksburg, Mississippi Technical Bull. No. 5.

Ivey, G., and Imberger, J. (1978). Field investigation of selective withdrawal. *J. Hydraul. Div. Am. Soc. Civ. Eng.* **104**, 1225–1237.

Jackman, A. P., and Yotsukura, N. (1977). Thermal Loading of Natural Streams. U. S. Geological Survey Professional Paper 991.

Jobson, H. E., and Sayre, W. W. (1970). Vertical transfer in open channel flow. *J. Hydraul. Div. Proc. Am. Soc. Civ. Eng.* **96**, 703–724.

John.Carollo Engineers (1970). Final Rep. on Ocean Outfall No. 2. County Sanitation District of Orange County.

Kalkanis, G. (1964). Transportation of Bed Material due to Wave Action. U.S. Army Corps of Engineers, Coastal Engineering Research Center, Technical Memorandum No. 2, Washington, D.C.

Keagy, W. R., and Weller, A. E. (1949). A study of freely expanding inhomogeneous jets. *Heat Transfer Fluid Mech. Inst. 2nd Berkeley, California* pp. 89–98.

Kirkland, W. B., Jr., and Fischer, H. B. (1976). Hydraulic Model Studies San Francisco Bay-Delta Model. U.S. Army Corps of Engineers, for East Bay Dischargers Authority. Waterfront Design Associates unpublished report.

Kizer, K. M. (1963). Material and momentum transport in axisymmetric turbulent jets of water. *Am. Inst. Chem. Eng. J.* **9**, 386–390.

Knudsen, M. (1901). "Hydrographical Tables." G. E. C. Gad, Copenhagen.

Koh, R. C. Y. (1973). Hydraulic Tests of Discharge Ports. W. M. Keck Laboratory of Hydraulics and Water Resources Technical Memorandum 73-4, California Institute of Technology, Pasadena, California.

Koh, R. C. Y. (1976). Buoyancy-driven gravitational spreading. *Proc. Int. Conf. Coastal Eng. 15th, Honolulu, Hawaii, July 11–17*, Vol. 4, pp. 2956–2975.

Koh, R. C. Y., and Brooks, N. H. (1975). Fluid mechanics of waste water disposal in the ocean. *Ann. Rev. Fluid Mech.* **7**, 187–211.

Koh, R. C. Y., and Fan, L. N. (1970). Mathematical Models for the Prediction of Temperature Distributions Resulting from the Discharge of Heated Water into Large Bodies of Water. Environ. Prot. Agency Rep. 16130 DWO 10/70, 219 pp. (Also Tetra Tech., Inc. Rep. TC-170.)

Koh, R. C. Y., Brooks, N. H., List, E. J., and Wolanski, E. J. (1974). Hydraulic Modeling of Thermal Outfall Diffusers for the San Onofre Nuclear Power Plant. W. M. Keck Laboratory of Hydraulics and Water Resources Tech. Rep. No. KH-R-30, California Institute of Technology, 118 pp.

Kolmogorov, A. N. (1931). Über die analytischen Methoden in der Wahrscheinlichkeitsrechnung. *Math. Ann* **104**, 415–458.

Kolmogorov, A. N. (1933). Zur Theorie der stetigen zufälligen Prozesse. *Math. Ann.* **108**, 149–160.

Kotsovinos, N. E. (1975). A Study of the Entrainment and Turbulence in a Plane Buoyant Jet. Ph.D. thesis, California Institute of Technology, Pasadena, California.

Kotsovinos, N. E., and List, E. J. (1977). Plane turbulent buoyant jets. Part 1. Integral properties. *J. Fluid Mech.* **81**, 25–44.

Kraus, E. B. (1977). "Modelling and Prediction of the Upper Layers of the Ocean." Pergamon, Oxford.

Kraus, E. B., and Turner, J. S. (1967). A one-dimensional model of the seasonal thermocline, II. The general theory and its consequences. *Tellus*. **13**, 98–106.

Kullenberg, G. (1974). An Experimental and Theoretical Investigation of the Turbulent Diffusion in the Upper Layer of the Sea. Dept. No. 25, Institute for Physical Oceanography, Univ. of Copenhagen.

Labus, T. L., and Symons, E. P. (1972). Experimental Investigation of an Axisymmetric Free Jet with an Initially Uniform Velocity Profile. NASA TN D-6783.

Lau, Y. L., and Krishnappan, B. G. (1977). Transverse dispersion in rectangular channels. *J. Hydraul. Div. Proc. Am. Soc. Civ. Eng.*, **103**, 1173–1189.

Laufer, J. (1950). Investigation of Turbulent Flow in a Two-dimensional Channel. NACA TN 2123.

Leendertse, J. J. (1967). Aspects of a Computational Model for Long-period Water-wave Propagation. Memorandum RM-5294-PR, The Rand Corp., Santa Monica, California.

Leendertse, J. J. (1970). "A Water-quality Simulation Model for Well-mixed Estuaries and Coastal Seas," Vol. 1, Principles of Computation. Rep. RM-6230-RC, Rand Corp., Santa Monica, California.

Leendertse, J. J., and Gritton, E. C. (1971). A Water-quality Simulation Model for Well Mixed Estuaries and Coastal Seas," Vol. II, Computation Procedures, Vol. III, Jamaica Bay Simulation. Rep. R-708-NYC and R-709-NYC, Rand Corp, Santa Monica, California.

Letter, J. J. Jr., and McAnally, W. H. Jr. (1975). Physical Hydraulic Models: Assessment of Predictive Capabilities, Rep. 1, Hydrodynamics of the Delaware River Estuary Model. U.S. Army Waterways Experiment Station, Vicksburg, Mississippi Rep. H-75-3.

Li, W. H. (1974). DO-sag in oscillating flow. *J. Environm. Eng. Div. Proc. Am. Soc. Civ. Eng.* **100**, 837–854.

Lin, C. C. (1960). On a theory of dispersion by continuous movements. *Proc. Nat. Acad. Sci. U.S.*, **46**, 566–570.

List, E. J., and Koh, R. C. Y. (1976). Variations in coastal temperatures on the southern and central California coast. *J. Geophys. Res.* **81**, (12), 1971–1979.

Liu, S. K., and Leendertse, J. J. (1978). Multidimensional numerical modeling of estuaries and coastal seas. *Adv. Hydrosci.*, **11**, 95–164.

Mackay, J. R. (1970). Lateral mixing of the Liard and Mackenzie Rivers downstream from their confluence. *Can. J. Earth Sci.* **7**, 111–124.

Markofsky, M., and Harleman, D. R. F. (1971). A Predictive Model for Thermal Stratification and Water Quality in Reservoirs. M. I. T. Hydrodynamics Laboratory Technical Rep. No. 134.

McBean, G. A., and Paterson, R. D. (1975). Variations of the turbulent fluxes of momentum, heat and moisture over Lake Ontario. *J. Phys. Oceanogr.* **5**, 523–531.

McGuirk, J. J., and Rodi, W. (1976). Calculation of three-dimensional heated surface jets. *Proc. Int. Conf. Heat Mass Transfer, 9th, Dubrovnik, Yugoslavia, August 29–September 4* pp. 275–287.

McNown, J. S. (1954). Mechanics of manifold flow. *Trans. Amer. Soc. Civil Eng.* **119**, 1103–1118.

McQuivey, R. S., and Keefer, T. (1974). Simple method for predicting dispersion in streams. *J. Environ. Eng. Div., Proc. Amer. Soc. Civil Eng.* **100**, 997–1011.

Miller, A. C., and Richardson, E. V. (1974). Diffusion and dispersion in open channel flow. *J. Hydraul. Div. Proc. Am. Soc. Civ. Eng.* **100**, 159–171.

Miller, D. R., and Comings, E. W. (1957). Static pressure distribution in the free turbulent jet. *J. Fluid Mech.* **3**, 1–16.

Monon, A. S., and Yaglom, A. M. (1965a). "Statistical Fluid Mechanics: Mechanics of Turbulence," Vol. 1. M.I.T. Press, Cambridge, Massachusetts (English transl., 1971).

Monin, A. S., and Yaglom, A. M. (1965b). "Statistical Fluid Mechanics: Mechanics of Turbulence," Vol. 2. M.I.T. Press, Cambridge, Massachusetts (English transl., 1975).

Mortimer, C. H. (1952). Water movements in lakes during summer stratification. Evidence from the distribution of temperature in Windermere. *Phil. Trans. Roy. Soc. London, Ser. B* **236**, 355–404.

Mortimer, C. H. (1953). The resonant response of stratified lakes to wind. *Schweiz. Z. Hydrol.* **15**, 94–151.

Mortimer, C. H. (1974). Lake hydrodynamics. *Int. Assoc. Theoret. Appl. Limnol. Mitteilungen* **20**, 124–197.

Morton, B., Taylor, G. I., and Turner, J. S. (1956). Turbulent gravitational convection from maintained and instantaneous sources. *Proc. R. Soc. London Ser. A* **234**, 1–23.

Moussa, Z. M., Trischka, J. W., and Eskinazi, S. (1977). The near field in the mixing of a round jet with a cross stream. *J. Fluid Mech.* **80**, 49–80.

Munk, W., and Anderson, E. R. (1948). Notes on a theory of the thermocline. *J. Mar. Res.* **7**, 276–295.

Murray, S., Conlon, D., Siripong, A., and Santoro, J. (1975). Circulation and salinity distribution in the Rio Guayas Estuary, Ecuador. *In* "Estuarine Research" (L. E. Cronin, ed.), Vol. 2, pp. 345–363. Academic Press, New York.

Murray, S. P., and Siripong, A. (1978). Role of lateral gradients and longitudinal dispersion in the salt balance of a shallow, well-mixed estuary. *In* "Estuarine Transport Processes" (B. Kjerfve, ed.), pp. 113–124. Univ. of South Carolina Press, Columbia, South Carolina.

Murthy, C. R. (1976). Horizontal diffusion characteristics in Lake Ontario. *J. Phys. Oceanogr.* **6**, 76–84.

National Academy of Sciences (1977). "Research and development in the Environmental Protection Agency," Vol. III, Analytical Studies for the U.S. Environmental Protection Agency, Washington, D.C.

Nelson, A. W., and Lerseth, R. J. (1972). A Study of Dispersion Capability of San Francisco Bay-Delta Waters. California Department of Water Resources, Sacramento, California.

Nihoul, J. C. J., and Ronday, F. C. (1976). Hydrodynamic models of the north sea, a comparative assessment. *Mem. Soc. R. Sci. Liege 6th Ser.* **10**, 61–96.

Niiler, P. P. (1975). Deepening of the wind-mixed layer. *J. Mar. Res.* **33**, 405–422.

Nordin, C. F., and Sabol, B. V. (1974). Empirical Data on Longitudinal Dispersion in Rivers. U.S. Geological Survey Water Resources Investigations 20–74, Open File Rep.

Obukhov, A. M. (1959). Description of turbulence in terms of Lagrangian variables. *Adv. Geophys.* **6**, 113–115.

Okoye, J. K. (1970). Characteristics of Transverse Mixing in Open-channel Flows. Rep. KH-R-23, California Institute of Technology, Pasadena, California.

Okubo, A. (1967). The effect of shear in an oscillatory current on horizontal diffusion from an instantaneous source. *Int. J. Oceanol. Limnol.* **1**(3), 194–204.

Okubo, A. (1973). Effect of shoreline irregularities on streamwise dispersion in estuaries and other embayments. *Neth. J. Sea Res.* **6**, 213–224.

Okubo, A. (1974). Some speculations on oceanic diffusion diagrams. *Rapp. P.-v. Réun. Cons. Int. Explor. Mer.* **167**, 77–85.

Okubo, A., and Ozmidov, R. V. (1970). Empirical dependence of the coefficient of horizontal turbulent diffusion in the ocean on the scale of the phenomenon question. *Izv. Atmos. Oceanic Phys.* **6**, 308–309.

Onsager, L. (1945). The distribution of energy in turbulence. *Phys. Rev.* **68**(2), 286 (Abstract).

Owens, M., Edwards, R. W., and Gibbs, J. W. (1964). Some reaeration studies in streams. *Air Water Pollut. Int. J.* **8**, 469–486.

OWQP (1972). Final Report, Water Quality Program for Oahu with Special Emphasis on Waste Disposal. Engineering Science, Inc., Sunn, Low, Tom and Hara, Inc., and Dillingham Environmental Company, March 1972.

Ozmidov, R. V. (1965a). Some features of the energy spectrum of oceanic turbulence. *Dokl. Acad. Nauk SSSR Earth Sci.* **160**, 11–18.

Ozmidov, R. V. (1965b). On the turbulent exchange in a stably stratified ocean. *Atmos. Oceanic Phys. Ser.* **1**, 853–860.

Partch, E. N., and Smith, J. D. (1978). Time dependent mixing in a salt wedge estuary. *Estuarine Coastal Mar. Sci.* **6**, 3–19.

Patterson, J., Loh, I., Imberger, J., and Hebbert, B. (1978). Management of a salinity affected reservoir. *Proc. Hydrol. Symp. I.E. Aust.* pp. 18–22.

Paulson, R. W. (1969). The longitudinal diffusion coefficient in the Delaware River estuary as determined from a steady state model. *Water Resour. Res.* **5**, 59–67.

Phillips, O. M. (1977). "The Dynamics of the Upper Ocean." Cambridge Univ. Press, London and New York.

Pollard, R. T., Rhines, P. B., and Thompson, R. O. R. Y. (1973). The deepening of the wind-mixed layer. *Geophys. Fluid Dynam.* **3**, 381–404.

Price, W. A., and Kendrick, M. P. (1963). Field and model investigations into the reasons for siltation in the Mersey Estuary. *Proc. Inst. Civ. Eng.* **24**, 473–518.

Priestley, C. H. B., and Ball, F. K. (1955). Continuous convection from an isolated source of heat. *Q. J. R. Meteorol. Soc.* **81**(348), 144–157.

Pritchard, D. W. (1967). Observations of circulation in coastal plain estuaries. *In* "Estuaries" (G. H. Lauff, ed.), pp. 37–44. AAAS Publ. No. 83, Washington, D.C.

Pritchard, D. W. (1971). Hydrodynamic models. *In* "Estuarine Modelling: An Assessment" (G. Ward and W. Espey, eds.), pp. 5–33. Environmental Protection Agency Rep. 16070DZV 02/71, Washington, D.C.

Proudman, J. (1953). "Dynamical Oceanography." Methuen, New York.

Prych, E. A. (1970). Effects of Density Differences on Lateral Mixing in Open Channel Flows. Rep. No. KH-R-21, California Institute of Technology, Pasadena, California.

Prych, E. A. (1972). A Warm Water Effluent Analyzed as a Buoyant Jet. Sveriges Meteorologiska Och Hydrologiska Institut, Ser. Hydrol. No. 21, Stockholm, Sweden.

Raphael, J. M. (1962). Prediction of temperature in rivers and reservoirs. *J. Power Div. Am. Soc. Civ. Eng.* **88**, 157–182.

Raudkivi, A. J. (1976). "Loose Boundary Hydraulics," 2nd ed., Pergamon, Oxford.

Rawn, A. M., Bowerman, F. R., and Brooks, N. H. (1961). Diffusers for disposal of sewage in sea water. *Trans. Am. Soc. Civ. Eng.* **126**, Part III, 344–88.

Richardson, L. F. (1926). Atmospheric diffusion shown on a distance-neighbour graph. *Proc. R. Soc. London, Ser. A* **110**, 709–737.

Ricou, F. P., and Spalding, D. B. (1961). Measurements of entrainment by axisymmetrical turbulent jets. *J. Fluid Mech.* **11**, 21–32.

Rigter, B. P. (1973). Minimum length of salt intrusion in estuaries. *J. Hydraul. Div. Proc. Am. Soc. Civ. Eng.* **99**, 1475–1496.

R. M. Towill Corp. (1972). Final Design Rep. Sand Island Ocean Outfall System, City and County of Honolulu.

Roberts, P. J. W. (1977). Dispersion of Buoyant Waste Water Discharged from Outfall Diffusers of Finite Length. Ph.D. thesis, California Institute of Technology, Pasadena, California.

Roberts, P. J. W. (1979). Line plume and ocean outfall dispersion. *J. Hydraul. Div., Proc. Amer. Soc. Civil Eng.* **105**, 313–331.

Robinson, R. M., and McEwan, A. D. (1975). Instability of a periodic boundary layer in a stratified fluid. *J. Fluid Mech.* **68**, 41–48.

Rosenweig, R. E., Hottel, H. C., and Williams, G. C. (1961). Smoke scattered light measurement of turbulent concentration fluctuations. *Chem. Eng. Sci.* **15**, 111–129.

Rosler, R. S., and Bankoff, S. G. (1963). Large scale turbulence characteristics of a submerged water jet. *AIChE J.* **9**(5), 672–676.

Rouse, H. (1938). "Fluid Mechanics for Hydraulic Engineers." McGraw-Hill, New York.

Rouse, H., Yih, C.-S., and Humphreys, H. W. (1952). Gravitational convection from a boundary source. *Tellus* **4**, 201–210.

Rozovskii, I. L. (1957). Flow of Water in Bends of Open Channels. Academy of Sciences of the Ukranian SSR, Kiev (translation No. OTS 60-51133), Office of Technical Services, U.S. Department of Commerce, Washington, D.C.

Ruden, P. (1933). Turbulente Ausbneitrorgönge im Freistahl. *Naturwissenschaften* **21**, 375–378.

Safaie, B. (1978). Mixing of Horizontal Buoyant Surface Jet Over Sloping Bottom. Rep. HEL-27-4, Hydraulic Engineering Laboratory, Univ. of California, Berkeley, California.

Saffman, P. G. (1962). The effect of wind shear on horizontal spread from an instantaneous ground source. *Q. J. Meteorol. Soc.* **88**, 382–393.

Sami, S., Carmody, T., and Rouse, H. (1967). Jet diffusion in the region of flow establishment. *J. Fluid Mech.* **27**, 231–252.

Sayre, W. W., and Caro-Cordero, R. (1977). Shore-attached thermal plumes in rivers. *Proc. Institute on River Mechanics—Modeling of Rivers. Colorado State University, July 5–15.*

Sayre, W. W., and Chang, F. M. (1968). A Laboratory Investigation of the Open Channel Dispersion Process for Dissolved, Suspended, and Floating Dispersants. U.S. Geological Survey Professional Paper 433-E.

Sayre, W. W., and Yeh, T. (1973). Transverse Mixing Characteristics of the Missouri River Downstream from the Cooper Nuclear Station. Iowa Institute of Hydraulic Research Rep. No. 145.

Schatzmann, M. (1976). Auftriebsstrahlen in natürlichen Strömungen-Entwicklung eines mathematischen Modells. SFB80/T/86, Sonderforschungsbereich 80, Univ. Karlsruhe, Germany.

Schatzmann, M., and Flick, W. (1977). Fluiddynamisches Simulations Modell zur Vorhesage von Ausbreitungsvorgängen in der Atmosphärischen Grenzschicht. SFB80/T/90 März, Sonderforschungsbereich 80, Univ. Karlsruhe, Germany.

Schijf, J. B., and Schönfeld, J. C. (1953). Theoretical considerations on the motion of salt and fresh water. *Proc. Minnesota Int. Hydraul. Conf., Minneapolis, Minnesota* pp. 321–333.

Schiller, E. J., and Sayre, W. W. (1973). Vertical Mixing of Heated Effluents in Open-channel Flow. Iowa Institute of Hydraulic Research Rep. No. 148.

Schuster, J. C. (1965). Canal discharge measurements with radioisotopes. *J. Hydraul. Div. Proc. Am. Soc. Civ. Eng.* **91**, 101–124.

Scorer, R. S. (1959). The behavior of chimney plumes. *Int. J. Air Pollut.* **1**, 198–220.

Serruya, S. (1974). The mixing patterns of the Jordan River at Lake Kinnert. *Limnol. Oceanogr.* **19**, 175–181.

Sherman, F. S., Imberger, J., and Corcos, G. M. (1978). Turbulence and mixing in stably stratified waters. *Ann. Rev. Fluid Mech.* **10**, 267–288.

Shirazi, M. A., and Davis, L. R. (1974). Workbook of Thermal Plume Predictions, Vol. 2, Surface discharges. Rep. No. EPA-R2-72-0056, U.S. Environmental Protection Agency, Corvallis, Oregon.

Simmons, H. B., and Bobb, W. H. (1965). Hudson River Channel, New York and New Jersey, Plans to Reduce Shoaling in Hudson River Channels and Adjacent Pier Slips, Hydraulic Model Investigation. Technical Rep. No. 2-694, U.S. Army Waterways Experiment Station, Vicksburg, Mississippi.

Slawson, P. R., and Csanady, G. T. (1971). The effect of atmospheric conditions on plume rise. *J. Fluid Mech.* **47**, 33–49.

Sotil, C. A. (1971). Computer Program for Slot Buoyant Jets into Stratified Ambient Environments. W. M. Keck Laboratory Technical Memorandum 71-2, California Institute of Technology, Pasadena, California.

Spigel, R. H. (1978). Wind Mixing in Lakes. Ph.D. Thesis, Univ. of California, Berkeley, California.

State of California (1962). Sacramento River Water Pollution Survey. Bull. No. 111, Department of Water Resources.

Stommel, H. (1953). Computation of pollution in a vertically mixed estuary. *Sewage Ind. Wastes* **24**, 1065–1071.

Stommel, H., and Farmer, H. G. (1952). On the Nature of Estuarine Circulation. Woods Hole Oceanographic Inst. References Nos. 52–51, 52–63, 52–88 (3 vols. containing chapters 1–4 and 7).

Stone, H. L., and Brian, P. T. (1963). Numerical solution of convective transport problems. *AICE J.* **9**, 681–688.

Streeter, V. L., and Wylie, E. B. (1967). "Hydraulic Transients." McGraw-Hill, New York.

Sullivan, P. J. (1968). Dispersion in a Turbulent Shear Flow. Ph.D. thesis, Univ. of Cambridge, Cambridge, England.

Sumer, S. M., and Fischer, H. B. (1977). Transverse mixing in partially stratified flow. *J. Hydraul. Div. Proc. Am. Soc. Chem. Eng.* **103**, 587–600.

Sunavala, P. D., Hulse, C., and Thring, M. W. (1957). Mixing and combustion in free and enclosed turbulent jet diffusion flames. *Combust. Flame* **1**, 79–193.

Taylor, G. I. (1921). Diffusion by continuous movements. *Proc. London Math. Soc. Ser. A* **20**, 196–211.

Taylor, G. I. (1953). Dispersion of soluble matter in solvent flowing slowly through a tube. *Proc. R. Soc. London Ser. A* **219**, 186–203.

Taylor, G. I. (1954). The dispersion of matter in turbulent flow through a pipe. *Proc. R. Soc. London Ser. A* **223**, 446–468; (1960). *Sci. Pap.* **2**, 466–488.

Tee, K. T. (1976). Tide-induced residual current, a 2-d nonlinear numerical tidal model. *J. Mar. Res.* **34**, 603–628.

Tee, K. T. (1977). Tide-induced residual current—verification of a numerical model. *J. Phys. Oceanogr.* **7**, 396–402.

Tennekes, H., and Lumley, J. L. (1972). "A First Course in Turbulence." M.I.T. Press, Cambridge, Massachusetts.

Thatcher, M., and Harleman, D. R. F. (1972). A Mathematical Model for the Prediction of Unsteady Salinity Intrusion in Estuaries. R. M. Parsons Laboratory Rep. No. 144, Massachusetts Institute of Technology, Cambridge, Massachusetts.

Thomas, I. E. (1958). Dispersion in Open-channel Flow. Ph.D. thesis. Northwestern Univ.

Thompson, E. F. (1969). The Mixing of a Layer of Fresh Water with Underlying Salt Water under the Influence of Wind. Master of Science Thesis, School of Engineering, University of California, Berkeley, California.

Thornton, K. W., and Lessen, A. S. (1976). Sensitivity Analysis of the Water Quality for River-reservoir Systems Model. Misc. paper Y-76-4, Environmental Effects Laboratory, U.S. Army Engineer Waterways Experiment Station.

Thorpe, S. A. (1971). Asymmetry of the internal seiche in Loch Ness. *Nature* **231**, 306–308.

Thorpe, S. A. (1974). Near resonant forcing in a shallow two layer fluid: A model for the internal surge in Loch Ness? *J. Fluid Mech.* **63**, 509–527.

Thorpe, S. A. (1977). Turbulence and mixing in a Scottish loch. *Philos. Trans. R. Soc. London* **A286**, 125–181.

Thorpe, S. A., and Hall, A. J. (1977). Mixing in the upper layer of a lake during heating cycle. *Nature* **265**, 719–722.

Tsai, Y. H., and Holley, E. R. (1978). Temporal moments for longitudinal dispersion. *J. Hydraul. Div. Proc. Am. Soc. Civil Eng.* **104**, 1617–1634.

Turner, J. S. (1960). A comparison between buoyant vortex rings and vortex pairs. *J. Fluid Mech.* **7**, 419–432.

Turner, J. S. (1966). Jets and plumes with negative or reversing buoyancy. *J. Fluid Mech.* **26**, 779–792.

Turner, J. S. (1969). A Note on Wind Mixing at the Seasonal Thermocline. *Deep-Sea Res. Suppl.* **16**, 297–300.

Turner, J. S. (1973). "Buoyancy Effects in Fluids." Cambridge Univ. Press, London and New York.

T.V.A. (1972). Heat and Mass Transfer between a Water Surface and the Atmosphere. T.V.A. Rep. No. 0-6803.

Uberoi, M., and Garby, L. C. (1967). Effect of density gradient on an air jet. *Phys. Fluids*, **10**, 200–202.

Uhlenbeck, G. E., and Ornstein, L. S. (1930). On the theory of the Brownian motion. *Phys. Rev.* **36**, 823–841.

U.S. Army Corps of Engineers (1956). Delaware River Model Study Rep. No. 1, Hydraulic and Salinity Verification. WEST.M. 2-337, Waterways Experiment Station, Vicksburg, Mississippi.

U.S. Army Corps of Engineers (1974). San Francisco Bay and Sacramento-San-Joaquin Delta Water Quality and Waste Disposal Investigation; Model Verification and Results of Sensitivity Tests. San Francisco Bay-Delta Model Technical Memo No. 1. U.S. Army Engineer District, San Francisco, California.

U.S. Army Corps of Engineers (1975). San Francisco Bay and Sacramento-San-Joaquin Delta Water Quality and Waste Disposal Investigation; Results of Model Tests for Peripheral Canal—Phase I. San Francisco Bay-Delta Model Technical Memo No. 3. U.S. Army Engineer District, San Francisco, California (2nd rough draft).

U.S. Army Corps of Engineers (1963). Comprehensive Survey of San Francisco Bay and Tributaries, California, Appendix " H " hydraulic Model Studies, Volume II—Plates: Verification and Tests of Barriers, to the Technical Rep. on San Francisco Bay barriers. U.S. Army Engineer District, San Francisco, California.

U.S. Navy Hydrographic Office (1952). Tables for Sea Water Density. H.O. Publication No. 615, U.S. Navy Hydrographic Office, Washington, D.C.

Valentine, E. M. (1978). Dispersion in Turbulent Flow. Ph.D. thesis, Univ. of Canterbury, Christchurch, New Zealand.

Valentine, E. M., and Wood, I. R. (1977). Longitudinal dispersion with dead zones. *J. Hydraul. Div. Proc. Am. Soc. Civ. Eng.* **103**, 975–990.

Vanoni, V. A. (ed.) (1975). "Sedimentation Engineering," Amer. Soc. Civil Eng. Manuals and Reports on Engineering Practice No. 54. American Society of Civil Engineers, New York.

Vanoni, V. A., and Brooks, N. H. (1957). Laboratory Studies of the Roughness and Suspended Load of Alluvial Streams. Rep. No. E-68, Sedimentation Laboratory, California Institute of Technology, Pasadena, California.

von Weisacker, C. F. (1948). Das Spectrum der Turbulenz bei grossen Reynoldsschen Zahlen. *Z. Phys.* **124**, 614–627.

Ward, P. R. B. (1974). Transverse dispersion in oscillatory channel flow. *J. Hydraul. Div. Proc. Am. Soc. Civ. Eng.* **100**, 755–772.

Ward, P. R. B. (1976). Measurements of estuary dispersion coefficient. *J. Environm. Eng. Div. Proc. Am. Soc. Civ. Eng.* **102**, 855–859.

Wedderburn, E. M. (1912). Temperature observations in Loch Earn with a further contribution to the hydrodynamical theory of temperature seiches. *Trans. Roy. Soc. Edinburgh* **48**, 629–295.

Wilson, R. A. M., and Danckwerts, P. V. (1964). Studies in turbulent mixing. II. A hot air jet. *Chem. Eng. Sci.* **19**, 885–895.

Wood, I. R., and Binney, P. (1976). Selective Withdrawal from a Two-Layered Fluid. Research Rep., Department of Civil Engineering, Univ. of Canterbury, Christchurch, New Zealand.

Wright, S. J. (1977). Effects of Ambient Crossflows and Density Stratification on the Characteristic Behavior of Round, Turbulent Buoyant Jets. Technical Rep. KH-R-36, W. M. Keck, Laboratory of Hydraulics and Water Resources, California Institute of Technology, Pasadena, California. 254 pp.

Wu, J. (1965). Collapse of Turbulent Wakes in Density-stratified Media. Hydronautics, Inc. Technical Rep. 231-4,

Wu, J. (1973). Wind induced entrainment across a stable density interface. *J. Fluid Mech.* **61**, 275–287.

Yotsukura, N., and Cobb, E. D. (1972). Transverse Diffusion of Solutes in Natural Streams. U.S. Geological Survey Professional Paper 582-C.

Yotsukura, N., and Sayre, W. W. (1976). Transverse mixing in natural channels. *Water Resour. Res.* **12**, 695–704.

Yotsukura, N., Fischer, H. B., and Sayre, W. W. (1970). Measurement of Mixing Characteristics of the Missouri River between Sioux City, Iowa and Plattsmouth, Nebraska. U.S. Geological Survey Water-Supply Paper 1899-G.

Zimmerman, J. T. F. (1978a). Topographic generation of residual circulation by oscillatory (tidal) currents. *Geophys. Astrophys. Fluid Dynam.* **11**, 35–47.

Zimmerman, J. T. F. (1978b). Dispersion by tide-induced residual current vortices. *In* "Hydrodynamics of Estuaries and Fjords" (J. C. J. Nihoul, ed.), pp. 207–216. Elsevier, Amsterdam.

Author Index

Subject Index

A

Advection
  definition, 7
  numerical model, 285, 290
Advective diffusion, *see* Diffusion, advective
Aris, dispersion analysis, 90–91

B

Baroclinic circulation, 243, 246, 257
Barotropic circulation, 243
Bend
  effect on longitudinal dispersion, 134–135
  effect on transverse mixing, 111
Bifurcation of buoyant plume, 359, 383
Biochemical oxygen demand, 4, *see also*
    Decaying substance
  standards for, 422
BOD, *see* biochemical oxygen demand
Buckingham–π theorem, 23
Buoyancy, definition, 20
Buoyancy flux
  definition, 319
  of effluent in river, 123
Buoyancy frequency, 157
Buoyant effluent, 123, *see also* Jet; Plume

C

California thermal plan, 430
Change of moment method, 136
Circulation, *see* Barotropic circulation;
    Baroclinic circulation; Residual
    circulation
Coliforms, 423
Complete mixing, 113–115, 118
Complimentary error function, 45
Concentration
  computation of in river, 112–121
  definition, 16
  from outfall in estuary, 268
Concentration moment method, 90
Continental shelf, 100
Continuous release, method for measurement
    of stream discharge, 142
Contraction coefficient, table of, 415
Convection, definition, 7
Cooling, once through, 426
Cumulative discharge method, 120
Curve, *see* Bend

D

Dead zone, 134, 136, 241, 257

478

ISBN 0-12-258150-4

90065